THE COUNTER-CREATIONISM HANDBOOK

THE COUNTER-CREATIONISM HANDBOOK

Mark Isaak

Greenwood Press
Westport, Connecticut • London

Library of Congress Cataloging-in-Publication Data

Isaak, Mark, 1959–
    The counter-creationism handbook / Mark Isaak.
        p. cm.
    Includes bibliographical references and index.
    ISBN 0–313–33305–X (alk. paper)
    1. Creationism.   2. Creation.   I. Title.
    BS651.I82   2005
    231.7'652—dc22   2005003394

British Library Cataloguing in Publication Data is available.

Library of Congress Catalog Card Number: 2005003394
ISBN 0–313–33305–X

First published in 2005

Greenwood Press, 88 Post Road West, Westport, CT 06881
An imprint of Greenwood Publishing Group, Inc.
www.greenwood.com

Printed in the United States of America

The paper used in this book complies with the Permanent Paper Standard issued by the National Information Standards Organization (Z39.48–1984).

10 9 8 7 6 5 4 3 2 1

CONTENTS

PREFACE

Science has made astounding progress toward explaining the world, and yet the world is still largely a mystery. This is partly because every discovery raises new questions about that discovery. It is partly because science has advanced at such a rate that education cannot possibly keep up. Even the most educated scientists would be at a loss outside their specialties. Paradoxically, the magnitude of unknowns in our lives is due largely to the success of scientific explanation.

And yet there are those who would use the unknowns to cast doubt on the scientific explanations, including some of the most successful explanations in the history of science. Most of the case for creationism relies on its arguments falling among the unknowns. Creationist claims are numerous and varied. They touch on diverse aspects of biology, geology, physics, astronomy, and more. Even an expert cannot be expected to know answers to all of them. And when people find they cannot answer all of the claims, they can begin to wonder whether the claims are valid after all. Much of the strength of creationism comes not from its having good arguments but from its creating so many arguments that educators cannot easily teach the answers to all of them.

This book replies to more than 400 of the most common claims that creationists make. Each creationist claim comes with a brief rebuttal showing faults with and, often, counterevidence against the claim. The scientific arguments are current, gathered from experts in the relevant fields. This book also covers a variety of philosophical and religious arguments that bear on creationism. Parts of rebuttals to these, inevitably, get somewhat subjective, but they still attempt to reflect mainstream views. A work like this cannot examine all of the claims in depth, so most of the rebuttals also come with references for where to look for further information.

Creationists themselves are a diverse group, including highly credentialed academics, charlatans even in the consideration of other creationists, and every class in between. They have different views about biblical interpretation, the age of the earth, and other issues. This book covers claims from the whole spectrum of creationist views. Claims are included if they are common or influential in creationist teachings, not on the basis of who made them or how credible the claim is.

This book is intended for anyone who is the least bit confused by or skeptical of what creationists say. Creationists themselves may want to read it to find weaknesses in their arguments. Some scientific literacy is expected of the reader but usually not much more than is needed to understand what the claim is saying in the first place. For the Further Reading sections, I have tried to find works that are written for a general audience; more technical works are so labeled. Many of the references are highly technical; readers who wish to delve into the details of a claim will, in some cases, need a good knowledge of the field, but no such special knowledge is necessary to understand the gist of the rebuttals.

Keep in mind that the body of literature that covers most of these claims is far, far greater than the few references I include. And more is being added every day. Interested readers may enjoy doing further research on their own. There will always be wonders to discover and mysteries to explore.

ACKNOWLEDGMENTS

This book would not exist without the contributions of many people who supplied information, explanation, correction, and encouragement. I am particularly indebted to those associated with the TalkOrigins.org Web site and talk.origins newsgroup and to the staff of the National Center for Science Education. I thank specifically the following people: Paul Barber, John Brawley, Ed Brayton, Troy Britain, Reed A. Cartwright, Craig Corson, Mike Dworetsky, Andrew Ellington, Wesley Elsberry, Jon Fleming, Jim Foley, Barbara Forrest, David Iain Greig, John Harshman, Paul Heinrich, Kevin R. Henke, Mike Hopkins, Gary Hurd, Sverker Johansson, David Ewan Kahana, Justin Kerk, Mark D. Kluge, Sherry Konkus, Don Lindsay, Andrew MacRae, Adam Marczyk, Nick Matzke, Mickey Mortimer, Glenn Morton, Ian Musgrave, Mark Perakh, Roger Scott, John Stear, Douglas Theobald, Tim Thompson, and Kari Tikkanen; and I apologize to those I have overlooked.

HOW TO USE THIS BOOK

For those looking for information on a specific claim, the best place to start is the index. Page numbers in boldface refer to the information most relevant to the entry.

The claims are grouped by topic, so if you are exploring a more general subject, the claims before or after a specific claim may be of interest, too. Cross-references to related issues are provided as well. For easier browsing, a brief version of the complete list of claims follows in the next section. For general background information as well as information that covers a topic in depth, readers are strongly encouraged to pursue the recommended readings and/or research the topic on their own.

Much of creationism is folklore; it survives and spreads mainly by word of mouth. In fact, some of the claims here, such as the "missing day" and "Lady Hope" stories, are explicitly folkloric and are included in collections of urban legends. Inspired by this aspect of creationism, I have created an indexing system patterned after that used by Stith Thompson in his *Motif-Index of Folk-Literature*. Thompson organized folklore motifs in an outline form. For example, his section D includes motifs about magic; motifs about transformation of person to animal start at D100; D150 is transformation of person to bird; D152.2 is man to eagle. I follow Thompson in using letters for the main sections, prefixing everything with "C" (for creationism) to distinguish this numbering from Thompson's. Sections CA through CG contain claims against mainstream science. Sections CH through CJ contain claims for creationism rather than against evolution. Specifically, the sections are as follows:

CA: Philosophy and Theology
CB: Biology (including Abiogenesis)
CC: Paleontology
CD: Geology
CE: Astronomy and Cosmology
CF: Physics and Mathematics

CG: Miscellaneous Antievolution
CH: Biblical Creationism
CI: Intelligent Design
CJ: Other Creationism

Claims are numbered within these sections. For example, the "missing day" story is CE010, and the "Lady Hope" story is CG001.

IF YOU CAN NOT FIND THE CLAIM HERE

A companion Web site, http://www.talkorigins.org/indexcc, contains some less common claims that did not make it into this book. The Web site is updated regularly, so new creationist claims might be addressed there already.

Still, some creationist claims will come up that are not included. How should one deal with them? The answer, in most cases, is to go to the source of the claim. Demand references and look them up. If the persons making the claim cannot supply references (and copies of the referenced articles, if necessary), that alone shows that they cannot support their claim. Once you have a reference, read it to see if it really says what is claimed of it.

One common type of claim is the out-of-context quote (see CA113). These alone could fill another book, and I could not include many here for lack of space. (Pieret 2003 has probably the most coverage of specific quotes.) Many quotes used by creationists, especially those most damning to evolution, have quite another message when read in context. Darwin's quote about the absurdity of eye evolution (CA113.1) is a classic example: Darwin presents a seeming problem and solves it; the creationist shows you only the problem. With this and other quotes, you need to see the surrounding context to understand all of what the author is saying.

Scientific advances give creationists fuel for more new claims. As scientists explore the cutting edge of their fields, they come up with unexpected discoveries and new uncertainties. Scientists themselves sometimes add to the appearance of problems by hyping their discoveries to make them sound more revolutionary. Creationists use the uncertainties for "God of the gaps" arguments (CA100), and they seize on unexpected discoveries as problems with the field as a whole. For example, when neutrinos from the sun were first measured, fewer were found than expected. Creationists took this as evidence against the standard, old-age model of the sun (see CE301). But our understanding of neutrino physics was still very incomplete at the time. Later experiments found that the number of neutrinos from the sun matches what is expected, but some of them change into a different "flavor" of neutrino that was not detected by the earlier instruments.

In truth, pioneering science does have problems by definition. Investigating these problems is what science is all about. In time, these problems get solved, and the cutting edge advances to other problems. The rest of science, where the problems have already been solved, is well founded and stable. Only very rarely do new discoveries make a dent in an established field of science. By looking at the reference that a new claim is based on, one can often see that the claim is about some cutting-edge science and that its implications to the established body of the field are nonexistent.

Occasionally, creationists will come up with a claim that is neither out of context nor about areas of science that are uncertain to begin with. Readers may be able to address such a claim by referring to a similar claim in this book. For example, a discrepant carbon-14 age may suffer the same problem as one of the listed claims about C-14 dating (see CD011). Other claims, however, may need examination by an expert. If that is the case, do not be afraid to ask an expert. Most scientists are pleased to answer questions from people who are honestly interested in the answers.

LIST OF CREATIONIST CLAIMS

This index briefly lists the claims covered by the book, using the indexing system described earlier.

CA: Philosophy and Theology

Ethics (see also *CH010*: Creationism is good)

Evolution is immoral, CA001. **Crime rates**, CA001.1.

Might makes right implied, CA002 . **Social Darwinism**, CA002.1. **Marx admired Darwin**, CA002.2, CA002.2.1.

Evolution is racist, CA005, CA005.1, CA005.2.

Evolution encourages eugenics, CA006, CA006.1.

Promiscuity and lust, CA008. **Animal behavior taught**, CA009.

Homosexuality approved, CA010. **Intellectual snobbery**, CA012.

Fairness and equal time, CA040. **Teach the controversy**, CA041. **Santorum Amendment**, CA041.1. **Biology taught without evolution**, CA042.

Epistemology

Foundation of knowledge

Argument from incredulity, CA100. **Lots unexplained**, CA100.1.

Evolution soon to be rejected, CA110. **Current scientists reject evolution**, CA111, CA112.

Quote mining, CA113. **Darwin on eye evolution**, CA113.1.

Famous scientists were creationists, CA114.

Are you qualified?, CA118. **An evolved mind is fallible**, CA120.

Theory of science

Evolution is only a theory, CA201. **It has not been proved**, CA202. **It does not make predictions**, CA210. **It can not be falsified**, CA211. **It is ambiguously defined**, CA212. **It cannot be replicated**, CA220. **Were you there?**, CA221.

Evolutionary algorithms smuggle in design, *CF011*. **WEASEL program prespecified**, *CF011.1*. **NFL theorems and blind search**, *CF011.2*.

First Law of Thermodynamics

The universe's energy cannot come from nothing, *CF101*. (see also CE440: Origin of everything; CI200: First cause)

Radiometric Decay (see also CD000: Radiometric dating; CE410: Physical constants only assumed constant)

Polonium halos indicate young earth, *CF201*.
Radiometric decay rates not constant, *CF210*.
Short-lived isotopes on the moon, *CF220*.

CG: Miscellaneous Antievolution

History

Darwin recanted, *CG001*. **Oldest living thing is young**, *CG010*.

Linguistics

Chinese glyph for "ark," *CG101*.
First languages are complex, *CG110*.

Folklore

There are flood myths from all over the world, *CG201*.

CH: Biblical Creationism

Biblical Creationism Generally

Creationism has explanatory power, *CH001*.
Creationism is good, *CH010, CH010.1*.
Genesis is foundational, *CH050*. **Noncreationist compromisers**, *CH055*.

Biblical Accuracy

Bible is ultimate authority, *CH100*.
Bible is inerrant, *CH101, CH101.1*.
Bible is literal, *CH102, CH102.1, CH102.2*.
Bible claims inspiration, *CH103*. **Prophecies prove Bible**, *CH110*.
Archaeology supports Bible, *CH120*.
Bible accurate on science, *CH130*. (see also CB621: Humanity traced to an African Eve)
Bible is harmonious, *CH190*.

Young-Earth Creationism

Age of the universe

Universe 6,000–10,000 years old, *CH200*. **Earth 6,000–10,000 years old**, *CH210*. **Apparent age**, *CH220*.

Death and the Fall

No death before the Fall, *CH301*.
Vapor canopy affected earth, *CH310*. **Extended lifetimes**, *CH311*.
Life is deteriorating, *CH320*. **Parasites are degenerate**, *CH321*.

Created kinds

Organisms are discrete kinds, *CH350*. **Stars, galaxies unchanging**, *CH370*.

Source of Flood

Flood from vapor canopy, *CH401*. **Hydroplate theory**, *CH420*.
Runaway subduction, *CH430*.

Meta-arguments

Science rules out considering design, *CI401*, *CI402*. (see also CA301.1: Science rules out supernatural explanations; CA230.1: Evolutionists interpret per their preconceptions)

Design requires a designer, *CI410*.

CJ: Other Creationism

Vedic Creationism

Mankind unchanged for billions of years, *CJ001*.

Native North American Creationism

Kennewick Man was Umatilla ancestor, *CJ311*.

Islamic Creationism

Qur'an accurate on science, *CJ530*.

INTRODUCTION

HOW TO ARGUE CREATIONISM (AND HOW NOT TO)

Perhaps you are using this book only to satisfy your own curiosity and have no interest in discussing it with others. But for those who are interested in communicating about a controversial subject (not just limited to creationism), knowing answers to claims is not enough. One must also know how and when to present them. Giving accurate information in a rude and hostile manner, for example, can make people less likely to consider it. How to argue depends on who you are arguing with and where. Different tactics are necessary when you are arguing personally one-on-one, in an ongoing public dialogue, in a public debate, or in a public presentation.

Personal One-on-One

The first thing to consider when arguing one-on-one is whether there is any point. If the person is hard-set in his or her ways and unwilling to consider contrary evidence, if the person is arguing only to try to convert you or to reinforce his or her own dogma, and if you do not have a great deal of patience, then it may be a waste of time to discuss the topic with such a person. This is a difficult call to make, though. The appearance of hardheadedness could come from the fact that the creationist had only encountered other creationists before; even if you do not see your arguments bear fruit, they may do so later.

On the other hand, if the other person is open to learning, supply some. Two essential qualities in such a dialogue are patience and politeness. Even if the other person's claim looks stupid to you, treat it seriously and answer politely, because it is a reasonable idea to the other person. If you do not know an answer, admit as much, do your research, and get back with the answer. Note that patience and politeness can be effective even against the dogmatist in certain situations, such as a family or work relationship, in which you have a chance to practice it for a long time.

If you can uncover fundamental motives, it helps to address these, too. Most creationists have the idea that evolution is incompatible with their religion. It may help, then, to show them other members of their religion who accept evolution.

Written Public Dialogue

In public dialogues, such as online forums and letters to the editor, it is important that all substantive claims be addressed. Space limitations (especially in letters) mean it is not always possible to do this as completely as one would like, but one should always at least say when claims are false and give pointers on where to find details.

A big problem with public dialogues is that the same claims get repeated over and over. It suffices most of the time simply to say, "that claim has already been dismantled; see here." For newcomers to the debate, the details should at least be sketched out occasionally. Patience and politeness are important here, too. It is especially important that polite queries and claims be met with polite replies. If someone treats you rudely, it is understandable to reply in kind but better to set the good example.

Teacher–Student

Biology teachers are likely to be challenged occasionally by students with creationist beliefs. Try to understand the point of view of the student. Usually, they are coming from a position of unshakeable religious belief, and trying to argue them out of their belief would be futile. Many teachers find it works well simply to tell the students that they do not need to believe evolution but that they are expected to understand it.

Biology teachers can arrange their curriculum to address common misconceptions about evolution. Even among noncreationists, many people do not know what evolution means. Other common misunderstandings are that evolution operates purely by chance, that it involves sudden transitions (such as a fish giving birth to a frog), that it cannot be studied because it happened in the past, and that it violates the laws of thermodynamics.

There are several resources available especially for teachers. Alters and Alters (2001) is particularly good. Consider contacting the National Center for Science Education (www.ncseweb.org), the National Science Teachers Association (www.nsta.org), and the Biological Sciences Curriculum Study (www.bscs.org).

Students receiving creationist arguments from their teacher are in a difficult position. In addition to being morally wrong, is unlawful in the United States for teachers to press religious views on their students, and the students have a right to make that stop, but the process may be uncomfortable. Whatever the problem, students should enlist the help of a parent, another teacher, or other adult. Even if a teacher is stopped from teaching creationism, do not expect this individual to start teaching evolution. It may be necessary to seek some education outside of the class (which is a useful skill to develop anyway).

Challenges and Public Debates

The first thing to consider in public debates is that the evolution side usually loses simply by showing up; the existence of the debate sends the message that the two sides are roughly of equal validity. Furthermore, during the debate, the creationist may make any number of wild, confabulated claims, more than anyone can counter in a limited time. And scientific acumen is not even a criterion for success; most people will judge the debater on how entertaining the presentation is. Finally, the audience is usually packed with people sympathetic to creationism who will ensure that the creationism side "wins" no matter what happens behind the podium. "The purpose of a debate is to rouse the local troops, to stir them to action, and inspire them to go forth and support the teaching of creationism. Why should we help?" (Scott 1994, 24) The serious debate about evolution has been occurring continuously in scientific meetings and publications; creationists have chosen not to participate.

If you do debate in a creationist forum, learn all you can about your opponent beforehand. Often, the creationist debaters have a set pattern for which you can prepare. Make sure you get some say in the topic of debate. Narrow the topic as much as possible. If possible, demand something that the creationists do not want to consider, such as the question "Is population genetics consistent with a global flood?"

Because creationists are so wedded to their prepackaged spiel, one of the primary tactics of the debater must be to force the creationist to depart from the script and wander into unrehearsed territory. The very best way to do this is to force the creationist to argue the merits of creation "science" rather than the more familiar (to him or her) territory of evolutionary science. In a debate with Canadian creationist Ian Taylor, Robert P. J. Day accomplished this with a simple and clever tactic:

> I compared two scenarios; what we have now, "Evolution in, Creationism out," with what the creationists, with their demand for "equal time," seem to be asking for, "Evolution in, Creationism in." I then pointed out that, in comparing the two scenarios, *there was no difference in the status of evolution*; that is, both evolutionists and creationists agree that evolution should be taught and evolution was therefore not the issue here. Rather, the controversy hinged on the inclusion of creation science in the public school curriculum; my task, in wanting to exclude it from science classes, would be to show that it did not qualify as science, while Taylor's job, in trying to include it, would be to defend it. I stated that any attacks on evolution by Taylor would be completely irrelevant, since evolution clearly was not an issue. In doing so, I deprived Taylor of his most effective weapon. . . . It left me free to use my entire presentation to eviscerate creation science. (R.P.J. Day 1990)

If the debate is held in a local space (e.g., a church auditorium rather than on radio), have some one-page fliers available for volunteers to hand out. This reinforces your main points and allows references to other resources.

Public Presentations

Perhaps the place where your voice can have the most impact is in public presentations, such as before school boards and in textbook hearings. Advanced preparation is important for such presentations.

Organize the community. Do not be a lone voice. Recruit scientists, clergy, teachers, and other members of the community to speak, too. The National Center for Science Education (www.ncseweb.org) is a good place to start for connecting with others in your community.

Having handouts ready is useful here, too. The people you are speaking to will forget much of what you say; they can see it again if it is summarized on paper.

General Considerations

Do not alienate your audience. As tempting as it may be at times to treat all your opponents with scorn, treating them thus will only reinforce their belief that evolutionists are evil scum.

Know your opposition. In particular, know what creationists believe. Demolishing a claim that your opponent does not believe in the first place may make you feel better, but it will leave your opponent unphased.

Do not argue beyond your expertise. If you do not know something, say so. If you make a mistake, admit it right away. One substantive mistake can destroy your credibility. Inviting your opponents to learn an answer with you may be a better learning experience for them than telling them the answer in the first place. In particular, be aware that this book will not make you an expert on creationist claims. It is merely an introduction.

A NOTE TO CREATIONISTS

This book is not written with the expectation that creationists will flock to it, but it can be of value to them in the following ways: First, it can help counter the arguments of other kinds of creationists. Biblical creationism alone includes young-earth creationism, old-earth creationism, day–age creationism (which says that the "days" of Genesis 1 refer to long stretches of time), and geocentrism. Then there are creationists from other religions, which, barring appeals to religious bigotry, have as much spiritual justification as biblical creationism. These various forms of creationism make particular claims with which other creationists disagree. This book can provide scientific and logical rebuttals to these claims. I expect every creationist will find at least some claims in this book that they already reject.

Second, this book can show creationists what arguments to avoid. Many arguments have been refuted so often that creationists lose credibility just by mentioning them. (Prime examples are the second law of thermodynamics [CF001] and moon dust [CE101].) At least one creationist group has recognized this problem and compiled a list of "arguments we think creationists should NOT use" (AIG, n.d.a.). This book is a more extensive version of such a compilation. Creationists may not accept that their favorite arguments are flawed just because they find them here, but they may at least begin to look at the evidence and see for themselves how credible the arguments are.

Finally, this book can show the point of view of those who oppose creationism. It does not show the evidence for evolution in detail (for that, see the further reading with CA202), but it does show why scientists reject essentially all of the creationists' arguments. Creationists may not like the rebuttals, but pretending that they do not count will not make it so. Understanding an issue requires looking at all sides of it. This is especially important for those who wish to discuss the issue with others. Anyone who wants to understand creationism fully should look at the side presented here.

Why should creationists (or anyone, for that matter) believe anything I say? They should not, at least not just because I say it. No single authority should be treated as definitive. However, there are some commonsense approaches one may take, which, I expect, will verify most of the information in this book.

Check things out for yourself. I have included both creationist and scientific references for just that purpose. Use them. And do further research on your own. In particular, it is useful to have a good broad but basic understanding of a field to give a framework for further understanding. There are several ways to acquire such a basic knowledge. Probably the most effective is to work in that field yourself, as a hobby if nothing else. Reading and taking introductory college courses are useful, too.

Consider all sides of any controversial issue. In particular, look beyond sources that feed your existing beliefs and preconceptions.

Consider motives (including your own). Creationists sometimes argue against evolution by saying, "Do you want to be related to a monkey?" They fail to appreciate that wanting evolution to be false is actually a strong argument that evolution is true. Claims are more trustworthy if they are supported by people who have no ulterior motive to accept them. Very few scientists want to be related to monkeys either, and yet they overwhelmingly accept evolution. Creationists who know few scientists may claim that scientists are motivated by their own "religion of evolution" (CA610) or commitment to materialism (CA601). Again, check it out for yourself. You will find that scientists come in all religions and spiritual persuasions. Many of them are as committed to God as any creationist. The value they all share is that they consider what the evidence actually shows to be more important than their wishes for what it should be. Scientists are also motivated by reputation and money to make sure their data and conclusions are sound, and they would get fame and fortune by finding support for a theory that overthrows conventional views. All of these motives suggest that scientists would not accept evolution unless the evidence left them no choice.

WHAT WE ARE ARGUING FOR

I hope that readers will not lose sight of the fact that the creation–evolution debate is about more than arguing against creationism. We who support quality science have valuable issues to argue *for*.

The benefits of good, accurate science should be obvious. Science has made possible our current technology and the standard of living that comes with it. Evolution has contributed to these benefits (CA201, section 2); creationism has not. And creationism opposes more than evolution; it also contradicts substantial parts of biology in general, paleontology, geology, astronomy, physics, and psychology, not to mention the scientific method itself. We benefit daily from the fruits of the science that creationism would deny. Accurate scientific knowledge is useful also to nonscientists. It can help directly as one applies the scientific principles, and it is part of the scientific literacy needed to support good policy decisions in a technological society.

Science has aesthetic value, too. Science opens up new realms—from distant galaxies to subatomic particles, from rain forest canopies to the depths of the ocean—to feed the curiosity with which we are born. We gain new ways to appreciate things we may never have even considered before. The rocks exposed by road cuts were just barren ground until I studied geology, and the night sky became more enchanting simply by learning the names of a few stars and constellations. What I study becomes more beautiful as I learn more about it. I know other scientists have similar feelings. Most fear is fear of the unknown; science replaces it with wonder.

Finally, although science is not a religion, there is a moral and philosophical dimension to it. Science is built upon accepting whatever the evidence shows. The scientist gets to design the experiments, but the universe gets the ultimate say in their outcome. It is humbling. Scientists who try to manipulate conclusions to their desired outcome come to grief. The acceptance of reality is the basis for honesty and effective action, in addition to being a positive value in its own right. There are few if any social institutions that encourage and enforce this value to the extent that science does.

Creationism is fought in the courts as a religious freedom issue. Those who support the teaching of evolution believe that the views of one particular religion should not be taught as fact to all children in public schools. More than that, we believe that a religion's opposition to prevailing scientific evidence and conclusions is not sufficient justification to keep the science from being taught. Most people support the idea of religious freedom. But religious freedom requires support of the practice, not just support of the idea. Unfortunately, most people are not motivated to get involved unless they think they are adversely affected themselves. I hope that they will reflect upon what else besides religious freedom is denied when evolution is rejected.

PHILOSOPHY AND THEOLOGY

ETHICS

CA001: Evolution is the foundation of an immoral worldview.

Evolution is the foundation of an immoral worldview. (Moon 1990)

1. Evolution is descriptive. It can be immoral only if attempting to accurately describe nature is immoral.

2. Any morals derived from evolution would have to recognize the fact that humans have evolved to be social animals. In a social setting, cooperation and even altruism lead to better fitness (Wedekind and Milinski 2000). Evolution leads naturally to ethical principles such as the Golden Rule.

3. Some bad morals, such as eugenics (see CA006) and social Darwinism (see CA002.1), are based on misunderstandings of evolution. Therefore, it is important that evolution be taught well to negate such misunderstandings.

4. Despite claims otherwise, creationism has its own problems (see CH010). For one thing, it is founded on religious bigotry, so the foundation of creationism, by most standards, is immoral.

5. Probably the most effective weapon against bad morals is exposure and publicity. Evolution (and science in general) is based on a culture of making information public.

6. Scientists are their own harshest critics. They have developed codes of ethical behavior for several circumstances, and they have begun to talk about a general ethics (Rotblat 1999). Creationists have nothing similar.

7. Some people feel better about themselves by demonizing others. Those people who are truly interested in morals begin by looking for immorality within themselves, not others.

Further Reading: Huxley, T.H.H. 1893. Evolution and ethics. http://aleph0.clarku.edu/huxley/CE9/index.html.

CA001.1: Crime rates and other social ills have increased since evolution began to be taught.

Since evolution began to be taught in public schools, crime rates and other social ills have increased. (Living Word n.d.)

1. Crime rates go up and down and are associated mostly with the age of the population. There does not appear to be any correlation between crime rates and teaching evolution. The United States was generally more violent in the years 1870–1910 before evolution was taught. In recent years, crime rates have been dropping since 1989.

Regional trends show a negative correlation between crime and teaching evolution. Other countries that teach evolution to a far greater extent than the United States have lower crime rates. In the United States, southern states tend to emphasize creationism more, but they also have generally higher crime rates.

2. Correlation does not imply causation. Since the teaching of evolution, death rates from most cancers have decreased, air travel has increased, and the earth's temperature has risen, but we do not attribute any of those to teaching evolution.

3. In the United States, at least, most people do not believe evolution. If social ills follow from belief about origins, creationists deserve more of the responsibility.

4. "Do not ask why the old days were better than the present; for that is a foolish question" (Ecclesiastes 7:10).

CA002: Survival of the fittest implies might makes right.

Survival of the fittest implies that "might makes right" is a proper guide to behavior. (Keyes 2001)

1. This claim exemplifies the naturalistic fallacy by arguing that the way things are implies how they ought to be. It is like saying that if someone's arm is broken, it should stay broken. But "is" does not imply "ought." Evolution is descriptive. It tells how things are, not how they should be.

2. Humans, being social, improve their fitness through cooperation with other people. Even if survival of the fittest were taken as a basis for morals, it would imply treating other people well.

Further Reading: Hume, D. (1779). 1947. *Dialogues Concerning Natural Religion*; Wilkins, J. 1997a. Evolution and philosophy: Does evolution make might right? http://www.talkorigins.org/faqs/evolphil/social.html.

CA002.1: Evolution leads to social Darwinism.
Darwinism leads to social Darwinism, the policy that the weak should be allowed to fail and die. (H. M. Morris 1985, 179)

1. This is an example of the naturalistic fallacy (see CA002)—the argument that how things are implies how they ought to be. But "is" does not imply "ought." Evolution only tells how things are; it does not say how they should be.

2. The source of social Darwinism was not Darwin but Herbert Spencer and the tradition of Protestant nonconformism going back to Hobbes via Malthus. Spencer's ideas of evolution were Lamarckian. The only real connection between Darwinism and social Darwinism is the name.

3. Diverse political and religious ideas draw upon evolutionary biology, including ideas advocating greater cooperation.

4. Evolutionary theory shows us that the long-term survival of a species is strongly linked with its genetic variability. All social Darwinist programs advocate minimizing genetic variability, thus reducing chances of long-term survival in the event of environmental change. An understanding of evolution should then rebuke any attempt at social Darwinism if the long-term survival of humanity is treated as a goal.

5. Eugenics and social Darwinian accounts are more often tied to the rise of the science of genetics than to evolutionary theory.

Further Reading: Bannister, R. C. 1979. *Social Darwinism: Science and Myth in Anglo-American Social Thought*; Bowler, P. J. 1993. *Biology and Social Thought, 1850–1914*; Hofstadter, R. 1944. *Social Darwinism in American Thought*; Kevles, D. 1995. *In the Name of Eugenics: Genetics and the Uses of Human Heredity*; Ruse, M. 2001. "Social Darwinism." Chap. 10 in *Can a Darwinian Be a Christian?*; Singer, P. 2000. *A Darwinian Left: Politics, Evolution, and Cooperation*; Wilkins, J. 1997a. Evolution and philosophy: Does evolution make might right? http://www.talkorigins.org/faqs/evolphil/social.html.

with Michael Suttkus

CA002.2: Marx admired and corresponded with Darwin.
Marx sent a personally inscribed copy of the second edition of Das Kapital *to Darwin and wanted to dedicate it to him, but Darwin wrote a letter politely declining.* (Humber 1987b)

1. Darwin wrote a letter declining the dedication of an unnamed book on atheism, but he wrote it to Edward Aveling. Aveling's common-law wife was Elanor Marx, Karl's daughter, and she inherited his papers. They got mixed up with Karl Marx's papers, and the letter was assumed to have been to Marx. This view found ideological favor in Russia, so it was widely repeated. Later, a letter from Aveling, requesting permission to dedicate his book *The Student's Darwin* to Darwin, was found among Darwin's papers. Darwin declined permission and argued that science should not address religious matters directly (Carter 2000; Colp 1982).

2. Darwin did have a copy of *Das Kapital*, but its pages were unseparated when he died, so he never read it.

3. None of this matters to the science of evolution.

Further Reading: Colp, R., Jr. 1982. The myth of the Darwin–Marx letter; Dawkins, R. 2000. There's more to books than titles. http://archive.workersliberty.org/wlmags/wl61/dawkins.htm.

CA005: Evolution is racist.
Evolution promotes racism. (H. M. Morris 1985, 179)

1. When properly understood, evolution refutes racism. Before Darwin, people used typological thinking for living things, considering different plants and animals to be their distinct "kinds." This gave rise to a misleading conception of human races, in which different races are thought of as separate and distinct. Darwinism helps eliminate typological thinking and with it the basis for racism.

2. Genetic studies show that humans are remarkably homogeneous genetically, so all humans are only one biological race. Evolution does not teach racism; it teaches the very opposite.

3. Darwin himself was far less racist than most of his contemporaries (see CA005.1).

4. Although creationism is not inherently racist, it is based upon and inseparable from religious bigotry, and religious bigotry is no less hateful and harmful than racism.

5. Racism historically has been closely associated with creationism, as is evident in the following examples:

- George McCready Price, who is to young-earth creationism what Darwin is to evolution, was much more racist than Darwin. In *The Phantom of Organic Evolution*, he referred to "Negroes" and "Mongolians" as degenerate humans. (Numbers 1992, 85)

- During much of the long history of apartheid in South Africa, evolution was not allowed to be taught. The Christian National Education system, formalized in 1948 and accepted as national policy from 1967 to 1993, stated, among other things,

    that white children should "receive a separate education from black children to prepare them for their respective superior and inferior positions in South African social and economic life, and all education should be based on Christian National principles." (Esterhuysen and Smith 1998)

  The policy excluded the concept of evolution, taught a version of history that negatively characterize nonwhites, and made Bible education, including the teaching of creationism, and religious assemblies compulsory (Esterhuysen and Smith 1998).

- The Bible Belt in the southern United States fought hardest to maintain slavery.

- Henry Morris, of the Institute for Creation Research, has in the past read racism into his interpretation of the Bible:

    Sometimes the Hamites, especially the Negroes, have even become actual slaves to the others. Possessed of a genetic character concerned mainly with mundane, practical matters, they have often eventually been displaced by the intellectual and philosophical acumen of the Japhethites and the religious zeal of the Semites. (H. M. Morris 1985, 241)

Further Reading: Mayr, E. 2000. Darwin's influence on modern thought; Trott, R., and J. Lippard. 2003. Creationism implies racism? http://www.talkorigins.org/faqs/racism.html.

CA005.1: Darwin himself was racist.

Charles Darwin was himself a racist, referring to native Africans and Australians, for example, as savages. (Humber 1987a; Weston-Broome 2001; see also CA005.2.)

1. Virtually all Englishmen in Darwin's time viewed blacks as culturally and intellectually inferior to Europeans. Some (such as Louis Agassiz, a staunch creationist) went so far as to say they were a different species. Charles Darwin was a product of his times and no doubt viewed non-Europeans as inferior in ways, but he was far more liberal than most: He vehemently opposed slavery (Darwin 1913 especially chap. 21), and he contributed to missionary work to better the condition of the native Tierra del Fuegans. He treated people of all races with compassion.

2. The views of Darwin, or of any person, are irrelevant to the fact of evolution. Evolution is based on evidence, not on people's opinions.

Further Reading: Britian, T. n.d. Darwin on race and slavery. http://home.att.net~troybritain/articles/darwin_on_race.htm.

CA005.2: Darwin's work refers to "preservation of favoured races."

The subtitle of Darwin's Origin of Species *refers to "the preservation of favoured races," showing the racist nature of Darwin's ideas.* (Weston 1998)

1. Race, as used by Darwin, refers to varieties, not to human races. It simply points out that some variations that occur naturally survive in greater numbers. *Origin of Species* hardly refers to humans at all.

2. Evolution is not racist (see CA005).

Further Reading: Mayr, E. 2000. Darwin's influence on modern thought.

CA006: Evolution encourages eugenics.

Evolution promotes eugenics. (DeWitt 2002)

1. Eugenics is based on genetic principles that are independent of evolution. It is just as compatible with creationism, and in fact at least one young-earth creationist (William J. Tinkle) advocated eugenics and selective human breeding (Numbers 1992, 222–223).

2. Many eugenics arguments, such as the expected effect of selective sterilization and the results of interracial mating, are based on bad biology. Better biology education, including the teaching of evolution, can only counter the assumptions on which eugenics is based.

Further Reading: Wilkins, J. 2000. Evolutionists against eugenics. http://www.talkorigins.org/origins/postmonth/nov00.html.

CA006.1: Hitler based his views on Darwinism.
Adolf Hitler exploited the racist ideas of Darwinism to justify genocide. (Weston-Broome 2001)

1. Hitler based his ideas not on Darwinism but on a "divine right" philosophy:

> Thus, it [the folkish philosophy] by no means believes in an equality of races, but along with their difference it recognizes their higher or lesser value and feels itself obligated, through this knowledge, to promote the victory of the better and stronger, and demand the subordination of the inferior and weaker in accordance with the eternal will that dominates this universe. (Hitler 1943, 383)

The first edition of *Mein Kampf* indicates that Hitler was a young-earth creationist at the time of its writing: "this planet will, as it did thousands of years ago, move through the ether devoid of men" (p. 65; the second edition substitutes "millions" for "thousands.") Other passages further support his creationist leanings:

> The undermining of the existence of human culture by the destruction of its bearer seems in the eyes of a folkish philosophy the most execrable crime. Anyone who dares to lay hands on the highest image of the Lord commits sacrilege against the benevolent Creator of this miracle and contributes to the expulsion from paradise. (Hitler 1943, 383)

Quotes from Hitler invoking Christianity as a basis for his actions could be multiplied ad nauseam. For example:

> Hence today I believe that I am acting in accordance with the will of the Almighty Creator: by defending myself against the Jew, I am fighting for the work of the Lord. (Hitler 1943, 65)

> [T]he task of preserving and advancing the highest humanity, given to this earth by the benevolence of the Almighty, seems a truly high mission. (Hitler 1943, 398)

> A campaign against the "godless movement" and an appeal for Catholic support were launched Wednesday by Chancellor Adolf Hitler's forces. (Associated Press 1933)

Of course, this does not mean that Hitler's ideas were based on creationism any more than they were based on evolution. Hitler's ideas were a perversion of both religion and biology.

2. Genocide and racism existed long before Darwin. Obviously, they did not need any contribution from Darwinism. In many instances, such as the Crusades and the Spanish conquest of Central America, religion was explicitly invoked to justify them.

3. Evolution does not promote social Darwinism (see CA002.1), racism (see CA005), or eugenics (see CA006).

Further Reading: Toland, J. 1976. *Adolf Hitler.*

CA008: Evolution encourages promiscuity and lust.

Evolutionists promote the concepts of promiscuity and lust, pointing out that the quest to produce many offspring is a main goal of organisms under Darwinism. (H. M. Morris 2000a)

1. Description does not imply promotion. Solving a problem works best if you first understand the source of the problem. Creationists, by denying sources of behavior, are less likely to deal with behavioral problems effectively.

Further Reading: Wright, R. 1994. *The Moral Animal.*

CA009: Evolution teaches that we are animals and to behave as such.

Evolution teaches that people are animals. We should not be surprised when people who are taught evolution start behaving like animals. (Rendle-Short 1980)

1. Evolution does not teach that humans are animals; biology in general does (and so does the Bible, in Ecclesiastes 3:18–21). More specifically, humans are a species of primate, which is a category of mammal, which is a category of vertebrate, which is a category of animal. This was known more than 2,000 years ago.

2. "Behaving like animals" does not mean anything, because different animals behave in different ways. A part of human behavior is the ability of people to learn and to modify their behavior according to cultural norms. Evolution teaches that people behave like humans.

3. Creationism teaches that similarities are designed, that God designed our bodies to be like animals. If God designed us to be like animals, then (creationism teaches) we should behave like animals.

CA012: Evolutionists are intellectual snobs.

Anticreationist complaints are a form of snobbery. There are many more important traits on which to judge people than whether they believe in evolution. (Derbyshire 2003)

1. Most anticreationists would be content to let creationists believe as they wish and not make an issue of it. However, creationists make creationism an important trait. They make a huge deal of it and want to impose it on others. If creationists did not believe that it is one of the most important traits on which to judge people, the creationism–evolution dispute would not exist.

2. The complaint of snobbery is based on the attitude that all opinions are equal. Although that attitude sounds democratic and fair, it is indefensible. Opinions have value to the extent that they are informed. If you are suffering serious stomach problems, would you give equal weight to opinions from a professional gastroenterologist and a supermarket bag-boy? When someone speaking on the subject of evolution is woefully uninformed on issues concerning evolution, it is entirely appropriate to point that out. And the claims

made by creationists show that almost all of them are woefully ignorant of evolution. There are exceptions (Kurt Wise, for example), but they are very few.

3. The resolution of creationism as a scientific proposition rests on what the facts indicate, and facts are not determined by the personalities of the people who talk about them. Creationists are free to avoid this issue entirely by approaching creationism as an entirely unscientific religious viewpoint, but they have chosen not to do so.

4. Many creationists have a literally holier-than-thou attitude. For example, they (falsely) claim that believers of evolution are atheistic (see CA602) and evil (see CA001). You cannot get any more snobbish or elitist than that.

Further Reading: Sonnert, G. 2002. *Ivory Bridges*.

CA040: Fairness demands that evolution and creation be given equal time.

In fairness, creation and evolution deserve equal time in science classes. (H. M. Morris 1985, 197–198)

1. The teaching of creationism does not belong in science classes because creationism has no science to teach. It is based on personal religious belief, not on evidence. For the most part, creationism can fit with anything we find, making it unscientific. Where creation models do make specific predictions that can be tested against evidence, they fail the tests. Asking for equal time is asking for nonscience to be taught in science classes.

A 1999 United States poll found that most people favor teaching evolution—and teaching it as science—and that when creationism is taught, most prefer that it be taught either in nonscience classes or as a religious belief (DYG 2000).

2. Equal time would open creationism, and by extension Christianity in general, to ridicule and attack. Saint Augustine recognized this in the fifth century:

> Usually, even a non-Christian knows something about the earth, the heavens, and the other elements of this world, about the motion and orbit of the stars and even their size and relative positions, . . . and this knowledge he holds to as being certain from reason and experience. Now, it is a disgraceful and dangerous thing for an infidel to hear a Christian, presumably giving the meaning of Holy Scripture, talking non-sense on these topics; and we should take all means to prevent such an embarrassing situation, in which people show up vast ignorance in a Christian and laugh it to scorn. (1982, 42–43)

3. Equal time would mean teaching

- other versions of creationism from other denominations of Christianity (including young-earth, old-earth, day–age, gap theory, geocentrism, and flat earth). All have equal basis for being taught, since they are all based on exactly the same Bible. All are mutually incompatible (DYG 2000; H. M. Morris 1985, 215–247; Watchtower 1985, 186).

- other versions of scientific creationism from other religions. Claims have been made for Muslim, Hindu, and Native American versions of creationism.

  The only legal precedent favoring creationism in the United States in the last fifty years was an Interior Department decision finding, on the basis of native creation and flood myths, that 9,400-year-old Kennewick Man was associated with present-day Native American tribes (Chatters 2001, 266; see CJ311).

- creation traditions from more than 300 other religions and cultures, from Abenaki to Zulu.
- other ideas for the origin of life and the universe, such as
  - solipsism
  - Last Thursdayism, the unfalsifiable view that the universe and everything in it was created last Thursday with only the appearance of earlier history
  - multiple designers (Hoppe 2004)
  - Raelianism or other extraterrestrial involvement
  - creation by time travelers.

Creationists do not want all of these taught in science class any more than science educators do. Clearly, creationism in school is an attempt to get greater time than all the opposing views, not equal time. That is not fair.

4. Creationists do not advocate equal time for evolutionary theory in church services. Why?

Further Reading: Edwords, F. 1981. Why creationism should not be taught as science; Part 2; Isaak, M. 2000. What is creationism? http://www.talkorigins.org/faqs/wic.html; Leeming, D., and M. Leeming. 1994. A *Dictionary of Creation Myths*; Sproul, B. 1991. *Primal Myths*.

CA041: Teach the controversy.
Students should be taught all sides of a controversial issue. Evolution should not be taught without teaching the controversy that surrounds it. (S.C. Meyer 2002)

1. On the fundamental issues of the theory of evolution, such as the facts of common descent and natural selection, there is no scientific controversy. The "teach the controversy" campaign is an attempt to get pseudoscience taught in classrooms. Lessons about the sociological issues of the evolution–creation controversy may be appropriate in history or other nonscience classes.

If the object is to keep bad science from the classroom, the same standards should be applied to the counterarguments from creationists, which are all bad science.

2. There are controversies over details of evolutionary theory, such as the relative contributions of sympatric versus allopatric speciation. These controversies require a great deal of background in biology even to understand what they are about. They should not be taught to beginning students. They should be taught to graduate-level students in biology, and they are.

Further Reading: Scott, E.C., and G. Branch. 2003. Evolution: What's wrong with 'teaching the controversy.'

CA041.1: Federal law (Santorum Amendment) supports teaching alternatives.
The Santorum Amendment, which was passed by the Senate and preserved in the conference committee report of the No Child Left Behind bill, gives federal legislative weight to the idea that alternatives to evolution should be taught in science classes. (Santorum 2002)

1. The legislative history of the Santorum Amendment actually opposes the idea of teaching intelligent design, because it was removed from the bill.

Federal laws in the United States must be passed by both the Senate and the House of Representatives. Often, a bill is submitted to both houses at the same time, and the two houses work on them independently. But before the bill goes to the president, there must be just one bill; both houses must vote on exactly the same language. To accomplish this, a conference committee, formed of members of both houses, hashes out the differences and makes a compromise bill. The committee produces an explanatory statement that tells the legislative history of the bill and any views about it that conference committee members want to include. Such a report may be taken into account if courts later need to consider the intent of the bill, but it has no legal force per se.

The Santorum Amendment was added to an early draft of HR-1 by Senator Rick Santorum. It was a recommendation that would have made no binding requirements anyway. It was not included in the House version of the bill. The conference committee for the bill considered the amendment and deliberately chose to omit it, and the Santorum Amendment does not exist in the law that was signed by the president (Public Law 107–110). The amendment still exists, in watered-down form, in the explanatory statement due to the special interest of one or more of the conference committee members.

As noted above, the explanatory statement can still be used to interpret the law. But what does it mean to interpret a nonbinding recommendation that was removed before the bill became law?

2. Even in its original form, the Santorum Amendment did not oppose teaching evolution or advocate teaching intelligent design. The language said, "where topics are taught that may generate controversy (such as biological evolution), the curriculum should help students to understand the full range of scientific views that exist." Evolution is used only as an example. It is a poor example, since there are no competing scientific views. In particular, intelligent design is not scientific (see CI001).

3. A governmental requirement for teaching intelligent design or other so-called alternatives to evolution would likely be found unconstitutional, given the precedent of past cases, particularly *Edwards v. Aguillard*, 393 U.S. 97 (1987).

4. Members of the U.S. Congress are not well qualified in educational matters. Rick Santorum in particular proposed and supported the amendment for religious reasons, not for pedagogy.

Further Reading: Branch, G., and E. Scott. 2003. The antievolution law that wasn't. Miller, K.R. 2002. The truth about the "Santorum Amendment" language on evolution. http://www.millerandlevine.com/km/evol/santorum.html; NCSE. 2002b. Is there a federal mandate to teach intelligent design creationism? http://www.ncseweb.org/resources/articles/ID-activists-guide-v1.pdf.

CA042: Biology can reasonably be taught without evolution.

Evolution does not need to be taught in science classes. The important parts of biology, such as how organisms function, how they are classified, and how they interact with one another, do not depend on evolution.

1. Biology without evolution is natural history, not biology. There is a great deal of important information in natural history that should be taught, but evolution is the unifying idea that ties it all together, allowing one not only to know the facts but to understand them and to know where the facts come from. Teaching biology without evolution would be like teaching chemistry without the periodic table of the elements.

2. To quote an anonymous writer to TalkOrigins.org (June 15, 2003 feedback):

> Evolution matters because science matters, and too many people (including some presidents) are willing to believe that science is something you can pick and choose from, with "good" science being anything that supports your own views and "bad" science being anything that doesn't. Physicists are great guys because they say nothing to offend us, biologists are mad scientists leading us down the path to perdition with their genetic meddling, evolutionists are self-delusional fools, and anyone studying environmental science is a left-wing tree-hugging extremist whose sole goal is to destroy the American economy and lead us to one-world government. If scientists in a given discipline argue about any conclusion, whoever says what you want to hear is the right one. Too many people can't accept that although scientists are not perfect, and do make mistakes (sometimes whoppers), science isn't something you can pick through like a buffet, accepting only what is to your "taste" and designating the rest inedible. If people feel free to reject the science of evolution, they feel free to reject any science on no better grounds. Whether my students accept evolution may have little direct effect on my future. Whether they understand biology, ecology, environmental geology (water is a big issue in my community), and other subjects and can make informed decisions regarding scientific issues *does* matter. If they feel free to reject evolution as part of a "buffet" approach to science, their other choices will be no better informed.

Further Reading: Dobzhansky, T. 1973. Nothing in biology makes sense except in the light of evolution.

EPISTEMOLOGY

Foundation of Knowledge

CA100: Argument from incredulity.
It is inconceivable that (fill in the blank) could have originated naturally. Therefore, it must have been created.

This argument, also known as the argument from ignorance or "god of the gaps," is implicit in a very many different creationist arguments. In particular, it is behind all arguments against abiogenesis and any and all claims of intelligent design.

1. Really, the claim is "I can't conceive that (fill in the blank)." Others might be able to find a natural explanation; in many cases, they already have. Nobody knows everything, so it is unreasonable to conclude that something is impossible just because you do not know it. Even a noted antievolutionist acknowledges this point: "The peril of nega-

tive arguments is that they may rest on our lack of knowledge, rather than on positive results" (Behe 2003).

2. The argument from incredulity creates a god of the gaps. Gods were responsible for lightning until we determined natural causes for lightning, for infectious diseases until we found bacteria and viruses, for mental illness until we found biochemical causes for them. God is confined only to those parts of the universe we do not know about, and that keeps shrinking.

Further Reading: Drummond, H. 1904. *The Lowell Lectures on the Ascent of Man*. Chap. 10. http://www.ccel.org/d/drummond/ascent/ascent14.htm; Van Till, H. J. 2002. Is the Creation a 'right stuff' universe?

CA100.1: Evolution leaves lots of things unexplained.
Evolution leaves lots of things unexplained, such as gravity, the origin of life, biological complexity, and morals. (Behe 1996b)

1. No theory explains everything, and evolution makes no pretense of being different. Evolution does not even apply to some areas, such as cosmology and physics. In biology, evolution is broadly applicable, and it explains a great deal (Theobald 2004), but it is not everything. Some explanations depend on other factors; some we simply have not found yet; and some may be beyond our ability to uncover or understand. It is silly to condemn evolution, despite its strengths, for not achieving godhood.

2. Evolution does explain some things that people claim it does not (morals, for example; see CB411), at least in broad outline. Sometimes the people making this claim simply have not done their homework.

CA110: Evolution will soon be widely rejected.
Evolution is a theory in crisis; it will soon be widely rejected. (see Morton 2002a)

1. Evolution is one of the most strongly supported theories in all of science. It is nowhere near a theory in crisis.

2. This claim has been made constantly since even before Darwin. In all that time, the theory of evolution has only gotten stronger. Prior to the development of evolutionary theory, almost 100 percent of relevant scientists were creationists. Now the number is far less than 1 percent (see CA111). The numbers continue to drop as the body of evidence supporting evolutionary theory continues to build. Thus, claims of scientists abandoning evolution theory for creationism are untrue.

3. This claim directly contradicts another common claim, that evolution cannot be falsified (see CA211).

Further Reading: Morton, G. R. 2002a. The imminent demise of evolution: The longest running falsehood in creationism. http://home.entouch.net/dmd/moreandmore.htm; Morton, G. R. 2002b. Morton's Demon. http://www.talkorigins.org/origins/postmonth/feb02.html.

CA111: Many current scientists reject evolution.

Many scientists reject evolution and support creationism. (ICR 1980)

1. Of the scientists and engineers in the United States, only about 5 percent are creationists, according to a 1991 Gallup poll (Robinson 1995). However, this number includes those working in fields not related to life origins (such as computer scientists, mechanical engineers, etc.). Taking into account only those working in the relevant fields of earth and life sciences, there are about 480,000 scientists, but only about 700 believe in "creation-science" or consider it a valid theory (Robinson 1995). This means that less than 0.15 percent of relevant scientists believe in creationism. And that is just in the United States, which has more creationists than any other industrialized country. In other countries, the number of relevant scientists who accept creationism drops to less than one tenth of 1 percent.

Additionally, many scientific organizations believe the evidence so strongly that they have issued public statements to that effect [NCSE n.d.]. The National Academy of Sciences, one of the most prestigious science organizations, devotes a Web site to the topic (NAS 1999). A panel of seventy-two Nobel Laureates, seventeen state academies of science, and seven other scientific organizations created an amicus curiae brief, which they submitted to the Supreme Court (Edwards v. Aguillard 1986). This report clarified what makes science different from religion and why creationism is not science. Note that there are no creationist Nobel Laureates.

2. One needs to examine not how many scientists and professors believe something, but what their conviction is based upon. Most of those who reject evolution do so because of personal religious conviction, not because of evidence. The evidence supports evolution. And the evidence, not personal authority, is what objective conclusions should be based on.

3. Often, claims that scientists reject evolution or support creationism are exaggerated or fraudulent. Many scientists doubt some aspects of evolution, especially recent hypotheses about it. All good scientists are skeptical about evolution (and everything else) and open to the possibility, however remote, that serious challenges to it may appear. Creationists frequently seize such expressions of healthy skepticism to imply that evolution is highly questionable. They fail to understand that the fact that evolution has withstood many years of such questioning really means it is about as certain as facts can get.

Further Reading: NAS. 1999. Science and creationism. http://www.nap.edu/html/creationism/; NCSE. 2003. Project Steve. http://www.ncseweb.org/article.asp?category=18; Schafersman, S. 2003. Texas Citizens for Science responds to latest Discovery Institute challenge. http://www.txscience.org/files/discovery-signers.htm.

with Derek Mathias

CA112: Many scientists find problems with evolution.

Many mainstream scientists point out serious problems with evolution, including problems with some of its most important points. (Discovery Institute 2001)

1. There are no known serious problems with the theory of evolution. Claims that there are fall into two (overlapping) categories:

- Some supposed problems are questions about details about the mechanisms of evolution. There are, and always will be, unanswered details in every field of science, and evolution is no exception. Creationists take controversies about details out of context to falsely imply controversy about evolution as a whole.

- Some supposed problems are misunderstandings, ignorance, or fraudulent claims about what the science says.

Further Reading: NCSE. 2002a, Analysis of the Discovery Institute's bibliography. http://www.ncseweb.org/resources/articles/3878_analysis_of_the_discovery_inst_4_5_2002.asp; Tamzek, N. 2002. Icon of obfuscation. http://www.talkorigins.org/faqs/wells.

CA113: Quote mining
Quotes from many noncreationist authorities show that evolutionists themselves find many various failures of evolution. (Watchtower 1985)

1. Quotes are very easy to misuse to give a false impression of what an author means. Many people develop their ideas over long passages, and no single quote can do justice to their argument. Many people, especially scientists, play devil's advocate with their own ideas, so some of their quotes will say exactly the opposite of the point they are supporting. In other cases, good summary quotes exist, but the quoter is either unable or unwilling to find and use them. It is extremely easy to find out-of-context quotes that do damage to a person's main ideas, even unintentionally. Quotes should probably be regarded with more skepticism than any other references.

2. Creationists use quotes as appeals to authority. They apparently see the printed word as a weighty authority. In science, though, the ultimate authority is the evidence itself, so that is what writers refer to. Quotes cannot substitute for evidence.

Appealing to authority is a misuse of quotations. The vast majority of good writing, when it refers to other people's work, summarizes the work and gives a reference to the original. In professional science writing, references are ubiquitous, but direct quotes are very rare.

3. Summarizing someone's work, rather than quoting it, shows understanding. Many creationists are limited to quoting because they have no idea what the author really means. In fact, most creationists probably repeat quotes without even having read the original author's work. For example, Darwin's quote about the eye (see CA113.1) would never be repeated in its usual abbreviated form by an honest person who has read the pages that follow it. If a person cannot understand a work well enough to summarize it, he or she should not be talking about it at all.

4. Even an accurate and in-context quote can be used to mislead. Many quotes are out of date, for example, and talk about our ignorance in areas of which we are no longer ignorant. Other quotes are from creationists, but they appear in a context that groups them with mainstream scientists.

Further Reading: Foley, J. 2002. Creationist arguments: Misquotes. http://www.talkorigins.org/faqs/homs/misquotes.html; Holloway, R. n.d. Evolution of a creationist quote. http://www.ntanet.net/quote.html; Hopkins, M. 2002. Quotations and misquotations. http://www.talkorigins.org/faqs/quotes; Ho-Stuart, C. 2003. Muller and mutations. http://www.talkorigins.org/faqs/quotes/muller.html; Lindsay, D. 2004. Famous quotes found in books. http://www.don-lindsay-archive.org/creation/quotes.html; Pieret, J. (ed.). 2003. The quote mine project. http://www.talkorigins.org/faqs/quotes/mine/project.html.

CA113.1: Darwin on evolution of the eye.

Charles Darwin acknowledged the inadequacy of evolution when he wrote, "To suppose that the eye, with all its inimitable contrivances for adjusting the focus to different distances, for admitting different amounts of light, and for the correction of spherical and chromatic aberration, could have been formed by natural selection, seems, I freely confess, absurd in the highest possible degree." (Huse 1983, 73)

1. The quote is taken out of its context. Darwin answered the seeming problem he introduced. The paragraph continues,

> Yet reason tells me, that if numerous gradations from a perfect and complex eye to one very imperfect and simple, each grade being useful to its possessor, can be shown to exist; if further, the eye does vary ever so slightly, and the variations be inherited, which is certainly the case; and if any variation or modification in the organ be ever useful to an animal under changing conditions of life, then the difficulty of believing that a perfect and complex eye could be formed by natural selection, though insuperable by our imagination, can hardly be considered real. How a nerve comes to be sensitive to light, hardly concerns us more than how life itself first originated; but I may remark that several facts make me suspect that any sensitive nerve may be rendered sensitive to light, and likewise to those coarser vibrations of the air which produce sound. (Darwin 1872, 143–144)

Darwin continues with three more pages describing a sequence of plausible intermediate stages between eyelessness and human eyes, giving examples from existing organisms to show that the intermediates are viable.

Further Reading: Babinski, E. n.d. An old, out of context quotation. http://www.talkorigins.org/faqs/ce/3/part8.html.

CA114: Many famous scientists were creationists.

There have been many famous scientists who believed in special creation in the past. In particular, the following scientists were creationists:

- Louis Agassiz (1807–1873; glacial geology)
- Charles Babbage (1792–1871; computer science)
- Robert Boyle (1627–1691; gas dynamics)
- David Brewster (1781–1868; optical mineralogy)
- Georges Cuvier (1769–1832; comparative anatomy)
- Leonardo da Vinci (1452–1519; hydraulics)
- Humphrey Davy (1778–1829; thermokinetics)
- Henri Fabre (1823–1915; entomology of living insects)

- Michael Faraday (1791–1867; electromagnetics)
- John Ambrose Fleming (1849–1945; electronics)
- William Herschel (1738–1822; galactic astronomy)
- James Joule (1818–1889; reversible thermodynamics)
- Lord Kelvin (1824–1907; energetics)
- Johann Kepler (1571–1630; celestial mechanics)
- Carolus Linnaeus (1707–1778; systematic biology)
- Joseph Lister (1827–1912; antiseptic surgery)
- Matthew Maury (1806–1873; oceanography)
- James Clerk Maxwell (1831–1879; electrodynamics)
- Gregor Mendel (1822–1884; genetics)
- Isaac Newton (1642–1727; calculus)
- Blaise Pascal (1623–1662; hydrostatics)
- Louis Pasteur (1822–1895; bacteriology)
- William Ramsay (1852–1916; isotopic chemistry)
- John Ray (1627–1705; natural history)
- Lord Rayleigh (1842–1919; dimensional analysis)
- Bernhard Riemann (1826–1866; non-Euclidean geometry)
- James Simpson (1811–1870; gynecology)
- Nicholas Steno (1631–1686; stratigraphy)
- George Stokes (1819–1903; fluid mechanics)
- Rudolph Virchow (1821–1902; pathology)
- John Woodward (1665–1728; paleontology)

Agassiz, Pasteur, Lord Kelvin, Maxwell, Dawson, Virchow, Fabre, and Fleming were strong opponents of evolution. (H. M. Morris 1982)

1. The validity of evolution rests on what the evidence says, not on what people say. There is overwhelming evidence in support of evolution and no valid arguments against it.

2. Many of the scientists in the above list lived before the theory of evolution was even proposed. Others knew the theory, but were not familiar with all the evidence for it. Evolution is outside the field of most of those scientists.

A couple hundred years ago, before the theory of evolution was developed and evidence for it was presented, virtually all scientists were creationists, including scientists in relevant fields such as biology and geology. Today, virtually all relevant scientists accept evolution. Such a turnabout could only be caused by overwhelming evidence. The alternative—that almost all scientists today are thoroughly incompetent—is preposterous.

3. Even if they did not believe in evolution, all these scientists were firmly committed to the scientific method, including methodological naturalism. They actually serve as counterexamples to the common creationist claim that a naturalistic practice of science is atheistic.

4. Evolution is entirely consistent with a belief in God (see CA602), including even "special creation." Special creation need not refer to the creation of every animal; it can refer simply to creation of the universe, of the first life, or of the human soul, for example. Many of the above scientists were not creationists in the sense that Henry Morris uses the term. For example, Lord Kelvin believed in evolution, but with intelligent guidance (Thomson 1871); Pasteur accepted some kind of evolution (Cuny 1965, 122); Linnaeus accepted speciation (Linne 1760).

CA118: Your arguments do not count because you are not qualified.

Many arguments may be discounted because they were put together by amateurs who are not scientifically qualified. ("Socrates" 2003)

1. A person's qualifications, although important, are not the only thing to consider. The ultimate authority for arguments about the world is the world itself. If the argument is logical and is based on reliable real-world data (for example, if it contains verifiable data or has reliable references), then the argument has authority regardless of who is giving it.

2. Qualifications consist of a lot more than letters after one's name. Perhaps the most important quality is how the person is regarded by others in the field. The soundness of the person's past work is another important consideration.

3. One must also consider the qualifications of others who approve or disapprove of the argument. When an argument withstands peer review, the authority of those who review it adds to the authority of the original author. Withstanding further exposure adds even more to the argument's reliability.

4. This argument about qualifications, if applied uniformly, would sink creationism in a second. For every creationist who claims one thing, there are dozens of scientists (probably more), all with far greater professional qualifications, who say the opposite.

Further Reading: Isaak, M. 2002b. Is that so? The art of evaluating information. http://home.earthlink.net~misaak/claims.html.

CA120: An evolved mind is fallible, its conclusions untrustworthy.

If our minds arose from lesser animals via natural processes, then our minds may be fallible. Then the conclusions that we come up with are subject to doubt, including the conclusion of evolution itself.

Darwin (1881) wrote in a letter, "With me the horrid doubt always arises whether the convictions of man's mind, which has been developed from the mind of the lower animals, are of any value or are at all trustworthy." (Plantinga 1991a)

1. It is well established that the mind is fallible. Ordinary memory and reasoning are mistaken surprisingly often (Gilovich 1991; Schacter 2001). Pathologies add further complications (Sacks 1970). This fallibility exists whatever the source of our minds may be.

Doubt exists in all areas of life. Nothing can be proven absolutely (see CA202). However, many things are certain enough that we call them facts and do not worry about the possibility that they are wrong until we see actual evidence that they are wrong. Without such an attitude, we would never be able to get on with our lives.

2. The fallibility of our minds argues more against creationism. Nobody can be certain of it either, and minds as imperfect as ours argue against their being divinely created.

3. Darwin only applied this argument to questions beyond the scope of science. He thought science was well within the scope of a modified monkey brain.

Further Reading: Gilovich, T. 1991. *How We Know What Isn't So*; Ruse, M. 2001. *Can a Darwinian Be a Christian?* 106–110; Sacks, O. 1970. *The Man Who Mistook His Wife for a Hat*; Schacter, D. L. 2001. *The Seven Sins of Memory.*

Theory of Science

CA201: Evolution is only a theory.

Evolution is only a theory. It is not a fact. (State of Oklahoma 2003)

1. The word theory, in the context of science, does not imply uncertainty. It means "a coherent group of general propositions used as principles of explanation for a class of phenomena" (Barnhart 1948). In the case of the theory of evolution, the following are some of the phenomena involved. All are facts:

- Life appeared on earth more than two billion years ago;
- Life forms have changed and diversified over life's history;
- Species are related via common descent from one or a few common ancestors;
- Natural selection is a significant factor affecting how species change.

Many other facts are explained by the theory of evolution as well.

2. The theory of evolution has proved itself in practice. It has useful applications in epidemiology, pest control, drug discovery, and other areas (Bull and Wichman 2001; Eisen and Wu 2002; Searls 2003).

3. Besides the theory, there is the fact of evolution, the observation that life has changed greatly over time. The fact of evolution was recognized even before Darwin's theory. The theory of evolution explains the fact.

4. If "only a theory" were a real objection, creationists would also be issuing disclaimers complaining about the theory of gravity, atomic theory, the germ theory of disease, and the theory of limits (on which calculus is based). The theory of evolution is no less valid than any of these. Even the theory of gravity still receives serious challenges (Milgrom 2002). Yet the phenomenon of gravity, like evolution, is still a fact.

5. Creationism is neither theory nor fact; it is, at best, only an opinion. Since it explains nothing, it is useless.

Further Reading: AIG. n.d.a Arguments we think creationists should NOT use. http://www.answersingenesis.org/home/area/faq/dont_use.asp; Gould, S. J. 1983. Evolution as fact and theory. In *Hen's Teeth and Horse's Toes*, pp. 253–262. http://www.stephenjaygould.org/library/gould_fact-and-theory.html. Isaak, M. 1995. Five major misconceptions about evolution. http://www.talkorigins .org/faqs/faq-misconceptions.html; Moran, L. 1993. Evolution is a fact and a theory. http://www .talkorigins.org/faqs/evolution-fact.html.

CA202: Evolution has not been proved.

Evolution has not been, and cannot be, proved. We cannot even see evolution (beyond trivially small change), much less test it experimentally. (H. M. Morris 1985, 4–6)

1. Nothing in the real world can be proved with absolute certainty. However, high degrees of certainty can be reached. In the case of evolution, we have huge amounts of data from diverse fields. Extensive evidence exists in all of the following different forms (Theobald 2004). Each new piece of evidence tests the rest.

- All life shows a fundamental unity in the mechanisms of replication, heritability, catalysis, and metabolism.

- Common descent predicts a nested hierarchy pattern, or groups within groups. We see just such an arrangement in a unique, consistent, well-defined hierarchy, the so-called tree of life.

- Different lines of evidence give the same arrangement of the tree of life. We get essentially the same results whether we look at morphological, biochemical, or genetic traits.

- Fossil animals fit in the same tree of life. We find several cases of transitional forms in the fossil record (see CC200).

- The fossils appear in a chronological order, showing change consistent with common descent over hundreds of millions of years and inconsistent with sudden creation.

- Many organisms show rudimentary, vestigial characters, such as sightless eyes or wings useless for flight.

- Atavisms sometimes occur. An atavism is the reappearance of a character present in a distant ancestor but lost in the organism's immediate ancestors. We only see atavisms consistent with organisms' evolutionary histories.

- Ontogeny (embryology and developmental biology) gives information about the historical pathway of an organism's evolution. For example, as embryos whales and many snakes develop hind limbs that are reabsorbed before birth.

- The distribution of species is consistent with their evolutionary history. For example, marsupials are mostly limited to Australia, and the exceptions are explained by continental drift. Remote islands often have species groups that are highly diverse in habits and general appearance but closely related genetically. This consistency still holds when the distribution of fossil species is included.

- Evolution predicts that new structures are adapted from other structures that already exist, and thus similarity in structures should reflect evolutionary history rather than function. We see this frequently. For example, human hands, bat wings, horse legs, whale flippers, and mole forelimbs all have similar bone structure despite their different functions.

- The same principle applies on a molecular level. Humans share a large percentage of their genes, probably more than 70 percent, with a fruit fly or a nematode worm.

- When two organisms evolve the same function independently, different structures are often recruited. For example, wings of birds, bats, pterosaurs, and insects all have different structures. Gliding has been implemented in many additional ways (see CB921.2). Again, this applies on a molecular level, too.

- The constraints of evolutionary history sometimes lead to suboptimal structures and functions. For example, the human throat and respiratory system make it impossible to breathe and swallow at the same time and make us susceptible to choking.

- Suboptimality appears also on the molecular level. For example, much DNA is nonfunctional.

- Some nonfunctional DNA, such as certain transposons, pseudogenes, and endogenous viruses, show a pattern of inheritance indicating common ancestry.

- Speciation has been observed (see CB910).

- The day-to-day aspects of evolution—namely heritable genetic change, morphological variation and change, functional change, and natural selection—are seen to occur at rates consistent with common descent.

Furthermore, the different lines of evidence are consistent; they all point to the same big picture. For example, evidence from gene duplications in the yeast genome shows that its ability to ferment glucose evolved about eighty million years ago. Fossil evidence shows that fermentable fruits became prominent about the same time. Genetic evidence for major change around that time also is found in fruiting plants and fruit flies (Benner et al. 2002).

The evidence is extensive and consistent, and it points unambiguously to evolution, including common descent, change over time, and adaptation influenced by natural selection. It would be preposterous to refer to these as anything other than facts.

Further Reading: Colby, C. 1993. Evidence for evolution: An eclectic survey. http://www .talkorigins.org/faqs/evolution-research.html; Moran, L. 1993. Evolution is a fact and a theory. http://www.talkorigins.org/faqs/evolution-fact.html; Theobald, D. 2004. 29+ evidences for macroevolution: The scientific case for common descent. http://www.talkorigins.org/faqs/comdesc.

CA210: Evolution does not make predictions.
A true science must make predictions. Evolution only describes what happened in the past, so it is not predictive. (Batten 2002)

1. The difference in predictive power between evolution and other sciences is one of degree, not kind. All theories are simplifications; they purposely neglect as many outside variables as they can. But these extraneous variables do affect predictions. For example, you can predict the future position of an orbiting planet, but your prediction will be off very slightly because you cannot consider the effects of all the small bodies in the solar system. Evolution is more sensitive to initial conditions and extraneous factors, so specific predictions about what mutations will occur and what traits will survive are impractical. It is still possible to use evolution to make general predictions about the future, though. For example, we can predict that diseases will become resistant to any new widely used antibiotics.

2. The predictive power of science comes from being able to say things we would not have been able to say otherwise. These predictions do not have to be about things hap-

pening in the future. They can be "retrodictions" about things from the past that we have not found yet. Evolution allows innumerable predictions of this sort.

3. Evolution has been the basis of many predictions. For example:

- Darwin predicted, based on homologies with African apes, that human ancestors arose in Africa. That prediction has been supported by fossil and genetic evidence (Ingman et al. 2000).

- Theory predicted that organisms in heterogeneous and rapidly changing environments should have higher mutation rates. This has been found in the case of bacteria infecting the lungs of chronic cystic fibrosis patients (Oliver et al. 2000).

- Predator–prey dynamics are altered in predictable ways by evolution of the prey (Yoshida et al. 2003).

- Ernst Mayr predicted in 1954 that speciation should be accompanied with faster genetic evolution. A phylogenetic analysis has supported this prediction (Webster et al. 2003).

- Several authors predicted characteristics of the ancestor of craniates. On the basis of a detailed study, they found that the fossil *Haikouella* "fit these predictions closely" (Mallatt and Chen 2003).

With predictions such as these and others, evolution can be, and has been, put to practical use in areas such as drug discovery and avoidance of resistant pests (Bull and Wichman 2001).

4. If evolution's low power to make future predictions keeps it from being a science, then some other fields of study cease to be sciences, too, especially archeology and astronomy.

Further Reading: Rainey, P. 2003. Evolution: Five big questions; Wilkins, J. 1997d. Evolution and philosophy: Predictions and explanations. http://www.talkorigins.org/faqs/evolphil/predict. html.

CA211: Evolution cannot be falsified.

Any fact can be fit into the theory of evolution. Therefore, evolution is not falsifiable and is not a proper scientific theory. (H.M. Morris 1985, 6–7)

1. There are many conceivable lines of evidence that could falsify evolution. For example:

- a static fossil record;

- true chimeras, that is, organisms that combined parts from several different and diverse lineages (such as mermaids and centaurs);

- a mechanism that would prevent mutations from accumulating;

- observations of organisms being created.

2. This claim, coming from creationists, is absurd, since almost all creationism is nothing more than (unsubstantiated) claims that evolution has been falsified (see CA110).

CA212: Evolution is ambiguously defined.

Evolution is defined ambiguously, and claims that it is fact are based on the ambiguity. It is usually defined as "change in heritable characteristics in a population over time" (often expressed as "change in allele frequencies"), which everyone accepts as fact, but that does not mean that macroevolution or common descent are fact. (POSH n.d.)

1. Language tends to be ambiguous at times (e.g., the entry for the word "set" covers more than twenty-two pages of the original Oxford English Dictionary). The word "evolution" is an unfortunate instance of that ambiguity; it is used for the fact of biological change over time; as shorthand for the theory of evolution, which encompasses a much broader range of observations and ideas; and for change generally, in any realm. The ambiguity can usually be resolved by the context in which the word is used, at least by people who know something about biological evolution.

Mixing contexts is indeed improper, and the fact of allele frequency change, by itself, does not establish the theory of evolution.

2. The soundness of the theory of evolution does not rest on ambiguity. On the contrary, scientific papers are written so other scientists can tell what the authors are talking about; they must be as unambiguous as possible. The evidence is overwhelming: evolution is not only a theory (see CA201); major aspects of it, such as common descent, are also facts.

3. Creationists sometimes misuse the ambiguity to their own advantage, trying, for example, to include cosmological change as part of the theory of evolution (Hovind n.d.b.). This is gross ignorance, deliberate dishonesty, or both.

Further Reading: Gould, S.J. 2002. "What does the dreaded 'E' word mean anyway?" 246; Wilkins, J. 2001. Defining evolution.

CA220: Evolution cannot be replicated.

Science requires experiments that can be replicated. Evolution cannot be replicated, so it is not science. (Morris 1985, 4)

1. Science requires that observations can be replicated. The observations on which evolution is based, including comparative anatomy, genetics, and fossils, are replicable. In many cases, you can repeat the observations yourself.

2. Repeatable experiments, including experiments about mutations and natural selection in the laboratory and in the field, also support evolution.

Further Reading: Weiner, J. 1994. *The Beak of the Finch.*

CA221: Were you there?

(In response to any claim about the history of life) Were you there? (Ham 1989)

1. Yes, because "there" is here. Events in the past leave traces that last into the present, and we can and do look at that evidence today.

2. If this response were a valid challenge to evolution, it would equally invalidate creationism and Christianity, since they are based on events that nobody alive today has witnessed.

3. A more useful and more general question is, "How do you know?" If the person making a claim cannot answer that question, you may consider the claim baseless (tentatively, as someone else may be able to answer). If the answer is subjective—for example, if it rests on the person's religious convictions—you know that the claim does not necessarily apply to anyone but that person. If you cannot understand the answer, you probably have some studying to do. If you get a good answer, you know to take the claim seriously.

CA230: Interpreting evidence is not the same as observation.

Evidence for evolution has not been observed. Claims that it has confuse observation with interpretation. What is observed has to be interpreted to fit the hypotheses. (T. Wallace 2002)

1. All observation requires interpretation. Even something as seemingly simple as seeing an object in front of you requires a great deal of interpretation to determine what it is, what properties it exhibits, how far away it is, and so forth (Sacks 1995). To dismiss absolutely everything we know because it is interpretation would be ludicrous.

2. Most of the evidence of evolution is not the sort about which interpretation is in question. The evidence consists of such things as the following:

- certain trilobite species are found in certain geological formations;
- many more varieties of marsupials are found in Australia than elsewhere;
- bacteria in test tubes have been seen to change in certain ways over time;
- flies share some traits that other insects do not;

and millions of other such facts, none of which are in dispute.

The sort of interpretation to which creationists object is how all the evidence fits together. They do not deny the evidence (not most of it, anyway); they deny that it is evidence *for evolution*.

However, a fact gets to be considered evidence for a theory if it fits that theory and does not fit or is not covered by competing theories. (Ideally, the theory should predict the fact before the fact is known, but that is not essential for the fact to be evidence.) The millions of facts referred to above fit this criterion, so they qualify as evidence for evolution.

3. The interpretation on which creationism depends, in contrast, is based only on highly questionable and subjective ideas that do not fit together into a coherent whole.

Further Reading: Wilkins, J. 1997c. Evolution and philosophy: Is evolution science and what does "science" mean? http://www.talkorigins.org/faqs/evolphil/falsify.html.

CA230.1: Evolutionists interpret evidence on the basis of their preconceptions.

The conclusions of scientists are based on their preconceptions. They prove only what they assume. (Oard 2003)

1. The conclusions of scientists are based on evidence, and the evidence remains for all to see. Scientists know that their ideas must stand the scrutiny of other scientists, who may not share their preconceptions. The best way to do this is to make the case strong enough on the basis of the evidence so that preconceptions do not matter. And scientists themselves condemn preconceptions when they see them. (Stephen J. Gould, the most vocal recent crusader against preconceptions, was vehemently anticreationism.)

The history of science is filled with scientists accepting ideas contrary to their preconceptions. Examples include the reality of extinctions, the reality of meteors, meteors as causes of mass extinctions, ice ages, continental drift, transposons, bacteria as the cause of ulcers, the nature of prions, and, of course, evolution itself. Scientists are not immune to being sidetracked by their preconceptions, but they ultimately go where the evidence leads.

Scientists make deliberate efforts to remove subjective influences from their evaluation of conclusions; they do a good job, on the whole, of reducing bias. They do such a good job, in fact, that what creationists really object to is the fact that scientists do not interpret evidence according to certain religious preconceptions.

2. The hypocrisy of this charge cannot be overstressed. Creationists state outright that they accept only what they already assume. Consider part of Answers in Genesis's statement of faith: "By definition, no apparent, perceived or claimed evidence in any field, including history and chronology, can be valid if it contradicts the Scriptural record" (AIG n.d.c.). The Institute for Creation Research has a similar statement of faith (ICR 2000). Creationists admit up front that their preconceptions, in the form of religious convictions, determine their conclusions.

CA240: Ockham's Razor says the simplest explanation (creation) is preferred.
Ockham's Razor says the simplest explanation should be preferred. That explanation is creation. (J. D. Morris 1999a)

1. Ockham's Razor does not say that the simplest explanation should be favored. It says that entities should not be multiplied beyond necessity (*non sunt multiplicanda entia praeter necessitatem*). In other words, new principles should not be invoked if existing principles already provide an explanation. If, however, the simpler explanation does not cover all the details, then additional "entities" are necessary.

2. Creationism is not an explanation. An explanation tells why something is one way instead of an alternative way. But creationism does not rule out any alternatives, since a creator God could have done anything. Because of this, creationism adds nothing to any argument. Thus, creationism is an unnecessary entity and, by Ockham's Razor, should be eliminated.

CA250: Scientific findings are always changing.
Scientific theories are always changing. You cannot trust what scientists say, since it may be different tomorrow. (Matthews 2003)

1. Science investigates difficult questions about unknown fields, and scientists are human, so it is inevitable that scientific findings will not be perfect. However, science works by investigating more and more, which means results get checked and rechecked with further findings. The reason some findings change is because they get corrected. This process of correction helps make science one of the most successful areas of human endeavor. The people who cannot be trusted are those who are always right.

2. As more evidence accumulates, scientific findings become more and more certain. Theories that have withstood several decades of study may undergo more refinement of details, but it is almost inconceivable that they would be overturned completely.

Scientific Method

CA301: Science is naturalistic.

Science is based on naturalism, the unproven assumption that nature is all there is. (Dembski 1996; P. E. Johnson 1990)

1. The naturalism that science adopts is methodological naturalism. It does not assume that nature is all there is; it merely notes that nature is the only objective standard we have. The supernatural is not ruled out a priori; when it claims observable results that can be studied scientifically, the supernatural is studied scientifically (e.g., Astin et al. 2000; Enright 1999). It gets little attention because it has never been reliably observed. Still, there are many scientists who use naturalism but who believe in more than nature.

2. The very same form of naturalism is used by everyone, including creationists, in their day-to-day lives. People literally could not survive without making naturalistic assumptions. Creationism itself is based on the naturalistic assumption that the Bible has not changed since the last time it was read.

3. Naturalism works. By assuming methodological naturalism, we have made tremendous advances in industry, medicine, agriculture, and many other fields. Supernaturalism has never led anywhere. Newton, for example, wrote far more on theology than he did on physics, but his theological work is largely forgotten because there has been no reason to remember it other than for historical curiosity.

4. Supernaturalism is contentious. Scientific findings are based on hard evidence, and scientists can point at the evidence to resolve disputes. People tend to have different and incompatible ideas of what form supernatural influences take, and all too often the only effective way they have found for reaching a consensus is by killing each other.

Further Reading: Isaak, M. 2002c. A philosophical premise of "naturalism"? http://www.talkdesign.org/faqs/naturalism.html.

CA301.1: Naturalistic science will miss a supernatural explanation.

If the correct explanation for a phenomenon happens to be supernatural, the naturalistic method of science will miss it. "With creationist explanations disqualified at the outset, it follows that the evidence will always support the naturalistic alternative." (P. E. Johnson 1990)

1. Nobody has ever come up with a useful definition of supernatural. By most definitions, something having an effect on nature makes that something a part of nature itself. So any explanation for something we see in nature can be considered natural by definition.

2. We cannot observe the supernatural, so the only way we could reach the supernatural explanation would be to eliminate all natural explanations. But we can never know that we have eliminated all possibilities. Even if a supernatural explanation is correct, we can never reach it.

3. Suppose we do conclude that a supernatural explanation is correct. It is impossible, even in principle, to distinguish one supernatural explanation from another. Many people, including many scientists, are willing to accept certain supernatural explanations on faith. There is nothing wrong with that as long as they do not claim special privilege for their faith. Some people, however, are not satisfied unless others believe as they do; this group includes all those who want to make the supernatural a part of science. Since they cannot make their case by using naturalistic evidence, they must resort to other means, such as force of arms. (This is not hyperbole. Such groups continually attempt to get political enforcement on their side.)

4. If we do miss a supernatural explanation, so what? Supernatural explanations cannot be generalized, so the explanation does not matter anywhere else. The usefulness of science comes from the ability to apply findings to different areas. Any supernatural explanation would be useless.

5. Creationist accounts of origins are not disqualified. People are free to believe whatever religion they choose. P. E. Johnson and others like him merely object to their religion not being taught as science to the exclusion of all other religious interpretations (not to mention to the exclusion of all of science).

CA310: Scientists find what they expect to find.

Scientists find what they expect to find.

1. Scientific results are tested. This has two very important consequences: First, the scientists know that their results will be subject to challenge, so they work harder to make sure the evidence really does support their results. Second, published ideas that the evidence does not support will get rejected, especially in times or places with different cultural biases.

2. Scientists more than most people are trained to be objective. Although expectations can affect their conclusions, they would not affect them to a large degree. Most certainly, they would not blind all biologists and geologists to all the evidence, as would be necessary if creationism were true.

3. At the start of the nineteenth century, scientists expected to find evidence for creation and a global flood. Instead, they found evidence for evolution, which is why evolution was the accepted theory by the end of the century.

4. Creationists find what they want to find. Since their entire worldview is threatened by finding disconfirming evidence, they are very highly motivated not to see it. Scientists, on the other hand, usually welcome disconfirming evidence when it comes along.

CA320: Scientists are pressured not to challenge established dogma.

Scientists are pressured not to challenge the established dogma. (Watchtower 1985, 182)

1. The pressures that science imposes do not weaken the validity of evolution—quite the contrary. Scientists are rewarded more for finding new things, not for supporting established principles. Thus, they tend to look more for novelties and for results that would overturn common beliefs. If a scientist found evidence that falsified evolution, he or she would be guaranteed world prestige and fame.

2. Creationists are under far more pressure than scientists. Since their entire worldview is threatened by finding disconfirming evidence, they are very highly motivated not to admit it. Many creationists have taken oaths saying that no evidence could change their dogma (AIG n.d.c). At least one admits that he became a scientist not to find the truth, but to destroy Darwinism (Wells n.d.). The commitment to established dogma is pretty well monopolized by creationists.

CA321: Scientists are motivated to support naturalism and reject creationism.

Scientists are motivated to support naturalism and reject nonnaturalistic ideas, such as creationism.

1. This claim is easily disproved by the fact that many scientists are strongly religious, having adopted nonnaturalistic ideas in their private lives.

2. Although motives in any large group are going to differ from person to person, the most common motive that makes people become scientists is curiosity. It has nothing to do with supporting naturalism.

3. Within the practice of science, there is not anything suggesting naturalism as a goal. The main motives are curiosity, professional pride, and material rewards. Pride enters because scientists must make their work available for all to see, so they want it to look good, and in particular they are motivated to do work that can withstand challenges. Material awards come mainly in the form of applying for funding, which means satisfying the funding agencies, which usually means the research must have some promise for practical value.

4. Although naturalism is not a motive for most scientists, its rejection is an explicit motive for most science pursued by antievolutionists. For example, the faculty and stu-

dents of the Institute for Creation Research Graduate School subscribe to a statement of faith in biblical inerrancy and antievolution (ICR 2000). Jonathan Wells pursued a biology degree in order to discredit evolution (Wells n.d.). He did so at the urging of Reverend Moon, whom Wells sees as the second coming of Christ (Wells 1991). William Dembski also sees religious motivation as paramount (Dembski and Richards 2001). The "overthrow of materialism" is the motivating basis for the Wedge Strategy, which is the operating principle for the intelligent design movement (CRSC 1998).

Perhaps when creationists claim that scientists are operating under ulterior motives, they are merely projecting how they themselves operate.

Further Reading: Isaak, M. 2002c. A philosophical premise of "naturalism"? http://www .talkdesign.org/faqs/naturalism.html.

CA325: Creationists are prevented from publishing in science journals.

Creationists cannot get their views accepted by mainstream science because they are prevented from publishing in mainstream scientific journals. (H. M. Morris 1998a)

1. The priorities of creationism are politics and religious evangelism. Science is not very important to creationists in the first place. The main reason that they do not get published in reputable science journals is that they do not try to publish there. In a survey of editors of sixty-eight journals, only eighteen out of an estimated 135,000 submissions were found that could be described as advocating creationism (Scott and Cole 1985).

In the *McLean v. Arkansas Board of Education* creationism trial, the creationists complained to the judge that the scientific journals refused to consider their articles, but they were unable to produce any articles that had been refused publication.

2. Creationists are free to publish in other venues, such as books and their own journals. These venues are as reputable as their authors and editors. Note that Darwin's major works were published in books.

3. Creationists do get published in reputable peer-reviewed science journals when they do real science. For example:

- Steven A. Austin, Gordon W. Franz, and Eric G. Frost, "Amos's Earthquake: An Extraordinary Middle East Seismic Event of 750 b.c." (*International Geology Review* 42: 657, 2000);

- Leonard Brand on the Flood deposition interpretation of Coconino Sandstone (*Palaeogeography, Palaeoclimatology, Palaeoecology* 28: 25–38, 1979; *Geology* 19: 1201–1204, 1991; *Journal of Paleontology* 70: 1004–1011, 1996);

- Harold G. Coffin on deposition environments of fossil trees (*Journal of Paleontology* 50: 539–543, 1976; *Geology* 11: 298–299, 1983);

- Robert Gentry on polonium haloes (*American Journal of Physics, Proceedings* 33: 878A, 1965; *Science* 184: 62–64, 1974; *Science* 194: 315–318, 1976);

- Grant Lambert on DNA error rates (*Journal of Theoretical Biology* 107: 387–403, 1984);

- Jan Peckzis on mass estimates of dinosaurs (*Journal of Theoretical Biology* 132: 509–510, 1988; *Journal of Paleontology* 63: 947–950, 1989; *Journal of Vertebrate Paleontology* 14: 520–533, 1995);

- Sigfried Scherer on ducks as a single kind (*Journal für Ornithologie* 123: 357–380, 1982; *Zeitschrift für zoologische Systematik und Evolutionsforschung* 24: 1–19, 1986).

In addition, many creationists have published science articles not related to creationism.

4. Scientists themselves are prevented from publishing in peer-reviewed journals when their science is not up to par. The peer-review process prevents lots of substandard work from being published, even from noncreationists such as myself. (The process, of course, is imperfect and produces a substantial borderline area, so some fairly good articles get rejected and some fairly poor ones get accepted. On the whole, however, it keeps quality up.) Creationists face no obstacles that mainstream scientists do not face themselves.

5. Creationists prevent others from publishing critical views in creationist journals. Glenn Morton, for example, has had papers rejected by the *Creation Research Society Quarterly* for violating their view that the Flood must be global and for criticizing Carl Froede's poor geology (Morton 1998c).

Further Reading: Flank, L. 1995a. Does science discriminate against creationists? http://www.geocities.com/CapeCanaveral/Hangar/2437/discrim.htm.

CA340: Evolutionists do not accept debate challenges.
Evolutionists are unwilling to debate creationists. (W. Brown 1995, 212)

1. The proper venue for debating scientific issues is at science conferences and in peer-reviewed scientific journals. In such a venue, the claims can be checked by anyone at their leisure. Creationists, with very rare exceptions, are unwilling to debate there.

2. Public debates are usually set up so that the winners are determined by public speaking ability, not by quality of material.

3. Debate formats, both spoken and written, usually do not allow space for sufficient examination of points. A common tactic used by some prominent creationists is to rattle off dozens of bits of misinformation in rapid succession. It is impossible for the responder to address each in the time or space allotted.

4. Notwithstanding the above points, there have been several debates, both live and online.

Further Reading: Talk.Origins Archive. n.d.a. Debates and gatherings. http://www.talkorigins.org/origins/faqs-debates.html.

CA341: Evolutionists have not met Hovind's challenge to prove evolution.
Evolutionists have been unable to claim $250,000 offered by Kent Hovind for proof of evolution. (Hovind n.d.b.)

1. The challenge is set up so that it is impossible to meet whether evolution is true or not. First, Hovind conflates many areas of science, including cosmology and abiogenesis, under his misuse of the word evolution. Second, he wants proof that the universe came from nothing, which is not known to be true (and which is not relevant to evolution). Third and most important, Hovind requires proof that "evolution . . . is the only possible way the observed phenomena could have come into existence" (Hovind n.d.b.). It is impossible to prove a universal negative. In fact, scientists already seriously consider alternatives for abiogenesis (namely panspermia).

2. The judging is likely to be unfair. The judges are all picked by Hovind, so they are probably biased, and Hovind has refused to let unbiased judges judge a challenge (Kolosick n.d.). Hovind's hand-picked judges may well be unqualified, too, since Hovind does not have the background to judge qualifications. There is even evidence that the judges do not exist: An advertisement in Pensacola headed "Attn: Hovind's Expert Committee" received no responses (Vlaardingerbroek n.d.).

Hovind himself says he will not accept important evidence. He will not accept macroevolution in the form of speciation as evidence for evolution.

3. Several people have tried to collect on his challenge, only to get a runaround or to be ignored:

- Lenny Flank received only nonanswers when he asked Hovind to clarify what "fundamentally different kind of animal" means (Flank n.d.).

- Kevin R. Henke called to inquire about the terms that would be necessary to win the award. Hovind told him that the award could be collected by recreating the Big Bang. One of Hovind's staff members agreed that the conditions were technically unfeasible and financially impossible. Hovind was willing to offer $2,000 for proof that a dog and a banana have a common ancestor, but he backed out of this when it was required that the judges be unbiased (Kolosick n.d.).

- Dr. Barend Vlaardingerbroek corresponded with Hovind concerning clarification of conditions and matters of fairness and got a runaround. He learned, however, that Hovind reserves the right to throw out any evidence he does not like before the judges see it (Vlaardingerbroke n.d.).

- Thomas, trying to meet the challenge, sent Hovind a list of evidences. He heard nothing back ("Thomas" n.d.).

- Ian Wood sent Hovind some evidence for evolution and found that Hovind lied about submitting evidence to a panel of judges (I. Wood n.d.).

In short, the challenge is a fraud.

Further Reading: Pieret, J. 2002. Kent Hovind's $250,000 offer. http://talkorigins.org/faqs/hovind.html.

CA342: Evolutionists do not accept Walt Brown's debate challenge.

A standing offer exists for a written debate between Walt Brown and an evolutionist, the debate to be published later. In over fifteen years, no evolutionist has accepted this offer. (W. Brown 1995, 212)

uced no publications in scientificuced no publications in scientificuced no publications in scientificuced no publications in scientific

1. Several aspects of Brown's twenty-two requirements may make his debate unattractive. The editor for the debate (who must be qualified, willing, and impartial) would be very difficult to find. Brown requires that his challenger must have a PhD, for no good reason. He stipulates that he will not allow theology into the debate, despite the fact that creationism is nothing but theology and that Brown himself uses theology as the basis for his conclusions.

2. People have attempted to debate Walt Brown, but Brown refuses. Joe Meert signed Brown's contract in 1996. He proposed (in accordance with the contract terms) that evidence regarding a global flood be a topic for discussion within the debate. Brown has steadfastly refused to debate Meert (Meert 2003b).

3. Brown himself refuses to debate in the proper venue for deciding scientific questions: scientific conferences and peer-reviewed literature (see CA340).

CA350: No gradual biochemical evolution models have been published.
Professional literature is silent on the subject of the evolution of biochemical systems. (Behe 1996a, 68, 72, 97, 114, 115–116, 138, 185–186)

1. The claim is simply false. Dozens of articles exist on the subjects for which Behe claims the literature is missing. David Ussery, for example, found 107 articles on cilia evolution, 125 on flagella evolution, 27 on the evolution of the entire coagulation system, 130 on the evolution of vesicle transport, and 84 on "molecular evolution of the immune system" (Ussery 1999).

2. Behe tries to make his claim appear more dramatic by overstating our understanding of the molecular workings of the cell. For example, he says, "Over the past four decades modern biochemistry has uncovered the secrets of the cell" (1996a, 232). But our understanding has only just begun. In the years since Behe wrote his book, journals have been filled with thousands of research articles uncovering new information, and much remains to be uncovered. When the complete *Escherichia coli* genome was sequenced in 1998, the functions of a third of its genes were still completely unknown, and *E. coli* is much simpler than human cells.

3. Behe's work on intelligent design theory has produced no publications in scientific literature. In fact, there have been no scientific publications on intelligent design by any of its proponents (Gilchrist 1997).

See for Yourself
You can do a search of biological and medical research yourself at PubMed (http://www.ncbi.nlm.nih.gov/PubMed). Try keywords such as flagella and evolution.

Further Reading: Catalano, J. (ed.). 1998. Publish or perish. http://www.talkorigins .org/faqs/behe/publish.html; Cavalier-Smith, T. 1997. The blind biochemist; Li, W.-H. 1997. *Molecular Evolution.*

CA500: "Survival of the fittest" is a tautology.

Natural selection, or "survival of the fittest," is tautologous (i.e., uses circular reasoning) because it says that the fittest individuals leave the most offspring, but it defines the fittest individuals as those that leave the most offspring. (Gish et al. 1981; H. M. Morris 1985, viii)

1. "Survival of the fittest" is a poor way to think about evolution. Darwin himself did not use the phrase in the first edition of *Origin of Species*. What Darwin said is that heritable variations lead to differential reproductive success. This is not circular or tautologous. It is a prediction that can be, and has been, experimentally verified. (Weiner 1994)

2. The phrase cannot be a tautology if it is not trivially true. Yet there have been theories proposing that the fittest individuals perish:

- Alpheus Hyatt proposed that lineages, like individuals, inevitably go through stages of youth, maturity, old age, and death. Toward the end of this cycle, the fittest individuals are more likely to perish than others (Hyatt 1866; Lefalophodon n.d.).

- The theory of orthogenesis says that certain trends, once started, keep progressing even though they become detrimental and lead to extinction. For example, it was held that Irish elks, which had enormous antlers, died out because the size increase became too much to support.

- The "fittest" individuals could be considered those that are ideally suited to a particular environment. Such ideal adaptation, however, comes at the cost of being more poorly adapted to other environments. If the environment changes, the fittest individuals from it will no longer be well adapted to any environment, and the less fit but more widely adapted organisms will survive.

3. The fittest, to Darwin, were not those that survived but those that could be expected to survive on the basis of their traits. For example, wild dogs selectively prey on impalas which are weaker according to bone marrow index (Pole et al. 2003). With that definition, survival of the fittest is not a tautology. Similarly, survival can be defined not in terms of the individual's life span but in terms of leaving a relatively large contribution to the next generation. Defined thus, survival of the fittest becomes more or less what Darwin said, and is not a tautology.

Further Reading: Lindsay, D. 1997b. Is "survival of the fittest" a tautology? http://www.don-lindsay-archive.org/creation/tautology.html; Wilkins, J. 1997b. Evolution and philosophy: A good tautology is hard to find. http://www.talkorigins.org/faqs/evolphil/tautology.html.

CA510: Creationism and evolution are the only two models.

Creation and evolution are the only two models of origins. (H. M. Morris 1985, 3, 8–10)

1. There are many mutually exclusive models of creation. Biblical creationism alone includes geocentrism, young-earth creationism, day–age creationism, progressive creationism, intelligent design creationism, and more. And then there are hundreds of very different varieties of creation from other religions and cultures. Some of the harshest criticism of creation models comes from creationists who believe other models.

2. Many noncreationist alternatives to Darwinian evolution, or significant parts of it, are possible and have received serious attention in the past. These include, among others,

- orthogenesis
- neo-Lamarckianism
- process structuralism
- saltationism

(See J. Wilkins 1998 for elaboration.)

3. Creation and evolution are not mutually exclusive. They coexist in models such as theistic evolution.

Further Reading: Isaak, M. 2000. What is creationism? http://www.talkorigins.org/faqs/wic.html; Kossy, D. 2001. *Strange Creations*; Wilkins, J. 1998. So you want to be an anti-Darwinian. http://www.talkorigins.org/faqs/anti-darwin.html.

CA510.1: Problems with evolution are evidence for creation.
Problems with evolution are evidence for creationism.

1. This claim assumes that creation and evolution are the only two possible models, which is very false (see CA510).

2. Even if the two-model idea were true, problems with one model do not imply that the other model is true. Another alternative is that another as-yet unknown model is correct.

THEOLOGY

CA601: Evolution requires naturalism.
Evolution is materialistic; it requires methodological naturalism. It irrationally rules out the possibility of any divine outside influence. (Dembski 1996; P. E. Johnson 1990)

1. The naturalism that science adopts is methodological naturalism (see CA301). It does not assume that nature is all there is; it merely notes that nature is the only objective standard we have. Supernaturalism is not ruled out a priori; it is left out because it has never been reliably observed. There are many scientists who use naturalism but who believe in more than nature.

2. Evolution does not in any way rule out the possibility of any outside influence, even divine influence. When evidence for outside influence has been observed, it has been included.

Science does not include anything that leaves no evidence that might be tested. Hypotheses that can be asserted but never supported are not part of science. However, these untestable phenomena are only removed from scientific consideration; they are not ruled out from life entirely. People are free to accept or reject them as they please, and science has absolutely nothing to say on the subject. Science not only rules out the acceptance of divine influence; it also rules out the rejection of divine influence.

3. Evolution is not alone in its naturalism. All science, all engineering, all manufacturing, and most other human endeavors are equally naturalistic. If we must discard evolution because of this philosophy, then we must also discard navigation, meteorology, farming, architecture, printing, law, and virtually all other subjects for the same reason.

4. Intelligent design implies philosophical naturalism. As noted above, all science, industry, agriculture, and so forth is based on nature. That does not stop evolutionists, other scientists, engineers, manufacturers, and farmers from being able to look beyond the materialism and find spirituality in their lives.

The intelligent design crowd, on the other hand, seems unable to make that step. They seem to require objective, material evidence to back up their spirituality. But that, of course, makes their spirituality naturalistic. For all their complaints about materialism, people like Dembski and Johnson are trying to expand materialism into the field of religion.

Further Reading: Isaak, M. 2002c. A philosophical premise of 'naturalism'? http://www .talkdesign.org/faqs/naturalism.html; Padgett, A. G. 2000. Creation by design. http://www.christian itytoday.com/bc/2000/004/13.30.html.

CA601.1: Evolution's materialism or naturalism denies a role for God.

Evolution's materialism or methodological naturalism denies a role for God. (P. E. Johnson 1999) *"Methodological naturalism asks us for the sake of science to pretend that the material world is all there is."* (Dembski 1996)

1. This is simply a lie. There are numerous scientists who embrace both evolution and God. They believe not only that God created a universe in which evolution occurs naturally but that God works constantly in their and other people's lives. The claim otherwise by Johnson, Dembski, and others is an overt rejection of other people's religious beliefs.

2. This claim applies not only to evolution; it logically should apply to people who believe in materialism or methodological naturalism in any science or any aspect of life. All people who believe that God does not intervene to keep planets rotating, cause winds, or make sodas fizz, according to this claim, must be atheists. It is obvious that they are not. Many famous scientists were and are devout Christians who use, in their work, exactly the same sort of naturalism that evolutionary science uses.

3. From a practical point of view, the people who make this claim are denying more than advocating a role for God. They say that theistic evolution is no different from atheistic evolution because it does not show God acting directly. For them, a God that does not act supernaturally is equivalent to no God at all. But nothing supernatural is happening around me right now; in fact, I have never seen anything supernatural occur. Does that mean God is not around? According to Johnson and others who believe his claim, God is irrelevant for almost all of our lives.

Further Reading: Isaak, M. 2002c. A philosophical premise of 'naturalism'? http://www .talkdesign.org/faqs/naturalism.html.

CA602: Evolution is atheistic.
Evolution is atheistic. (H. M. Morris 1985, 215)

1. For a claim that is so obviously false, it gets repeated surprisingly often. Evolution does not require a God, but it does not rule one out either. In that respect, it is no different from almost all other fields of interest. Evolution is no more atheistic than biochemistry, farming, engineering, plumbing, art, law, and so forth.

2. Many, perhaps most, evolutionists are not atheists. If you take the claim seriously, you must claim that the following people are atheists, to give just a few examples:

- Sir Ronald Fisher—the most distinguished theoretical biologist in the history of evolutionary thought. He was also a Christian (a member of the Church of England) and a conservative whose social views were somewhere to the right of Louis XIV.

- Pope John Paul II—a social conservative.

- Pierre Teilhard de Chardin—a paleontologist and priest who taught that God guided evolution.

- President Jimmy Carter—a devout and active Southern Baptist.

3. Anyone worried about atheism should be more concerned about creationism. Creationism can lead to a crisis of faith when people discover that its claims are false and its tactics frequently dishonest. This has led some people to abandon religion altogether (Greene n.d.). It has led others to a qualitatively different understanding of Christianity (Morton 2000c).

4. By saying that only one religious interpretation is correct and universal, creationism typically is a rejection of every other religious interpretation. For example, young-earth creationists reject the religious interpretation that the universe is more than 10,000 years old (Sarfati 2004), and design theorists reject the idea that God has guided evolution (Dembski 1996). For people whose beliefs about God differ from those of a creationist, that creationism might just as well be atheistic.

Further Reading: NCSE. n.d. Voices for evolution. http://www.ncseweb.org/resources/articles/ 5025_statements_from_religious_orga_12_19_2002.asp; Ruse, M. 2001. *Can a Darwinian Be a Christian?*

CA602.1: Darwin made it easy to become an intellectually fulfilled atheist.
By providing a naturalistic explanation of biological origins, evolution promotes atheism (Berlinski 1996): *"Although atheism might have been logically tenable before Darwin, Darwin made it possible to be an intellectually fulfilled atheist."* (Dawkins 1986, 6)

1. Naturalistic explanations of origins are not necessary for atheism. Nobody in the world can explain the origin of everything anyway. Leaving one more thing unexplained does not much matter. Dawkins's claim is the flip side of the "God of the Gaps" fallacy.

2. Naturalistic explanations of origins do not make atheism mandatory. If God is the creator, it would make sense that he would be responsible for creating everything, including evolution and the laws that make it operate.

3. Darwin was not alone in providing naturalistic explanations. Many people before him removed God from explanations for parts of the universe. Pierre-Simon Laplace provided a natural explanation for the origin and stability of the solar system. Friedrich Wöhler synthesized urea, showing that there was no "vital" element in organic material. David Hume argued that design was not necessary for the origin of life. Darwin, by providing the mechanism, merely filled in one of the last gaps. It was possible to be an intellectually fulfilled atheist even before Darwin (Gliboff 2000).

4. There is nothing wrong with being an atheist if you want to be an atheist. That some people disapprove only shows that there is something wrong with religious bigots.

5. Due mainly to its being rife with intellectual dishonesty creationism also drives some people to atheism (Babinski 1995).

CA602.2: Scientists aim to make God unnecessary.

The goal of many scientists, especially evolutionists and cosmologists, is to explain the universe without God. They want to make God unnecessary. (H. M. Morris 2003a)

1. The goal of scientists is to explain the universe, period. If that could best be done by including God in the equations, it would be. However, God is inscrutable, even according to creationists, so God is useless as an explanation.

2. Explaining things without God does not make God unnecessary. Many people believe that God is necessary in their personal lives, regardless of his scientific implications. Morris's claim, in contrast, implies that God has no value as a personal God.

3. Many scientists are firmly committed to God. Many interpret their findings as glorifying God (Campagna 2002; Livingstone 1984).

Further Reading: Livingstone, D. N. 1984. *Darwin's Forgotten Defenders.*

CA603: Naturalistic evolution rules out all but a Deist god.

"Naturalistic evolution is consistent with the existence of 'God' only if by that term we mean no more than a first cause which retires from further activity after establishing the laws of nature and setting the natural mechanism in motion." (P. E. Johnson 1990, 17)

1. This claim logically applies not only to evolution but to everything that naturalism applies to, including electricity, ecology, gravity, weather, optics, and very nearly everything else. Johnson effectively sets up a false dichotomy of rejecting all of nature or all of God.

The claim has been rejected by serious Christians (and devout members of other religions) since evolution was first proposed. They believe that God and nature are not incompatible, that God can work in ways consistent with evolution. For example, some people believe that God provides strength and inspiration on a personal level.

2. Johnson's view of God effectively rejects God. He says that supernaturalism is an essential aspect of God: Since nothing supernatural is happening around me, God is not part of my life. If I were to adopt Johnson's view of God, I would call myself an atheist.

3. A God that is active supernaturally brings the problem of evil to the forefront. It means that God created suffering and could eliminate it if he wanted to. Through his inaction, God becomes responsible for evil.

Further Reading: Ruse, M. 2001. *Can a Darwinian be a Christian?*

CA610: Evolution is a religion.
Evolution is a religion because it encompasses views of values and ultimate meanings. (H. M. Morris 1985, 196–200)

1. Evolution merely describes part of nature. The fact that that part of nature is important to many people does not make evolution a religion. Consider some attributes of religion and how evolution compares:

- Religions explain ultimate reality. Evolution stops with the development of life (it does not even include the origins of life).

- Religions describe the place and role of humans within ultimate reality. Evolution describes only our biological background relative to present and recent human environments.

- Religions almost always include reverence for and/or belief in a supernatural power or powers. Evolution does not.

- Religions have a social structure built around their beliefs. Although science as a whole has a social structure, no such structure is particular to evolutionary biologists, and one does not have to participate in that structure to be a scientist.

- Religions impose moral prescriptions on their members. Evolution does not. Evolution has been used (and misused) as a basis for morals and values by some people, such as Thomas Henry Huxley, Herbert Spencer, and E. O. Wilson (Ruse 2000), but their view, although based on evolution, is not the science of evolution; it goes beyond that.

- Religions include rituals and sacraments. With the possible exception of college graduation ceremonies, there is nothing comparable in evolutionary studies.

- Religious ideas are highly static; they change primarily by splitting off new religions. Ideas in evolutionary biology change rapidly as new evidence is found.

2. How can a religion not have any adherents? When asked their religion, many, perhaps most, people who believe in evolution will call themselves members of mainstream religions, such as Christianity, Buddhism, and Hinduism. None identify their religion as evolution. If evolution is a religion, it is the only religion that is rejected by all its members.

3. Evolution may be considered a religion under the metaphorical definition of something pursued with zeal or conscientious devotion. This, however, could also apply to

stamp collecting, watering plants, or practically any other activity. Calling evolution a religion makes religion effectively meaningless.

4. Evolutionary theory has been used as a basis for studying and speculating about the biological basis for morals and religious attitudes (Sober and Wilson 1998). Studying religion, though, does not make the study a religion. Using evolution to study the origins of religious attitudes does not make evolution a religion any more than using archaeology to study the origins of biblical texts makes archaeology a religion.

5. Evolution as religion has been rejected by the courts:

> Assuming for the purposes of argument, however, that evolution is a religion or religious tenet, the remedy is to stop the teaching of evolution, not establish another religion in opposition to it. Yet it is clearly established in the case law, and perhaps also in common sense, that evolution is not a religion and that teaching evolution does not violate the Establishment Clause. (*McLean v. Arkansas Board of Education* 1982)

The court cases *Epperson v. Arkansas*, *Willoughby v. Stever*, and *Wright v. Houston Indep. School Dist.* are cited as precedent.

Further Reading: VonRoeschlaub, W.K. 1998. God and evolution. http://www.talkorigins.org/faqs/faq-god.html.

CA611: Evolutionary theory has become sacrosanct.

Evolutionary theory, for a variety of nonscientific reasons, has obtained the status of sacred revelation. To express doubts by bringing up the counterevidence to the theory is to brand oneself an intellectual infidel. (Wiker 2003)

1. Evolution is far from sacrosanct. Since Darwin's formulation of it, there have been several significant revisions of important aspects of it:

- Mendelian heredity: Darwin thought genes were both blending (not particulate) and influenced by the environment of the organism, a kind of Lamarckian inheritance he called "pangenesis."

- Speciation: For a long while Darwin's own view on what caused new species to rise (natural selection) was rejected by most biologists in favor of geographical isolation. Only recently has Darwin's view come back into favor as one cause among many.

- Jumping genes: Barbara McClintock won the Nobel Prize for showing that genes can move from one place to another within the genome.

- Symbiotic origins of organelles: Lynn Margulis proposed that the ancestors of eukaryotic cells arose from prokaryote cells joined together in "symbiotic consortiums" (Margulis 1981).

- Genetic drift: This idea from Sewall Wright says that much genetic change in populations is due to random drift rather than natural selection.

- Neutral theory, proposing that most generic variation is neutral, not subject to selection (or nearly neutral, in Ohta's extension of the theory; Kimura 1983; Ohta 1992).

- Prions: The discovery of an entirely new kind of "life" form that replicates without genetic material via a catalytic change of molecular configuration. This also yielded a Nobel prize for Stanley B. Prusiner.
- Lateral gene transfer: Some genetic material is not inherited from an immediate ancestor but from distantly related organisms (e.g., Woese 2000).

Challenges to parts of evolutionary theory continue today. However, they are the sort of thing one rarely encounters below the graduate level.

Evolution has undergone a tremendous amount of testing, some of which has shown that correction is necessary. Correcting a scientific theory makes the (corrected) theory stronger. The testing and correction account for evolution's strong reputation today. If evolution were sacrosanct, it would not undergo testing and revision, and it would lose its reputation among scientists.

2. Critics of evolution are treated as intellectual outcasts not because they criticize evolution but because they do not know what they are talking about. Answers in Genesis (AIG) recognizes the problem of poorly educated creationists doing more harm than good to the reputation of creationists, so they devote a page to arguments creationists should not use (AIG n.d.a). Still, it is extremely common to hear creationists speak with ignorance about the second law of thermodynamics (see CF001), no transitional fossils (see CC200), irreducible complexity (see CB200), and other subjects, and AIG's list of bad arguments barely scratches the surface. The real infidels of evolution, such as Barbara McClintock and Stanley Prusiner, win acclaim.

3. Creationist works almost invariably cite mainstream science in their attempts to discredit evolution. If evolution is sacrosanct, how can creationists so readily find science articles to use against it?

with John Wilkins

CA612: Evolution requires as much faith as creationism.

Because evolution has never been observed, the theory of evolution requires as much faith as creationism does. (H. M. Morris 1985, 4)

1. The theory of evolution is based on evidence that has been observed. There is a great amount of this evidence (see CA202). When evidence is found to contradict previous conclusions, those conclusions are abandoned, and new beliefs based on the new evidence take their place. This "seeing is believing" basis for the theory is exactly the opposite of the sort of faith implied by the claim.

2. The claim implicitly equates faith with believing things without any basis for the belief. Such faith is better known as gullibility. Equating this sort of belief with faith places faith in God on exactly the same level as belief in UFOs, Bigfoot, and modern Elvis sightings.

A truly meaningful faith is not simply about belief. Belief alone does not mean anything. A true faith implies acceptance and trust; it is the feeling that whatever happens, things will somehow be okay. Such faith is not compatible with most creationism. Creationism usually demands that God acts according to peoples' set beliefs, and anything else is simply wrong (e.g., ICR 2000). It cannot accept that whatever God has done is okay.

CA620: If man comes from random causes, life has no purpose or meaning.

If man arose by chance, life would have no purpose or meaning. (H.M. Morris 1985, 178)

1. Purpose can come from anyone. The same object can have different purposes to different people or to the same person at different times. If you, God, or anyone else, want to do something with your life, then your life has purpose. Nothing else is relevant.

2. Purpose is not determined by origins. Things can have purpose even if their origin is due to chance. The North Star, for example, came to its position by chance, but people still find a purpose for it.

3. Like most people, virtually all creationists already acknowledge that people arise by chance. In the process of sexual recombination, it is chance that determines which genes come from each parent; thus chance determines the genetics that make us who we are.

4. The theory of evolution most emphatically does not say that humans arose purely by chance (see CB940).

CA622: Without a literal Fall, there is no need for Jesus and redemption.

Without a literal Fall, there is no need for redemption and thus no need for Jesus or Christianity. (Grant et al. n.d.; H.M. Morris 1998b)

1. It is sin in general, and not merely one particular instance of sin, that makes redemption necessary. If you can find any sin in the world, then the claim is baseless. Proof of this is given by the fact that many Christians feel the need for redemption but do not believe in a literal Fall.

2. This claim implies that sin and redemption are about things that happened thousands of years ago, not about anything happening to us today. It makes religion less relevant to people's lives.

3. Origins are not determined by our personal decisions of what religion to follow.

CA630: Animals are not moral, aesthetic, idealistic, or religious.

Evolution says humans are animals, but humans are moral, aesthetic, idealistic, and religious, and animals are not. (H.M. Morris 1985, 196)

1. It is not just evolution that says humans are animals. Any basic biology says that.

2. Who says animals are not moral, aesthetic, idealistic, or religious? Many animals show altruistic behavior. Others show respect for their dead. Some writers have claimed, not entirely in jest, that animals are more moral than humans.

CA640: Do you want to be descended from a monkey?

Evolution says you are descended from a monkey. Do you want to be descended from a monkey? (Moon 1976)

1. What we want does not determine what is. To believe that reality will bend to our wishes is a form of hubris.

2. What is wrong with being related to a monkey? In Yoruba tradition, it is believed that twins are descended directly from monkeys, and twins are considered good fortune (Courlander 1996, 233–238). Being related to a monkey does not make you a monkey. It does not make you any different from what you are.

CA650: Death and suffering before humanity implies an unmerciful God.

Evolution implies that animals suffered and died for billions of years before human beings appeared. This implies that God is not really a God of mercy and grace, because the suffering was not necessary to create human beings for fellowship with himself. This, in turn, implies that the loving God of the Bible does not exist. (H. M. Morris 1998b; 2000b)

1. Evolution assumes that the conditions of death and suffering in the past were much like they are now. If death and suffering mean God was not a god of grace and mercy before humanity, then the same conditions today mean he still is not such a god.

2. It is not for us to tell God how to do his job. There is no justification to claim that from God's point of view, death and suffering are not the best way to get things done.

3. Creationism does not solve the problem. It still proposes a god that allowed 2,000 years or more of suffering and dying before redemption came. If God is cruel for allowing billions of years of suffering, he is still cruel for allowing thousands of years of suffering.

4. Origins are not determined by our personal decisions of what religion to follow.

CA651: God would have pronounced death and suffering "very good."

If evolution is true, then God is directly responsible for death. God pronounced his original creation "very good" (Gen. 2:1), which implies that God is sadistic, taking pleasure in watching the suffering and dying. A God of grace and mercy would not use the principle of survival of the fittest. (H. M. Morris 2000b)

1. Humans do not get to decide what God would and would not do. To make such a decision is to place oneself above God.

2. If a God of grace and mercy would not allow natural selection, then a God of grace and mercy does not exist. Extinctions and selective individual deaths continue today. If a creator God exists, then God, at the very least, is responsible for creating the circumstances that cause extinction and death and for allowing them to continue. This is true

even if you believe that death is the result of the Fall, which God implemented in response to sin.

If creationism is true, then God is directly responsible for death. Blaming Adam, not God, for all suffering is scapegoating. Adam was human, so he did not have that much power. Even if Adam motivated God to cause suffering, God was the one responsible for bringing it to pass.

3. There would be little good about a world without death. Either there would be no reproduction, in which case none of us would exist, or animals would be many layers deep and would have no room to move around. Death is a necessary part of life.

4. The attitude behind the claim not only separates one from God but encourages ingratitude with the world, saying we cannot see it as "very good" as it is. Suffering is a state of mind and is largely a function of the person's attitude more than of external circumstances. As Abraham Lincoln said, "People are about as happy as they make up their mind to be." To blame suffering on others is to look for scapegoats.

5. Creationists link death and decay with the second law of thermodynamics, with the consequence that no decay before the Fall means the second law of thermodynamics was not in effect before the Fall. However, the second law of thermodynamics is intimately connected with the flow of time. Since the Bible says that time was established before the Fall, thermodynamics and therefore decay must have existed then, too.

6. Origins are not determined by one's personal decision of what religion to follow.

Further Reading: Larson, G. 1998. *There's a Hair in My Dirt!*

CA652: Christ's death was unjust if physical death was not the penalty for sin.

"[I]f physical human death was not really an important part of the penalty for sin, then the agonizingly cruel physical death of Christ on the cross was not necessary to pay that penalty, and thus would be gross miscarriage of justice on God's part." (H. M. Morris 2000b)

1. The claim is a non sequitur. According to creationists, God instituted the penalties for sin. If God has any power to speak of, he could have withdrawn the penalties whenever he wanted. Neither Christ's life nor his death would have been necessary. This is true whether or not the penalty for sin included physical death. It is not for us to say what is necessary for God to do.

2. If physical death was part of the penalty for sin, then Christ did not pay that penalty because physical death is still with us.

3. Origins are not determined by one's personal decision of what religion to follow.

BIOLOGY

ABIOGENESIS

CB000: Pasteur proved life only comes from life (law of biogenesis).

Pasteur and other scientists disproved the concept of spontaneous generation and established the "law of biogenesis"—that life comes only from previous life. (Watchtower 1985, 38)

1. The spontaneous generation that Pasteur and others disproved was the idea that life forms such as mice, maggots, and bacteria can appear fully formed. They disproved a form of creationism. There is no law of biogenesis saying that very primitive life cannot form from increasingly complex molecules.

CB010: The odds of life forming are incredibly small.

The proteins necessary for life are very complex. The odds of even one simple protein molecule forming by chance are 1 in 10^{113}, and thousands of different proteins are needed to form life. (Watchtower 1985, 44; see also CB010.2)

1. The calculation of odds assumes that the protein molecule formed by chance. However, biochemistry is not chance, making the calculated odds meaningless. Biochemistry produces complex products, and the products themselves interact in complex ways. For example, complex organic molecules are observed to form in the conditions that exist in space, and it is possible that they played a role in the formation of the first life (Spotts 2001).

2. The calculation of odds assumes that the protein molecule must take one certain form. However, there are innumerable possible proteins that promote biological activity. Any calculation of odds must take into account all possible molecules (not just proteins) that might function to promote life.

3. The calculation of odds assumes the creation of life in its present form. The first life would have been very much simpler.

4. The calculation of odds ignores the fact that innumerable trials would have been occurring simultaneously.

Further Reading: Musgrave, I. 1998a. Lies, damned lies, statistics, and probability of abiogenesis calculations. http://www.talkorigins.org/faqs/abioprob/abioprob.html; Stockwell, J. 2002. Borel's Law and the origin of many creationist probability assertions. http://www.talkorigins.org/faqs/abioprob/borelfaq.html.

CB010.1: Even the simplest life is incredibly complex.
Even the simplest, most primitive forms of life—bacteria—are incredibly complex, much too complex to have arisen by chance. (Sherwin 2001)

1. There is no reason to think that the life around today is comparable in complexity to the earliest life. All of the simplest life would almost certainly be extinct by now, outcompeted by more complex forms.

2. Self-replicators can be incredibly simple, as simple as a strand of six DNA nucleotides (Sievers and von Kiedrowski 1994). This is simple enough to form via prebiotic chemistry. Self-replication sets the stage for evolution to begin, whether or not you call the molecules "life."

3. Nobody claims the first life arose by chance (see CB010.2). To jump from the fact that the origin is unknown to the conclusion that it could not have happened naturally is the argument from incredulity (see CA100).

Further Reading: Musgrave, I. 1998a. Lies, damned lies, statistics, and probability of abiogenesis calculations. http://www.talkorigins.org/faqs/abioprob/abioprob.html.

CB010.2: First cells could not come together by chance.
The most primitive cells are too complex to have come together by chance. (H. M. Morris 1985, 59–69; Watchtower 1985, 44; see also CB010)

1. Biochemistry is not chance. It inevitably produces complex products. Amino acids and other complex molecules are even known to form in space.

2. Nobody knows what the most primitive cells looked like. All the cells around today are the product of billions of years of evolution. The earliest self-replicator was likely very

much simpler than anything alive today; self-replicating molecules need not be all that complex (D.H. Lee et al. 1996) and protein-building systems can also be simple (Ball 2001; Tamura and Schimmel 2001).

3. This claim is an example of the argument from incredulity (see CA100). Nobody denies that the origin of life is an extremely difficult problem. That it has not been solved, though, does not mean it is impossible. In fact, there has been much work in this area, leading to several possible origins for life on earth:

- Panspermia, which says life came from someplace other than earth. This theory, however, still does not answer how the first life arose.

- Proteinoid microspheres (Fox 1960, 1984; Fox and Dose 1977; Fox et al. 1995; Pappelis and Fox 1995): This theory gives a plausible account of how some replicating structures, which might well be called alive, could have arisen. Its main difficulty is explaining how modern cells arose from the microspheres.

- Clay crystals (Cairn-Smith 1985): This says that the first replicators were crystals in clay. Though they do not have a metabolism or respond to the environment, these crystals carry information and reproduce. Again, there is no known mechanism for moving from clay to DNA.

- Emerging hypercycles: This proposes a gradual origin of the first life, roughly in the following stages: (1) a primordial soup of simple organic compounds. This seems to be almost inevitable; (2) nucleoproteins, somewhat like modern tRNA (de Duve 1995b) or peptide nucleic acid (K.E. Nelson et al. 2000), and semicatalytic; (3) hypercycles, or pockets of primitive biochemical pathways that include some approximate self-replication; (4) cellular hypercycles, in which more complex hypercycles are enclosed in a primitive membrane; (5) first simple cell. Complexity theory suggests that the self-organization is not improbable. This view of abiogenesis is the current front-runner.

- The iron–sulfur world (Russell and Hall 1997; Wächtershäuser 2000): It has been found that all the steps for the conversion of carbon monoxide into peptides can occur at high temperature and pressure, catalyzed by iron and nickel sulfides. Such conditions exist around submarine hydrothermal vents. Iron sulfide precipitates could have served as precursors of cell walls as well as catalysts (W. Martin and Russell 2003).

- Polymerization on sheltered organophilic surfaces (Smith et al. 1999): The first self-replicating molecules may have formed within tiny indentations of silica-rich surfaces so that the surrounding rock was its first cell wall.

- Something that no one has thought of yet.

Further Reading: Cohen, P. 1996. Let there be life. http://www.newscientist.com/hottopics/astrobiology/letthere.jsp; de Duve, C. 1995a. The beginnings of life on earth. http://www.americanscientist.org/template/AssetDetail/assetid/21438?fulltext-true; de Duve, C. 1995b. *Vital Dust*; Fox, S. 1988. *The Emergence of Life*; Fry, I. 2000. *The Emergence of Life on Earth*; Lacey, J.C., et al. 1992. Experimental studies on the origin of the genetic code and the process of protein synthesis; Lewis, R. 1997. Scientists debate RNA's role at beginning of life on earth. http://www.mhhe.com/biosci/genbio/life/articles/article28.mhtml; McClendon, J.H. 1999. The origin of life; Orgel, L.E. 1994. The origin of life on the earth; Pigliucci, M. 1999. Where do we come from?; Russell, M. 2003. Evolution: Five big questions; Willis, P. 1997. Turning a corner in the search for the origin of life. http://www.santafe.edu/sfi/publications/Bulletins/bulletin-summer97/turning.html.

CB015: DNA needs proteins to form; proteins need DNA.

DNA needs certain proteins in order to replicate. Proteins need DNA to form. Neither could have formed naturally without the other already in existence. (H.M. Morris 1985, 47–48; Watchtower 1985, 45)

1. DNA could have evolved gradually from a simpler replicator; RNA is a likely candidate, since it can catalyze its own duplication (Jeffares et al. 1998; Leipe et al. 1999; Poole et al. 1998). The RNA itself could have had simpler precursors, such as peptide nucleic acids (Böhler et al. 1995). A deoxyribozyme can both catalyze its own replication and function to cleave RNA—all without any protein enzymes (Levy and Ellington 2003).

CB020: Why is new life not still being generated today?

Abiogenesis assumes life was created by processes still operating today, so new life should still be appearing today. (H.M. Morris 1985, 46)

1. Conditions today are different from conditions in the past in two important ways: First, there was little or no molecular oxygen in the atmosphere or oceans when life first appeared. Free oxygen is reactive and would likely have interfered with the formation of complex organic molecules. More importantly, there was no life around before life appeared. The life that is around today would scavenge and eat any complex molecules before they could turn into anything approaching new life.

CB025: Not all amino acids needed for life have been formed experimentally.

Stanley Miller's original abiogenesis experiment produced only four of the twenty amino acids from which proteins are built, and later experiments still have not produced all twenty amino acids under plausible conditions. (Watchtower 1985, 40)

1. Miller's experiments produced thirteen of the twenty amino acids used in life (Henahan 1996). Others may have formed via other mechanisms. For example, they may have formed in space and been carried to earth on meteors (Pizzarello and Weber 2004).

2. It is not known which amino acids are needed for the most primitive life. It could be that the amino acids that form easily were sufficient and that life later evolved to produce and rely on others.

CB026: Abiogenesis experiments produce toxins, such as cyanide and formaldehyde.

Miller–Urey type experiments produce toxic chemicals, such as cyanide and formaldehyde, but not amino acids. (Discovery Institute 2003, 5)

1. Cyanide and formaldehyde are necessary building blocks for important biochemical compounds, including amino acids (Abelson 1996). They are not toxins in this context.

2. Miller–Urey experiments produce amino acids among other chemical compounds (Kawamoto and Akaboshi 1982; Schlesinger and Miller 1983).

Further Reading: Ellington, M.D., and M. Levy. 2003. Gas, discharge, and the Discovery Institute.

CB030: Early molecules would have decayed.
Complex organic molecules, such as the bases in RNA, are very fragile and unstable, except at low temperatures. They would not hold together long enough to serve as the first self-replicating proto-life. (Bergman 2000)

1. The source Bergman cites for the fragility of RNA bases (Levy and Miller 1998) disputes abiogenesis only at high temperatures, around 100 degrees Celsius. They also conclude, "At 0 degrees C, A, U, G, and T appear to be sufficiently stable (half life greater than or equal to a million years) to be involved in a low-temperature origin of life." They also say that cytosine is unstable enough at 0 degrees Celsius (half life of 17,000 years) that it may not have been involved in the first genetic material. The discovery of a ribozyme without C-G bases shows that genetic material without cytosine is plausible (Reader and Joyce 2002).

2. If synthesis of nucleo-bases is catalyzed and hydrolysis is not, we expect the nucleo-bases to accumulate. Formamide, which can form under prebiotic conditions, has been found to catalyze the formation of nucleo-bases (Saladino et al. 2001; Saladino et al. 2003).

RNA degrades quickly today because there are enzymes (RNAses) to chew it up. Those enzymes would not have evolved if RNA degraded quickly on its own. If complex organic molecules were so fragile, life itself would be impossible. In fact, life exists even in boiling temperatures or at very high acidity.

3. Life need not have begun with highly stable molecules. Eigen and Schuster developed a notion of chemical hypercycles, in which many chemical components coexist; each component of the reaction leads to other components, which eventually reform the original one (Eigen and Schuster 1977). Chemicals involved in such a cycle need not persist longer than the duration of the hypercycle itself.

4. Organic molecules may have grown in association with stabilizing templates, such as clay templates (Ertem and Ferris 1996), or parts of the hypercycles mentioned above.

CB030.1: Early molecules would have been destroyed by ultraviolet light.
Since the early atmosphere had no ozone layer, ultraviolet (UV) light would have irradiated organic molecules that formed in the atmosphere, destroying complex molecules needed to form life. (Watchtower 1985, 41)

1. When simple organic molecules are held together in a fairly concentrated area, such as stuck to a dust or ice grain, the UV light actually enhances the formation of more complex molecules by breaking some bonds and allowing the molecules to recombine (Bernstein et al. 1999b; Cooper et al. 2001). DNA and RNA are relatively resistant to UV light, because some parts of the molecules shelter others and damage to the bases can provide the materials to repair the backbone. UV light gives nucleic acids a selective advantage and may in fact have been an essential ingredient for abiogenesis (Mulkidjanian et al. 2003; Mullen 2003).

2. The molecules need not all have stayed exposed to UV for long. Some would have dissolved in oceans and lakes. In one proposed scenario, the complex organic molecules form in the deep ocean around geothermal vents, well away from ultraviolet light.

Further Reading: Bernstein, M. P., et al. 1999a. Life's far-flung raw materials.

CB035: Miller's experiments had an invalid assumption of the type of atmosphere.

In Miller's experiment demonstrating the formation of complex organics from simple compounds, the atmospheric composition used was a reducing atmosphere, with no free oxygen. The early earth probably had a more oxidizing atmosphere. (Watchtower 1985, 40–41)

1. Since his first experiment, Miller and others have experimented with other atmospheric compositions, too (Chang et al. 1983; S. L. Miller 1987; Schlesinger and Miller 1983; Stribling and Miller 1987). Complex organic molecules form under a wide range of prebiotic conditions.

2. It is possible that life arose well away from the atmosphere—for example, around deep-sea hydrothermal vents (see CB010.2). This could make the atmospheric content largely irrelevant.

3. The early atmosphere, even if it was oxidizing, was nowhere near as oxidizing as it is today (see CB035.1).

Further Reading: Gishlick, A. D. n.d. Icons of evolution? http://www.ncseweb .org/icons/icon1millerurey.html; Tamzek, N. 2002. Icon of obfuscation. http://www.talkorigins .org/faqs/wells/iconob.html#Miller-Urey.

CB035.1: Earth's early atmosphere had abundant oxygen.

Free oxygen is fatal to abiogenesis scenarios, such as those that Stanley Miller experimented with. Evidence indicates that the early earth had significant oxygen. (Ankerberg et al. 1990)

1. There is a variety of evidence that the early atmosphere did not have significant oxygen (Turner 1981):

- Banded iron formations are layers of hematite (Fe_2O_3) and other iron oxides deposited in the ocean 2.5 to 1.8 billion years ago. The conventional interpretation is that oxygen was introduced into the atmosphere for the first time in significant quantities beginning about 2.5 billion years ago when photosynthesis evolved. This caused the free iron dissolved in the ocean water to oxidize and precipitate. Thus, the banded iron formations mark the transition from an early earth with little free oxygen and much dissolved iron in water to present conditions with lots of free oxygen and little dissolved iron.

- In rocks older than the banded iron formations, uranite and pyrite exist as detrital grains, or sedimentary grains that were rolling around in stream beds and beaches. These minerals are not stable for long periods in the present high-oxygen conditions.

- "Red beds," which are terrestrial sediments with lots of iron oxides, need an oxygen atmosphere to form. They are not found in rocks older than about 2.3 billion years, but they become increasingly common afterward.

- Sulfur isotope signatures of ancient sediments show that oxidative weathering was very low 2.4 billion years ago (Farquhar et al. 2000).

The dominant scientific view is that the early atmosphere had 0.1 percent oxygen or less (Copely 2001).

2. Free oxygen in the atmosphere today is mainly the result of photosynthesis. Before photosynthetic plants and bacteria appeared, we would expect little oxygen in the atmosphere for lack of a source. The oldest fossils (over a billion years older than the transition to an oxygen atmosphere) were bacteria; we do not find fossils of fish, clams, or other organisms that need oxygen in the oldest sediments.

Further Reading: Tamzek, N. 2002. Icon of obfuscation. http://www.talkorigins.org/faqs/wells/iconob.html#Miller-Urey.

CB035.2: Earth's early atmosphere had no reducing gases.
Reducing gases likely were not present in the early prebiotic earth's atmosphere. (Discovery Institute 2003; J. Wells 2000)

1. The claim is false. Current evidence indicates that the early earth had a mildly reducing atmosphere (Kasting 1993).

2. Even if the earth's overall atmosphere were neutral, there would have been many local areas that were reducing, such as areas near active volcanism (Delano 2001; Kasting 1993).

Further Reading: Ellington, A.D., and M. Levy. 2003. Gas, discharge, and the Discovery Institute.

CB035.3: Amino acids are not generated from just CO_2, nitrogen, and water.
When the Miller–Urey experiment is run with an atmosphere consisting only of carbon dioxide, nitrogen, and water vapor, no amino acids are produced. (Discovery Institute 2003, 5)

1. The claim is false. Such an atmosphere does give rise to amino acids (Schlesinger and Miller 1983).

Further Reading: Ellington, A.D., and M. Levy. Gas, discharge, and the Discovery Institute; Gishlick, A.D. n.d. Icons of evolution? http://www.ncseweb.org/icons/icon1millerurey.html.

CB040: Life uses only left-handed amino acids.

The twenty amino acids used by life are all the left-handed variety. This is very unlikely to have occurred by chance. (Watchtower 1985, 43)

1. The amino acids that are used in life, like most other aspects of living things, are very likely not the product of chance. Instead, they likely resulted from a selection process. A simple peptide replicator can amplify the proportion of a single handedness in an initially random mixture of left- and right-handed fragments (Saghatelian et al. 2001; TSRI 2001). Self-assemblies on two-dimensional surfaces can also amplify a single handedness (Zepik et al. 2002). An excess of handedness in one kind of amino acid catalyzes the handedness of other organic products, such as threose, which may have figured prominently in proto-life (Pizzarello and Weber 2004).

2. Amino acids found in meteorites from space, which must have formed abiotically, also show significantly more of the left-handed variety, perhaps from circularly polarized UV light in the early solar system (Cronin and Pizzarello 1999). The weak nuclear force, responsible for beta decay, produces only electrons with left-handed spin, and chemicals exposed to these electrons are far more likely to form left-handed crystals (Service 1999). Such mechanisms might also have been responsible for the prevalence of left-handed amino acids on earth.

3. The first self-replicator may have had eight or fewer types of amino acids (Cavalier-Smith 2001). It is not all that unlikely that the same handedness might occur so few times by chance, especially if one of the amino acids was glycine, which has no handedness.

4. Some bacteria use right-handed amino acids, too (McCarthy et al. 1998).

Further Reading: Clark, S. 1999. Polarized starlight and the handedness of life. http://www.amsci.org/amsci/articles/99articles/clark.html.

CB050: Abiogenesis is speculative without evidence.

Abiogenesis is speculative without evidence. Since it has not been observed in the laboratory, it is not science. (Watchtower 1985, 50–52)

1. There is a great deal about abiogenesis that is unknown, but investigating the unknown is what science is for. Speculation is part of the process. As long as the speculations can be tested, they are scientific. Much scientific work has been done in testing different hypotheses relating to abiogenesis, including the following:

- research into the formation of long proteins (J.P. Ferris et al. 1996; Orgel 1998; Rode et al. 1999);

- synthesis of complex molecules in space (Kuzicheva and Gontareva 1999; Schueller 1998; see also CB030);

- research into molecule formation in different atmospheres (see CB035); and

- synthesis of constituents in the iron–sulfur world around hydrothermal vents (Cody et al. 2000; Russell and Hall 1997).

2. See also the references and suggested readings with

- CB010.2: Primitive cells are too complex
- CB035: Abiogenesis experiments assume a reducing atmosphere
- CB015: DNA needs proteins to form, proteins need DNA
- CB040: Amino acids are left-handed

Further Reading: Deamer, D. W., and J. Ferris. 1999. The origins and early evolution of life. http://www.chemistry.ucsc.edu/~deamer/home.html; RESA. n.d. Origins of life. http://www.resa.net/nasa/origins_life.htm; Wächtershäuser, G. 2000. Life as we don't know it.

CB090: Evolution is baseless without a theory of abiogenesis.

Evolution is baseless without a good theory of abiogenesis, which it does not have. (Mastropaolo 1998)

1. The theory of evolution applies as long as life exists. How that life came to exist is not relevant to evolution. Claiming that evolution does not apply without a theory of abiogenesis makes as much sense as saying that umbrellas do not work without a theory of meteorology.

2. Abiogenesis is a fact. Regardless of how you imagine it happened (note that creation is a theory of abiogenesis), it is a fact that there once was no life on earth and that now there is. Thus, even if evolution needs abiogenesis, it has it.

GENETICS

CB100: Mutations are rare.

Evolution requires mutations, but mutations are rare. (H. M. Morris 1985, 55)

1. Very large mutations are rare, but mutations are ubiquitous. There is roughly 0.1 to 1 mutation per genome replication in viruses and 0.003 mutations per genome per replication in microbes. Mutation rates for higher organisms vary quite a bit between organisms, but excluding the parts of the genome in which most mutations are neutral (the junk DNA), the mutation rates are also roughly 0.003 per effective genome per cell replication. Since sexual reproduction involves many cell replications, humans have about 1.6 mutations per generation. This is likely an underestimate, because mutations with very small effect are easy to miss in the studies. Including neutral mutations, each human zygote has about 64 new mutations (Drake et al. 1998). Another estimate concludes 175 mutations per generation, including at least 3 deleterious mutations (Nachman and Crowell 2000).

Further Reading: Drake, J. W., et al. 1998. Rates of spontaneous mutation; Harter, R. 1999. Are mutations harmful? http://www.talkorigins.org/faqs/mutations.html.

CB101: Most mutations are harmful.

Most mutations are harmful, so the overall effect of mutations is harmful. (H. M. Morris 1985, 55–57; Watchtower 1985, 100)

1. Most mutations are neutral. Nachman and Crowell estimate around 3 deleterious mutations out of 175 per generation in humans (2000). Of those that have significant effect, most are harmful, but a significant fraction are beneficial. The harmful mutations do not survive long, and the beneficial mutations survive much longer, so when you consider only surviving mutations, most are beneficial.

2. Beneficial mutations are commonly observed. They are common enough to be problems in the cases of antibiotic resistance in disease-causing organisms and pesticide resistance in agricultural pests (e.g., Newcomb et al. 1997; these are not merely selection of preexisting variation; see CB110). They can be repeatedly observed in laboratory populations (Wichman et al. 1999). Other examples include the following:

   - Mutations have given bacteria the ability to degrade nylon (Prijambada et al. 1995).
   - Plant breeders have used mutation breeding to induce mutations and to select the beneficial ones (FAO/IAEA 1977).
   - Certain mutations in humans confer resistance to AIDS (Dean et al. 1996) or to heart disease (Long 1994; Weisgraber et al. 1983).
   - A mutation in humans makes bones strong (Boyden et al. 2002).
   - Transposons are common, especially in plants, and help to provide beneficial diversity (Moffat 2000).
   - In vitro mutation and selection can be used to evolve substantially improved function of RNA molecules, such as a ribozyme (M. C. Wright and Joyce 1997).

3. Whether a mutation is beneficial or not depends on environment. A mutation that helps the organism in one circumstance could harm it in another. When the environment changes, variations that once were counteradaptive suddenly become favored. Since environments are constantly changing, variation helps populations survive, even if some of those variations do not do as well as others. When beneficial mutations occur in a changed environment, they generally sweep through the population rapidly (Elena et al. 1996).

4. High mutation rates are advantageous in some environments. Hypermutable strains of *Pseudomonas aeruginosa* are more commonly found in the lungs of cystic fibrosis patients, where antibiotics and other stresses increase selection pressure and variability, than in patients without cystic fibrosis (Oliver et al. 2000).

5. Note that the existence of any beneficial mutations is a falsification of the young-earth creationism model (H. M. Morris 1985, 13).

Further Reading: Harter, R. 1999. Are mutations harmful? http://www.talkorigins.org/faqs/ mutations.html; Peck, J. R., and A. Eyre-Walker. 1997. The muddle about mutations; Williams, R. n.d.a. Examples of beneficial mutations and natural selection. http://www.gate.net/~rwms/EvoMu

tations.html; Williams, R. n.d.b. Examples of beneficial mutations in humans. http://www
.gate.net/~rwms/EvoHumBenMutations.html.

CB101.1: Mutations are accidents; things do not get built by accident.

Mutations are accidents, and things do not get built by accident. (H.M. Morris 1985, 55;
Watchtower 1985, 102)

1. There is more to evolution than mutation. A small percentage of mutations are
beneficial, and selection can cause the beneficial mutations to persist and the harmful
mutations to die off. The combination of mutation and selection can create new useful
adaptations.

Sometimes things do get built by accident. Many discoveries started out as accidents
that people recognized uses for. Many other designs (accidental or not) have been selected against, that is, discarded. Design itself is an evolutionary process.

2. Experiments and genetic analysis show that mutations (plus selection) do account
for new adaptations (Max 1999).

Further Reading: Max, E.E. 1999. The evolution of improved fitness by random mutation plus
selection. http://www.talkorigins.org/faqs/fitness.

CB101.2: Mutations do not produce new features.

Mutations only vary traits that are already there. They do not produce anything new. (H.M.
Morris 1985, 51; Watchtower 1985, 103)

1. Variation of traits is production of novelty, especially where there was no variation
before. The accumulation of slight modifications is a basis of evolution.

2. Documentation of mutations producing new features includes the following:
- the ability of a bacterium to digest nylon (Negoro et al. 1994; "Thomas" n.d.;
  Thwaites 1985);
- adaptation in yeast to a low-phosphate environment (Francis and Hansche 1972,
  1973; Hansche 1975);
- the ability of *E. coli* to hydrolyze galactosylarabinose (Hall 1981; Hall and Zuzel
  1980);
- evolution of multicellularity in a unicellular green alga (Boraas 1983; Boraas et al.
  1998);
- modification of *E. coli*'s fucose pathway to metabolize propanediol (Lin and Wu
  1984);
- evolution in *Klebsiella* bacteria of a new metabolic pathway for metabolizing 5-car-
  bon sugars (Hartley 1984).

There is evidence for mutations producing other novel proteins:

- Proteins in the histidine biosynthesis pathway consist of beta/alpha barrels with a twofold repeat pattern. These apparently evolved from the duplication and fusion of genes from a half-barrel ancestor (Lang et al. 2000).

3. For evolution to operate, the source of variation does not matter; all that matters is that heritable variation occurs. Such variation is shown by the fact that selective breeding has produced novel features in many species, including cats, dogs, pigeons, goldfish, cabbage, and geraniums. Some of the features may have been preexisting in the population originally (see CB110), but not all of them were, especially considering that creationism requires the animals to originate from a single pair.

Further Reading: Max, E. E. 1999. The evolution of improved fitness by random mutation plus selection. http://www.talkorigins.org/faqs/fitness; Musgrave, I., et al. 2003a. Apolipoprotein AI mutations and information. http://www.talkorigins.org/faqs/information/apolipoprotein.html; Thomas, D. n.d. Evolution and information. http://www.nmsr.org/nylon.htm.

CB102: Mutations do not add information.
Mutations are random noise; they do not add information. Evolution cannot cause an increase in information. (AIG n.d.b.)

1. It is hard to understand how anyone could make this claim, since anything mutations can do, mutations can undo. Some mutations add information to a genome; some subtract it. Creationists get by with this claim only by leaving the term "information" undefined, impossibly vague, or constantly shifting. By any reasonable definition, increases in information have been observed to evolve. We have observed the evolution of

- increased genetic variety in a population (Lenski 1995; Lenski et al. 1991)
- increased genetic material (Alves et al. 2001; C. J. Brown et al. 1998; Hughes and Friedman, 2003; Lynch and Conery 2000; Ohta 2003)
- novel genetic material (Knox et al. 1996; Park et al. 1996)
- novel genetically regulated abilities (Prijambada et al. 1995).

If these do not qualify as information, then nothing about information is relevant to evolution in the first place.

2. A mechanism that is likely to be particularly common for adding information is gene duplication, in which a long stretch of DNA is copied, followed by point mutations that change one or both of the copies. Genetic sequencing has revealed several instances in which this is likely the origin of some proteins. For example:

- Two enzymes in the histidine biosynthesis pathway that are barrel-shaped, structural and sequence evidence suggests, were formed via gene duplication and fusion of two half-barrel ancestors (Lang et al. 2000).
- RNASE1, a gene for a pancreatic enzyme, was duplicated, and in langur monkeys one of the copies mutated into RNASE1B, which works better in the more acidic small intestine of the langur (Zhang et al. 2002).
- Yeast was put in a medium with very little sugar. After 450 generations, hexose transport genes had duplicated several times, and some of the duplicated versions had mutated further (Brown et al. 1998).

The biological literature is full of additional examples. A PubMed search (at http://www.ncbi.nlm.nih.gov/entrez/query.fcgi) on "gene duplication" gives more than 3,000 references.

3. According to Shannon-Weaver information theory, random noise maximizes information. This is not just playing word games. The random variation that mutations add to populations is the variation on which selection acts (see CB110). Mutation alone will not cause adaptive evolution, but by eliminating nonadaptive variation, natural selection communicates information about the environment to the organism so that the organism becomes better adapted to it. Natural selection is the process by which information about the environment is transferred to an organism's genome and thus to the organism (Adami et al. 2000).

4. The process of mutation and selection is observed to increase information and complexity in simulations (Adami et al. 2000; T. D. Schneider 2000).

Further Reading: Adami, C., et al. 2000. Evolution of biological complexity. http://www.pnas .org/cgi/content/full/97/9/4463; Hillis, D. M., et al. 1992. Experimental phylogenetics; Max, E. E. 1999. The evolution of improved fitness by random mutation plus selection. http://www.talkorigins .org/faqs/fitness; Musgrave, I. 2001. The *Period* gene of Drosophila. http://www.talkorigins.org/ origins/postmonth/apr01.html.

CB102.1: Dawkins could not give an example of increasing information.

In an interview in 1997, Richard Dawkins was asked to "give an example of a genetic mutation or an evolutionary process which can be seen to increase the information in the genome." Apparently unable to answer, he paused a long time and finally responded by changing the subject. (AIG 1998)

1. Dawkins paused because the question revealed that the interviewers were creationists, that he had been duped about their motives. He paused to think about how to handle them, and the change of subject occurred due to the several minutes when he confronted them being omitted from the video.

2. The question is equivalent to asking how complexity could evolve, which Dawkins has covered in at least four books (*The Blind Watchmaker*, *River Out of Eden*, *Climbing Mount Improbable*, and *A Devil's Chaplain*). He has answered the question at great length.

3. The ability of a single person to answer a question is largely irrelevant. The scientific literature is rife with examples of information increasing (see CB102).

Further Reading: Dawkins, R. 2003. The information challenge. http://www.skeptics.com .au/journal/dawkins1.htm; Williams, B. 1998. Creationist deception exposed. http://www.tccsa.tc/ video/creationist_deception_exposed.pdf.

CB110: Microevolution selects only existing variation.

Microevolution (for example, the development of insecticide resistance) merely selects preexisting variation. It does not demonstrate that mutations create new variation. (T. Wallace 2002)

1. In experiments with bacteria, variation (including beneficial mutations) arises in populations that are grown from a single individual (Lederberg and Lederberg 1952). Since the population started with just one chromosome, there was no variation in the original population; all variation must have come from mutations.

Furthermore, disease organisms and insect pests have developed resistance to a variety of antibiotics and pesticides, many of them artificial and unlike anything in nature. It is highly improbable that all insects were created with resistance to all pesticides.

2. Mutation is the only natural process that adds variation to populations. Selection and genetic drift remove variation. If mutations did not create new variation, there would now be little or no variation to select from. In particular, reducing populations to a single pair of individuals, as Noah's Flood requires, would have removed very nearly all variation from the world's wildlife in one stroke.

3. It is true that much microevolution selects from preexisting variation. In animals, that kind of microevolution occurs much faster than waiting for certain mutations to occur, so we often see artificial selection programs stall when they have selected among all the variation that was there to begin with. However, if the selection is maintained, change should continue, albeit at a much slower rate.

Further Reading: True H. L., and S. L. Lindquist. 2000. A yeast prion provides a mechanism for genetic variation and phenotypic diversity.

CB120: Genetic load from mutations would make populations unviable.
The overall effect of mutations is to lower the viability of populations, due to the "genetic load," or genetic burden, that they add to the gene pool. (H. M. Morris 1985, 56–57)

1. As new harmful mutations enter the population, selection removes existing harmful traits. The genetic load of a stable population is an equilibrium between the two.

2. Bacteria mutate much faster than plants and animals do, yet their populations are not becoming less viable.

CB121: The cost of natural selection is prohibitive (Haldane's dilemma).
J.B.S. Haldane calculated that new genes become fixed only after 300 generations due to the cost of natural selection (Haldane 1957). Since humans and apes differ in 4.8×10^7 genes, there has not been enough time for difference to accumulate. Only 1,667 gene substitutions could have occurred if their divergence was ten million years ago. (ReMine 1993)

1. Haldane's "cost of natural selection" stemmed from an invalid simplifying assumption in his calculations. He divided by a fitness constant in a way that invalidated his as-

sumption of constant population size, and his cost of selection is an artifact of the changed population size. With corrected calculations, the cost disappears (B. Wallace 1991; R. Williams n.d.c).

Haldane's paper was published in 1957, and Haldane himself said, "I am quite aware that my conclusions will probably need drastic revision" (Haldane 1957, 523). It is irresponsible not to consider the revision that has occurred in the forty years since his paper was published.

2. ReMine (1993), who promotes the claim, makes several invalid assumptions:

- The vast majority of differences would probably be due to genetic drift, not selection.

- Many genes would have been linked with genes that are selected and thus would have hitchhiked with them to fixation.

- Many mutations, such as those due to unequal crossing over, affect more than one codon.

- Human and ape genes both would be diverging from the common ancestor, doubling the difference.

- ReMine's computer simulation supposedly showing the negative influence of Haldane's dilemma assumed a population size of only six (Musgrave 1999).

Further Reading: Williams, R. n.d.c. Haldane's dilemma. http://www.gate.net/~rwms/haldane1 .html.

CB130: "Junk" DNA is not really junk.
So-called junk DNA is not really junk. Functions have been found for noncoding DNA, which was previously thought to be junk, and we cannot be sure that the rest of the junk DNA is not functional as well. (Behe 2003)

1. It has long been known that some noncoding DNA has important functions. (This was known even before the phrase "junk DNA" was coined.) However, there is good evidence that much DNA has no function:

- Sections of DNA can be cut out or replaced with randomized sequences with no apparent effect on the organism (Nóbrega et al. 2004).

- Some sections of DNA are corrupted copies of functional coding DNA, but mutations in them, such as stop codons early in the sequence, show that they cannot have retained the same function as the coding copy.

- The fugu fish has a genome that is about one third as large as its close relatives.

- Mutations in functional regions of DNA show evidence of selection—nonsilent changes occur less often that one would expect by chance. In other sections of DNA, there is no evidence that any changes are selected against.

Further Reading: EvoWiki. 2004. Junk DNA. http://www.evowiki.org/wiki.phtml?title=Junk_ DNA; Knight, J. 2002. Evolutionary genetics.

CB141: Chromosome counts differ greatly and unsystematically between species.
DNA and chromosome counts differ widely between different organisms. This dissimilarity contradicts the similarity we expect from common descent. Chromosome counts should be either the

same because the different forms of life descended from a common ancestor (Pathlights n.d.c), *or more complex as organisms get more complex.* (B. Thompson and Butt 2001) *Neither is the case. For example, humans have 46 chromosomes, some ferns have 512, and some gulls have 12.*

1. Chromosome counts are poor indications of similarity; they can vary widely within a single genus or even a single species. The plant genus *Clarkia*, for example, has species with chromosome counts of n = 5, 6, 7, 8, 9, 12, 14, 17, 18, and 26 (H. Lewis 1993). Chromosome counts in the house mouse species (*Mus domesticus*) range from 2n = 22 to 40 (Nachman et al. 1994).

Chromosomes can split or join with little effect on the genes themselves. One human chromosome, for example, is very similar to two chimpanzee chromosomes laid end to end; it likely formed from the joining of two chromosomes (Yunis and Prakash 1982; see Figure CB1). Because the genes can still align, a change in chromosome number does not prevent reproduction. Chromosome counts can also change through polyploidy, where the entire genome is duplicated. Polyploidy, in fact, is a common mechanism of speciation in plants.

CB150: Functional genetic sequences are too rare to evolve from one to another.

Evolution requires that protein sequences change to very different sequences, with all the intermediate sequences staying functional. But out of all possible sequences, functional sequences are extremely rare, so most functional sequences are highly isolated from each other. Using language as an analogy, one sentence cannot be changed to another by gradual changes such that all the intermediate changes are meaningful. It is highly improbable that random mutations could change one functional sequence to another. (S. C. Meyer 2004)

1. Functional sequences are not so rare and isolated. Experiments show that roughly 1 in 10^{11} of all random-sequence proteins have ATP-binding activity (Keefe and Szostak 2001), and theoretical work by H. P. Yockey (1992, 326–330) showed that at this density all functional sequences are connected by single amino acid changes. Furthermore, there are several kinds of mutations that change multiple amino acids at once.

2. There is a great deal of evidence showing that novel genes with novel functions can and do evolve (see CB101.2 and CB904). Even an arbitrary genetic sequence can evolve to acquire functionality (Hayashi et al. 2003). Directed evolution in vitro is a powerful and increasingly popular method of producing new genes and useful gene products (Joyce 2004; Schmidt-Dannert 2001; Tao and Cornish 2002). Directed evolution can work even when starting from random sequences. Evolution of novel sequences cannot be very improbable if it happens so easily and so often.

3. The analogy to language is flawed. Proteins are far more flexible. They can differ greatly in their sequence similarity, even 70 to 80 percent or more, and still have the same function.

Further Reading: Gishlick, A., et al. 2004. Meyer's hopeless monster. http://www.pandasthumb .org/pt-archives/000430.html.

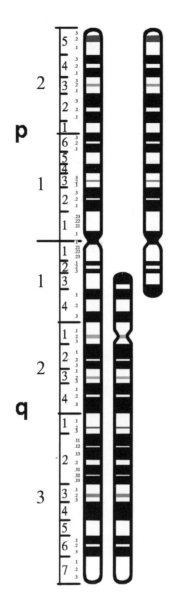

Figure CB1
Human chromosome 2 (*left*) and two corresponding chimpanzee chromosomes. The human chromosome likely formed from the joining of two smaller chromosomes. From Yunis and Prakash 1982.

MOLECULAR BIOLOGY

CB200: Some systems are irreducibly complex.

Some biochemical systems are irreducibly complex, meaning that the removal of any one part of the system destroys the system's function. Irreducible complexity rules out the possibility of a system having evolved, so it must be designed. (Behe 1996a)

1. Irreducible complexity can evolve. It is defined as a system that loses its function if any one part is removed, so it only indicates that the system did not evolve by the addition of single parts with no change in function. That still leaves several evolutionary mechanisms:

- deletion of parts
- addition of multiple parts; for example, duplication of much or all of the system (Pennisi 2001)
- change of function
- gradual modification of parts.

All of these mechanisms have been observed in genetic mutations. In particular, deletions and gene duplications are fairly common (S. D. Hooper and Berg 2003; Lynch and Conery 2000), and together they make irreducible complexity not only possible but expected. In fact, it was predicted as early as 1939 (Muller 1939).

Evolutionary origins of some irreducibly complex systems have been described in some detail. For example, the evolution of the Krebs citric acid cycle has been well studied; irreducibility is no obstacle to its formation (Meléndez-Hevia et al. 1996).

2. Even if irreducible complexity did prohibit Darwinian evolution, the conclusion of design does not follow (see CI102). Other processes might have produced it. Irreducible complexity is an example of a failed argument from incredulity (see CA100).

3. Irreducible complexity is poorly defined. It is defined in terms of parts, but it is far from obvious what a "part" is. Logically, the parts should be individual atoms, because they are the level of organization that does not get subdivided further in biochemistry, and they are the smallest level that biochemists consider in their analysis. Behe, however, considered sets of molecules to be individual parts, and he gave no indication of how he made his determinations.

4. Systems that have been considered irreducibly complex might not be. For example:

- The mousetrap that Behe used as an example of irreducible complexity can be simplified by bending the holding arm slightly and removing the latch.
- The bacterial flagellum (see CB200.1) is not irreducibly complex because it can lose many parts and still function, either as a simpler flagellum or a secretion system. Many proteins of the eukaryotic flagellum (also called a cilium or undulipodium) are known to be dispensable, because functional swimming flagella that lack these proteins are known to exist.
- In spite of the complexity of Behe's protein transport example (see CB200.3), there are other proteins for which no transport is necessary (see Ussery 1999 for references).
- The immune system example that Behe includes (see CB200.4) is not irreducibly complex because the antibodies that mark invading cells for destruction might themselves hinder the function of those cells, allowing the system to function (albeit not as well) without the destroyer molecules of the complement system.

Further Reading: Gray, T. M. 1999. Complexity—yes! Irreducible—maybe! Unexplainable—no! http://tallship.chm.colostate.edu/evolution/irred_compl.html; Lindsay, D. 1996. Review: "Darwin's black box, the biochemical challenge to evolution" by Michael Behe. http://www.don-lind say-archive.org/creation/behe.html; Miller, K.R. 1999. *Finding Darwin's God*, Chap. 5; Shanks, N., and K.H. Joplin. 1999. Redundant complexity. http://www.asa3.org/ASA/topics/Apologetics/ POS6-99ShenksJoplin.html; TalkOrigins Archive. n.d.b. Irreducible complexity and Michael Behe. http://www.talkorigins.org/faqs/behe.html; Ussery, D. 1999. A biochemist's response to "The bio-chemical challenge to evolution." http://www.cbs.dtu.dk/staff/dave/Behe.html.

CB200.1: Bacterial flagella are irreducibly complex.

Bacterial flagella and eukaryotic cilia are irreducibly complex (see CB200). *Since nonfunctional intermediates cannot be preserved by natural selection, these systems can only be explained by intelligent design.* (Behe 1996a, 59–73)

1. This is an example of argument from incredulity (see CA100) because irreducible complexity can evolve naturally (see CB200). Many of the proteins in the bacterial fla-gellum or eukaryotic cilium are similar to each other or to proteins for other functions. Their origins can easily be explained by a series of gene duplication events followed by modification and/or co-option, proceeding gradually through intermediate systems differ-ent from and simpler than the final flagellum.

One plausible path for the evolution of flagella goes through the following basic stages (keep in mind that this is a summary, and that each major co-option event would be fol-lowed by long periods of gradual optimization of function):

a. A passive, nonspecific pore evolves into a more specific passive pore by addition of gating protein(s). Passive transport converts to active transport by addition of an AT-Pase that couples ATP hydrolysis to improved export capability. This complex forms a primitive type III export system.

b. The type III export system is converted to a type III secretion system (T3SS) by addition of outer membrane pore proteins (secretin and secretin chaperone) from the type II secretion system. These eventually form the P- and L-rings, respectively, of modern flagella. The modern type III secretory system forms a structure strikingly similar to the rod and ring structure of the flagellum (Blocker et al. 2003; Hueck 1998).

c. The T3SS secretes several proteins, one of which is an adhesin (a protein that sticks the cell to other cells or to a substrate). Polymerization of this adhesin forms a prim-itive pilus, an extension that gives the cell improved adhesive capability. After the evolution of the T3SS pilus, the pilus diversifies for various more specialized tasks by duplication and subfunctionalization of the pilus proteins (pilins).

d. An ion pump complex with another function in the cell fortuitously becomes as-sociated with the base of the secretion system structure, converting the pilus into a primitive protoflagellum. The initial function of the protoflagellum is improved dis-persal. Homologs of the motor proteins MotA and MotB are known to function in di-verse prokaryotes independent of the flagellum.

e. The binding of a signal transduction protein to the base of the secretion system regulates the speed of rotation depending on the metabolic health of the cell. This imposes a drift toward favorable regions and away from nutrient-poor regions, such as those found in overcrowded habitats. This is the beginning of chemotactic motility.

f. Numerous improvements follow the origin of the crudely functioning flagellum. Notably, many of the different axial proteins (rod, hook, linkers, filament, caps) originate by duplication and subfunctionalization of pilins or the primitive flagellar axial structure. These proteins end up forming the axial protein family.

The eukaryotic cilium (also called the eukaryotic flagellum or undulipodium) is fundamentally different from the bacterial flagellum. It probably originated as an outgrowth of the mitotic spindle in a primitive eukaryote (both structures make use of sliding microtubules and dyneins). Cavalier-Smith (1987; 2002) has discussed the origin of these systems on several occasions.

2. The bacterial flagellum is not even irreducible. Some bacterial flagella function without the L- and P-rings. In experiments with various bacteria, some components (e.g., FliH, FliD [cap], and the muramidase domain of FlgJ) have been found helpful but not absolutely essential (Matzke 2003). One third of the 497 amino acids of flagellin have been cut out without harming its function (Kuwajima 1988). Furthermore, many bacteria have additional proteins that are required for their own flagella but that are not required in the "standard" well-studied flagellum found in *E. coli*. Different bacteria have different numbers of flagellar proteins (in *Helicobacter pylori*, for example, only thirty-three proteins are necessary to produce a working flagellum), so Behe's favorite example of irreducibility seems actually to exhibit quite a bit of variability in terms of numbers of required parts (Ussery 1999).

Eukaryotic cilia are made by more than 200 distinct proteins, but even here irreducibility is illusive. Behe (1996a) implied and Denton (1986) claimed explicitly that the common 9 + 2 tubulin structure of cilia could not be substantially simplified. Yet functional 3 + 0 cilia, lacking many microtubules as well as some of the dynein linkers, are known to exist (K. R. Miller 2003, 2004).

3. Eubacterial flagella, archebacterial flagella, and cilia use entirely different designs for the same function. That is to be expected if they evolved separately, but it makes no sense if they were the work of the same designer.

Further Reading: Dunkelberg, P. 2003. Irreducible complexity demystified. http://www.talkdesign.org/faqs/icdmyst/ICDmyst.html; Matzke, N. J. 2003. Evolution in (brownian) space. http://www.talkdesign.org/faqs/flagellum.html; Musgrave, I. 2000. Evolution of the bacterial flagella. http://www.health.adelaide.edu.au/Pharm/Musgrave/essays/flagella.htm; Ussery, D. 1999. A biochemist's response to "The biochemical challenge to evolution." http://www.cbs.dtu.dk/staff/dave/Behe.html.

with Nicholas Matzke

CB200.2: Blood clotting is irreducibly complex.
The biochemistry of blood clotting is irreducibly complex, indicating that it must have been designed. (Behe 1996a, 74–97)

1. The blood clotting systems appears to be put together by using whatever long polymeric bridges are handy. There are many examples of complicated systems made from components that have useful but completely different roles in different components. The co-opting of parts with different functions gets around the "challenge" of irreducible complexity evolving gradually.

2. Irreducible complexity is not an obstacle to evolution (see CB200) and does not imply design (see CI102).

Further Reading: Acton, G. 1997. Behe and the blood clotting cascade. http://www.talk origins.org/origins/postmonth/feb97.html; Doolittle, R. F. 1997. A delicate balance. http://bostonreview.net/BR22.1/doolittle.html; Dunkelberg, P. 2003. Irreducible complexity demystified. http://www.talkdesign.org/faqs/icdmyst/ICDmyst.html; Ussery, D. 1999. A biochemist's response to "The biochemical challenge to evolution." http://www.cbs.dtu.dk/staff/dave/Behe.html.

CB200.3: Protein transport within a cell is irreducibly complex.
The biochemistry of protein transport within a cell is irreducibly complex, indicating that it must have been designed. (Behe 1996a, 98–116)

1. Despite the complexity of the system that Behe describes, protein transport need not be that complex. Some proteins direct their own secretion so that no transport mechanism is necessary (see Ussery 1999 for references). Certainly, other simple systems that could serve as precursors to vesicular transport should be possible.

2. Many of the proteins involved in transport in eukaryote cells have molecular "ancestors" in bacteria. These molecules, the ABC transporters, serve in a much simpler system. If Behe is interested in the simplest system that accomplishes a function, why does he not even mention them?

3. Irreducible complexity is not an obstacle to evolution (see CB200) and does not imply design (see CI102).

Further Reading: Ussery, D. 1999. A biochemist's response to "The biochemical challenge to evolution." http://www.cbs.dtu.dk/staff/dave/Behe.html.

CB200.4: The immune system is irreducibly complex.
The human immune system is irreducibly complex, indicating that it must have been designed. (Behe 1996a, 117–139)

1. The complement system of the human immune system is not irreducibly complex. Urochordates have a functional complement system, yet they lack a component of the cascade.

2. Common mechanisms, such as gene duplication and co-option of molecules with other roles, allow the immune system to evolve naturally. Much has been written on the subject (Kasahara et al. 1997; Lindsay 1999b; J. Travis 1998).

3. Behe gets some of the basic biology wrong. Bacteria are not destroyed, as Behe says (1996a, 134), by water rushing in when the cell membrane is punctured but because their chemical gradients have been destroyed (Ussery 1999).

4. Irreducible complexity is not an obstacle to evolution (see CB200) and does not imply design (see CI102).

Further Reading: Coon, M. 1998. Is the complement system irreducibly complex? http:// www.talkorigins.org/faqs/behe/icsic.html; Inlay, M. 2002. Evolving immunity. http://www.talk design.org/faqs/Evolving_Immunity.html; Ussery, D. 1999. A biochemist's response to "The bio- chemical challenge to evolution." http://www.cbs.dtu.dk/staff/dave/Behe.html.

CB200.5: The metabolic pathway for AMP synthesis is too complex to have evolved.

The metabolic pathway for AMP synthesis is too complex to have evolved. It requires several intermediate steps, and it is highly improbable that all of the steps could have evolved simulta- neously. (Behe 1996a, 140–160)

1. Although AMP synthesis is done in several steps in modern life, several steps are not required. In fact, adenosine can and does form entirely outside of life, both in aque- ous solution (Ferris et al. 1984) and in space (Kuzicheva and Gontareva 2002). The ear- liest life could have used prebiotic AMP and later have gradually developed and refined mechanisms for synthesizing AMP itself.

Irreducible complexity itself is no obstacle to evolution (see CB200).

PHYSIOLOGY AND ANATOMY

CB300: Complex organs could not have evolved.

Complex organs and biological functions could not have evolved. (Kofahl 1977, chap. 5)

1. This is an example of the argument from incredulity (see CA100). In fact, several complex organs, which have previously been claimed unevolvable, have plausible means of evolving, including the eye (see CB301), the bombardier beetle defense mechanism (see CB310), the woodpecker tongue (see CB326), and more.

2. Evolutionary mechanisms do account for the evolution of complex organs. The ab- stract of Lenski et al. (2003, 139) is worth quoting in full:

> A long-standing challenge to evolutionary theory has been whether it can explain the origin of complex organismal features. We examined this issue using digital or- ganisms—computer programs that self-replicate, mutate, compete and evolve. Pop- ulations of digital organisms often evolved the ability to perform complex logic functions requiring the coordinated execution of many genomic instructions. Com- plex functions evolved by building on simpler functions that had evolved earlier, provided that these were also selectively favoured. However, no particular inter- mediate stage was essential for evolving complex functions. The first genotypes able to perform complex functions differed from their non-performing parents by only one or two mutations, but differed from the ancestor by many mutations that were also crucial to the new functions. In some cases, mutations that were delete- rious when they appeared served as stepping-stones in the evolution of complex features. These findings show how complex functions can originate by random mu- tation and natural selection.

Further Reading: National Science Foundation. 2003. Artificial life experiments show how complex functions can evolve. http://www.sciencedaily.com/releases/2003/05/030508075843.htm.

CB301: The eye is too complex to have evolved.

The eye is too complex to have evolved. (Hitching 1982, 66–68)

1. This is the quintessential example of the argument from incredulity (see CA100). The source making the claim usually quotes Darwin saying that the evolution of the eye seems "absurd in the highest degree" (see CA113.1). However, Darwin follows that statement with a three-and-a-half-page proposal of intermediate stages through which eyes might have evolved via gradual steps (Darwin 1872). These stages include the following:

- photosensitive cell
- aggregates of pigment cells without a nerve
- an optic nerve surrounded by pigment cells and covered by translucent skin
- pigment cells forming a small depression
- pigment cells forming a deeper depression
- the skin over the depression taking a lens shape
- muscles allowing the lens to adjust.

All of these steps are known to be viable because all exist in animals living today. The increments between these steps are slight and may be broken down into even smaller increments. Natural selection should, under many circumstances, favor the increments. Since eyes do not fossilize well, we do not know that the development of the eye followed exactly that path, but we certainly cannot claim that no path exists.

Nilsson and Pelger (1994) calculated that if each step were a 1 percent change, the evolution of the eye would take 1,829 steps, which could happen in 364,000 generations.

Further Reading: Dawkins, R. 1996. *Climbing Mount Improbable.* Chap. 5; Land, M. F., and D. E. Nilsson. 2001. *Animal Eyes*; Lindsay, D. 1998a. How long would the fish eye take to evolve? http://www.don-lindsay-archive.org/creation/eye_time.html.

CB302: The ear is too complex to have evolved.

The ear is too complex to have evolved. (W. Brown 1995, 7)

1. Not much complexity is needed for a functional ear. All that is necessary is a nerve connected to something that can vibrate. Insects have evolved "ears" on at least eleven different parts of their bodies, from antennae to legs (Hoy and Robert 1996). Even humans detect very low frequencies via tactile sensation, not through their ears.

2. The transition from reptile to mammal (see CC215) shows some of the intermediate stages in human hearing. Jaw bones, which likely helped the hearing of therapsid reptiles, became co-opted exclusively for hearing in the middle ear.

3. This is an example of the argument from incredulity (see CA100). That one does not know how something happened does not mean it cannot have happened.

CB303: The brain is too complex to have evolved.

The brain is too complex to have evolved. (W. Brown 1995, 7; Watchtower 1985, 168–178)

1. This is an argument from incredulity (see CA100). Complexity only indicates that something is difficult to understand, not that it is difficult to evolve. Evolution, unlike design, is not constrained by requirements for simplicity.

2. Brains come in many different sizes. The sea slug (*Aplysia*), for example, has only about 20,000 neurons in its entire nervous system. Coelenterates have an even simpler nervous system consisting of a nerve net and nothing even close to a brain. There are innumerable intermediate forms of brains between humans and brainless animals; gradual evolution of the brain presents no challenge.

CB310: The bombardier beetle is too complex to have evolved.

The bombardier beetle's complex defense mechanism (see CB310.1) *cannot be explained by evolution. It must have been designed.* (AIG 1990a; Gish 1977, 51–55)

1. This is an argument from incredulity (see CA100). It is based in part on an inaccurate description of how the beetle's bombardier mechanism works (see CB310.1), but even then the argument rests solely on the lack of even looking for evidence. In fact, an evolutionary pathway that accounts for the bombardier beetle is not hard to come up with (Isaak 1997).

One plausible sequence (much abbreviated) is thus:

a. Insects produce quinones for tanning their cuticle. Quinones make them distasteful, so the insects evolve to produce more of them and to produce other defensive chemicals, including hydroquinones.

b. The insects evolve depressions for storing quinones and muscles for ejecting them onto their surface when threatened with being eaten. The depression becomes a reservoir with secretory glands supplying hydroquinones into it. This configuration exists in many beetles, including close relatives of bombardier beetles (Forsyth 1970).

c. Hydrogen peroxide becomes mixed with the hydroquinones. Catalases and peroxidases appear along the output passage of the reservoir, ensuring that more quinones appear in the exuded product.

d. More catalases and peroxidases are produced, generating oxygen and producing a foamy discharge, as in the bombardier beetle *Metrius contractus* (Eisner et al. 2000).

e. As the output passage becomes a hardened reaction chamber, still more catalases and peroxidases are produced, gradually becoming today's bombardier beetles.

All of the steps are small or can be easily broken down into smaller ones, and all are probably selectively advantageous. Several of the intermediate stages are known to be viable by the fact that they exist in other living species.

2. Bombardier beetles illustrate other aspects of life that look undesigned:

- With design, we expect similar forms to be created for similar functions and different forms for different functions (H. M. Morris 1985, 70). However, what we see is different forms for similar functions. Many ground beetles have very similar habits and habitats as centipedes, but their forms differ greatly. Different groups of bombardier beetles use very different mechanisms for the same function of aiming their spray (Eisner 1958; Eisner and Aneshansley 1982).

- Some forms have no function. Some bombardier beetles have vestigial flight wings (Erwin 1970, 46, 55, 91, 114–115, 119).

- If bombardier beetles have a purpose, then death is an integral part of it, since the beetles are predators (some, as larvae, are parasitoids, gradually eating pupae of other beetles [Erwin 1967]), and their spray is a defense against other predators. Many creationists claim that death was not part of God's design.

Further Reading: Isaak, M. 1997. Bombardier beetles and the argument of design. http://www.talkorigins.org/faqs/bombardier.html; Weber, C. G. 1981. The bombardier beetle myth exploded.

CB310.1: Bombardier beetle chemicals would explode if mixed without an inhibitor.

The bombardier beetle would explode if the hydrogen peroxide and hydroquinone that produce their ejecta were mixed without a chemical inhibitor. Such a combination of chemicals could not have evolved. (Gish 1977, 51–52; Hitching 1982, 68)

1. That description of bombardier beetles' physiology is inaccurate. It is based on a sloppy translation of a 1961 German article by Schildknecht and Holoubek (Kofahl 1981). Hydrogen peroxide and hydroquinone do not explode when mixed (Dawkins 1986). What actually happens is this: Secretory cells produce a mixture of hydroquinones and hydrogen peroxide (and perhaps other chemicals), which collects in a reservoir. To produce the blast, the beetle releases some of this mixture into a reaction chamber, where catalases and peroxidases cause the mixture to oxidize in chemical reactions that generate enough heat to vaporize about a fifth of the mixture. The pressure of the released gasses causes the heated mixture to be expelled explosively from the beetle's abdomen (Aneshansley and Eisner 1969; Aneshansley et al. 1983; Eisner et al. 1989).

See for Yourself

Hydroquinone is available from photography shops and hydrogen peroxide, from supermarkets and drug stores. You can mix them yourself to see that they do not explode.

Further Reading: Isaak, M. 1997. Bombardier beetles and the argument of design. http://www.talkorigins.org/faqs/bombardier.html; Weber, C. G. 1981. The bombardier beetle myth exploded.

CB311: Butterfly metamorphosis is too complex to have evolved.

Butterfly metamorphosis is too complex to have evolved. (Poirier and Cumming 1993)

1. This is an argument from incredulity (see CA100). Because one does not understand how butterfly metamorphosis evolved does not mean it is too complex to have evolved.

Growth patterns intermediate to full metamorphosis already exist, ranging from growth with no metamorphosis (such as with silverfish) to partial metamorphosis (as with true bugs and mayflies), complete metamorphosis with relatively little change in form (as with rove beetles), and the metamorphosis seen in butterflies. It is surely possible that similar intermediate stages could have developed over time to produce butterfly metamorphosis from an ancestor without metamorphosis. In fact, an explanation exists for the evolution of metamorphosis based largely on changes in the endocrinology of development (Truman and Riddiford 1999).

CB325: The giraffe neck could not evolve without a special circulatory system.

A giraffe's heart must be quite large (it is over 24 lbs) to pump blood to the giraffe's head. A series of special one-way valves in the neck regulates blood flow, and there is a special net of elastic blood vessels at the base of the brain. Without these valves and elastic blood vessels, the blood pressure in the giraffe's head would be immense when it bends over, enough to cause brain damage. All of these features—large heart, valves in the jugular vein, and wondernet of vessels—must be in place simultaneously or the giraffe would die. They could not have evolved gradually. (P. Davis and Kenyon 1989, 69–72; Setterfield 1998)

1. Darwin answered this claim in 1868 (206). The claim assumes that "gradually" must mean "one at a time." Not so. The different features could have (and almost certainly would have) evolved both simultaneously and gradually. Partial valves would have been useful for reducing blood pressure to a degree. An intermediate heart would have produced enough pressure for a shorter neck. A smaller net of blood vessels in the head could have handled the lesser pressure. As longer necks were selected for, all of the other components would have been modified bit by bit as well. In other words, for each inch that the neck grew, the giraffe's physiology would have evolved to support such growth before the next inch of neck growth.

Further Reading: Gould, S. J. 1998a. The tallest tale.

CB326: The woodpecker tongue could not have evolved.

The unique arrangement of the woodpecker's tongue could not have evolved. If the tongue started anchored to the back of the beak, it would require a large sudden change to get to its present configuration.

1. The woodpecker's tongue (and hyoid apparatus, a rigid cartilage and bone skeleton of the tongue) is unusually long. However, it is simply an elongation of the same basic anatomy found in all birds. Like other birds, the main attachments are to the mandible, the cartilage of the throat, and the base of the skull. All that is required for the woodpecker's tongue to evolve is for it to grow longer, which could easily happen gradually.

The creationist claim stems from a mistaken understanding of the tongue anatomy. Creationists think the tongue is anchored in the nostril and grows backwards out of it. Although the back of the tongue in some species is long enough to extend to the nasal cavity, it is not anchored there.

Further Reading: Ryan, R. 2003. Anatomy and evolution of the woodpecker's tongue. http://www.talkorigins.org/faqs/woodpecker/woodpecker.html.

CB340: Organs and organ systems would have been useless until all parts were in place.
Organs and organ systems would have been useless until all the parts were in place. The coordinated innovation that they require is too improbable for evolution to create in one step. (Plantinga 1991b)

1. This claim is an instance of the argument from incredulity (see CA100). In all specific instances of this claim, there are ways for the organs and organ systems to evolve gradually. The idea that they could not evolve usually involves one or more of the following errors in thinking:

a. that organ parts appear suddenly. This seems to be an artifact of creationist thinking. Evolution, however, is not creationism; parts change gradually.

b. that organs less developed than what exists now must be completely useless. This is nonsense. A light-sensitive patch on the skin may not be as useful as the eyes we have now, but it is better than nothing. And just a little bit better is all that is required for the trait to evolve.

c. that parts must evolve separately. Coordinated innovation between parts of an organ or organ system is possible. Indeed, if the parts evolve gradually, it is inconceivable that parts that interact would not coevolve in such a way that changes are coordinated via natural selection.

d. that parts do not change function. Many organs do not start from nothing. Rather, they start as a part that serves a different function and gradually gets co-opted for a new function. For example, tetrapod legs evolved from fins.

The invalidity of specific examples of this claim are seen in the the bombardier beetle (see CB310), the giraffe neck (see CB325), and the woodpecker tongue (see CB326).

2. Sometimes multiple coordinated changes can occur when there is a mutation in a regulatory gene.

CB350: Sex cannot have evolved.
Sex is too complex for its origin to be explained by evolution. Males and females would have to evolve independently, and any incompatibility in any of the physical, chemical, or behavioral components would have caused extinction. Furthermore, evolutionary theory predicts that asexual reproduction would be favored because asexual species can reproduce faster. (W. Brown 1995, 14–15)

1. The variety of life cycles is very great. It is not simply a matter of being sexual or asexual. There are many intermediate stages. A gradual origin, with each step favored by natural selection, is possible (Kondrashov 1997). The earliest steps involve single-celled organisms exchanging genetic information; they need not be distinct sexes. Males and females most emphatically would not evolve independently. Sex, by definition, depends on both males and females acting together. As sex evolved, there would have been some incompatibilities causing sterility (just as there are today), but these would affect indi-

viduals, not whole populations, and the genes that cause such incompatibility would rapidly be selected against.

2. Many hypotheses have been proposed for the evolutionary advantage of sex (Barton and Charlesworth 1998). There is good experimental support for some of these, including resistance to deleterious mutation load (Davies et al. 1999) and more rapid adaptation in a rapidly changing environment, especially to acquire resistance to parasites (Sá Martins 2000).

Further Reading: Judson, O. 2002. *Dr. Tatiana's Sex Advice to All Creation*; Margulis, L., and D. Sagan. 1990. *The Origins of Sex*; Wuethrich, B. 1998. Why sex?

CB360: Vestigial organs may have functions.

Practically all "vestigial" organs in man have been shown to have definite uses and not to be vestigial at all. (H. M. Morris 1985, 75–76)

1. "Vestigial" does not mean an organ is useless. Per Merriam-Webster's dictionary, a vestige is a "trace or visible sign left by something lost or vanished." Examples from biology include leg bones in snakes, eye remnants in blind cave fish (Yamamoto and Jeffery 2000), extra toe bones in horses, wing stubs on flightless birds and insects, and molars in vampire bats. Whether these organs have functions is irrelevant. They obviously do not have the function that we expect from such parts in other animals, for which creationists say the parts are "designed."

Vestigial organs are evidence for evolution because we expect evolutionary changes to be imperfect as creatures evolve to adopt new niches. Creationism cannot explain vestigial organs. They are evidence against creationism if the creator follows a basic design principle that form follows function, as H. M. Morris himself expects (1985, 70). They are compatible with creation only if anything and everything is compatible with creation, making creationism useless and unscientific.

2. Some vestigial organs can be determined to be useless if experiments show that organisms with them survive no better than organisms without them.

Further Reading: Theobald, D. 2004. 29+ evidences for macroevolution. http://www.talkorigins.org/faqs/comdesc/section2.html#morphological_vestiges.

CB360.1: The human appendix is functional, not vestigial.

The human appendix is not really vestigial. It has an immunological function as part of the lymphatic system. Its lymphoid follicles produce antibodies. (Ham and Wieland 1997)

1. Vestigial does not mean functionless (see CB360). The appendix appears as part of the tissues of the digestive system; it is homologous to the end of the mammalian caecum. Since it does not function as part of the digestive system, it is a vestigial part of that system, no matter what other functions it may have.

2. The human appendix may not be functional. Its absence causes no known harmful effects (other than surgical complications from removing it). When it is present, there is

a 7 percent lifetime risk of acute appendicitis, which is usually fatal without modern surgical techniques (Hardin 1999).

3. Co-opting a part for an entirely different function, such as turning part of the intestines into part of the lymphatic system, is entirely compatible with, and even expected from, evolution. However, it argues against design because

1. it rarely occurs with known (human) designs, and

2. it invalidates design arguments, such as irreducible complexity (see CB200).

Further Reading: Theobald, D. 2003. The vestigiality of the human vermiform appendix. http://www.talkorigins.org/faqs/vestiges/appendix.html.

CB361: Vestigial organs are just evidence of decay, not evolution.

Vestigial organs (if any really exist; see CB360) are not evidence of evolution. They just show decay consistent with the second law of thermodynamics (H. M. Morris 1985, 75–76)

1. Vestigial organs include more than atrophied organs. The bones of the middle ear, for example, are vestiges of jaw bones of ancestral tetrapods.

2. Loss of organs is sometimes an advantage. For example, loss of legs is adaptive in whales. Thus, losses of organs often are evolution driven by natural selection. They are evidence of evolution when their vestigial forms show similarities to earlier nonvestigial forms.

3. The second law of thermodynamics allows for more than decay (see CF001).

CB370: Endorphins at death indicate a beneficent creator, not evolution.

The body releases endorphins at death, making dying less traumatic. Evolution cannot explain this, because there is no selection pressure for a biological mechanism to help people die peacefully. But it does show evidence of a merciful creator. (Easterbrook 2003)

1. Endorphin levels are elevated even by relatively minor trauma. Alleviating the pain and stress of a person who has suffered a trauma can keep them functional enough to take life-saving actions, such as staunching the bleeding or seeking help. That endorphins are released also during fatal trauma is a side effect.

2. Release of endorphins during trauma is not entirely merciful. If the person survives the trauma, the withdrawal from the endorphins can produce emotional distress that can contribute to post-traumatic stress disorder or alcohol addiction (Volpicelli et al. 1999).

BEHAVIOR AND COGNITION

CB400: Evolution cannot explain consciousness.

Evolution cannot explain the origin of consciousness or free will. (P. E. Johnson 1990)

1. This is an argument from ignorance (see CA100). Not knowing an explanation does not mean an explanation is impossible. And since we are barely beginning to understand what consciousness is, it is not surprising that we would not have its origin worked out yet.

In fact, preliminary explanations for the origin of consciousness have been proposed, although they are too complicated to try to summarize here (see Dennett 1991; Minsky 1985). Much more experimentation and refinement is needed before we have a full-fledged theory of the origin of consciousness, but we have more than enough to know that such a theory is possible.

2. A factor that likely contributes to the claim of consciousness's inexplicability is the fact that many people do not want a naturalistic explanation of consciousness, since a natural consciousness does not fit easily with a divine soul. This threatens people's desire for a divine origin and immortality (but see Dennett 1991, 430, for immortality of a naturalistic consciousness). An examination of this point alone could fill a book. However, suffice it to say,

1. There is much evidence—from genetic predispositions of behavior and personality, from brain injury studies, from brain imaging of healthy people—that consciousness is naturalistic now. A natural origin would not matter much beyond that.

2. What we want has no bearing on what really is.

CB400.1: Evolution does not explain human intelligence.
Evolution does not explain how humans became so intelligent. (Yahya 2004)

1. Intelligence has obvious advantages that can help with survival, so it is consistent with evolutionary theory. What remains to be explained is why human brains are significantly larger (relative to body size) than brains of other animals. Several hypotheses have been proposed, including the following:

- Greater intelligence allows more effective foraging, especially for learning and remembering where and when fruits ripen (Kaplan et al. 2000).
- Greater intelligence allows more effective tool invention and use.
- Greater intelligence allows more successful functioning in the complex social groups that primates form. This could be intertwined with the development of language (see CB402).

The last hypothesis is probably the most likely, although they are not mutually exclusive.

Much of how larger brains evolved may be explained by neoteny, the prolonging of immature periods of development. Evolving a larger brain does not require large genetic change, just a relatively small change to keep the brain growing for a longer time (S.J. Gould 1977a).

Further Reading: Cosmides, L., and J. Tooby. 1997. Evolutionary psychology. http://www .psych.ucsb.edu/research/cep/primer.html.

CB401: Instincts are too complex to have evolved.

Instincts are too complex to have evolved. Scientists have no way of explaining complex instincts, such as migrations and nest building. (Watchtower 1985, 160–167)

1. This claim is an example of the argument from incredulity (see CA100). One's inability to envision circumstances that lead to complex instincts does not preclude such circumstances.

2. Not all instincts are complex. Some phobias, for example, are no more than a basic emotional response to a simple stimulus, such as loud sudden noise. And there is nothing to prevent the complex instincts from arising gradually. For example, some bees only communicate information about flower species to others in the hive (Dornhaus and Chittka 1999). Complex instincts could arise via small steps such as this.

Further Reading: Gould, J.L., and C.G. Gould. 1995. *The Honey Bee.* Weiner, J. 1999. *Time, Love, Memory.*

CB402: Evolution does not explain language ability.

Evolution does not explain the evolution of language ability. (Watchtower 1985, 174–175; Yahya 2004)

1. We do not know the definitive explanation for how language ability arose, but there are plausible hypotheses. Intermediate stages may be reached by gradual changes. For example:

- The larger brain size of primates arose before language; it may have come from adaptation for functioning in social groups or for finding food.

- Human brain size has evolved to be unusually large through neoteny; the rapid brain growth of childhood is maintained for a longer time. It may have arisen in conjunction with the evolution of language or before it.

- Before language, some communication already was done via gestures (which itself probably arose as mimetic imitation) and via vocalizations. Spoken language ability may have built on those vocalizations, or it may have been transferred from gestural language, or both.

- Once vocal protolanguage began, the vocal tract and neural connections for producing and controlling speech would have begun evolving. Speech probably began in *Homo erectus* more than a million years ago.

- Once language becomes a central part of human behavior, language acquisition comes more readily because of the Baldwin effect; the plasticity of learning allows natural selection for language ability to proceed more quickly (Baldwin 1896).

2. Language is obviously useful, so it is not hard to see how it could have provided selective advantages. In fact, the bigger problem is explaining why chimpanzees did not evolve language. Language was probably primarily an adaptation to a social structure that was (and is) far more complex than that of other primates.

Further Reading: Deacon, T. W. 1998. *The Symbolic Species.* Johansson, S. 2002. The evolution of human language capacity. http://home.hj.se/~lsj/langevod.pdf (esp. pp. 65ff); Pinker, S. 1994. *The Language Instinct;* Tattersall, I. 2001. How we came to be human.

CB403: Evolution does not explain homosexuality.

Evolution does not explain homosexuality. Traits evolve as a result of greater reproductive success, and homosexuals are less likely to reproduce. (Macks 2000)

1. There are several possible explanations for this:

- Although homosexuality probably has a genetic component, much of its cause, perhaps most of it, appears to be nongenetic (Haynes 1995; Kendler et al. 2000; Kirk et al. 2000). To the extent it is not genetic, selection would not affect it.

- Homosexuals still have children. Sexual orientation is not an either–or trait but exists as a continuum (Haynes 1995). Those with some heterosexual orientation can still contribute homosexual genes (to the extent it is genetic; see above). And even the most extreme homosexuals sometimes have children.

  The most manifest heterosexuals may have homosexual tendencies, too. Homophobic male heterosexuals showed more arousal to homosexual images than did nonhomophobic heterosexuals (Adams et al. 1996). Societal condemnation of homosexuality may contribute to its genes being propagated by causing latent homosexuals to behave heterosexually.

- Genes for homosexuality could be beneficial on the whole. In bonobo chimpanzees, homosexual interactions are a form of social cement. It is possible that homosexuality evolved to serve social functions in humans, too (Kirkpatrick 2000). After all, social cohesion is still a main function of sex in humans.

  The genetic etiology of homosexuality may come from a collection of traits that, when expressed strongly and in concert, result in homosexuality; expressed less strongly or without supporting traits, these traits contribute to the robust nature of our species. The genes for these traits persist because they usually combine to make us better at survival and reproduction.

- Genes for homosexuality could be spread through kin selection, if the homosexuals care for their siblings' offspring. However, this explanation is unlikely (Kirkpatrick 2000).

It should be noted that the question of explaining homosexuality is not limited to humans. Homosexuality exists in hundreds of animal species (Bagemihl 1998).

Further Reading: Kirkpatrick R. C. 2000. The evolution of human homosexual behavior; Wright, R. 1994. *The Moral Animal*, pp. 384–386.

with Elf M. Sternberg

CB411: Evolution does not explain morals, especially altruism.

Evolution cannot explain moral behavior, especially altruism. Evolutionary fitness is selfish; individuals win only by benefitting themselves and their offspring. (Watchtower 1985, 177)

1. The claim ignores what happens when organisms live socially. In fact, much about morals can be explained by evolution. Since humans are social animals and they benefit from interactions with others, natural selection should favor behavior that allows us to better get along with others.

Fairness and cooperation have value for dealing with people repeatedly (Nowak et al. 2000). The emotions involved with such justice could have evolved when humans lived in small groups (Sigmund et al. 2002). Optional participation can foil even anonymous exploitation and make cooperation advantageous in large groups (Hauert et al. 2002).

Kin selection can explain some altruistic behavior toward close relatives; because they share many of the same genes, helping them benefits the giver's genes, too. In societies, altruism benefits the giver because when others see someone acting altruistically, they are more likely to give to that person (Wedekind and Milinski 2000). In the long term, the generous person benefits from an improved reputation (Wedekind and Braithwaite 2002). Altruistic punishment (punishing another even at cost to yourself) allows cooperation to flourish even in groups of unrelated strangers (Fehr and Gächter 2002).

Finally, evolution does not require that all traits be adaptive 100 percent of the time. The altruism that benefits oneself most of the time may contribute to life-risking behavior in some infrequent circumstances.

Further Reading: Netting, J. 2000. Model of good (and bad) behaviour. http://www.nature.com/nsu/001026/001026-2.html; Sigmund, K., et al. 2002. The economics of fair play; Wright, R. 1994. *The Moral Animal.*

CB440: Evolution does not explain religion.
Evolution does not explain religion.

1. Religion fits comfortably with evolutionary theory. Some of its important components probably even predate humanity. For example:

- A fear of death has obvious survival advantages and is probably as old as emotions. When intellect evolved to the point that imagination became possible, we could start thinking about alternatives.

- Humans and other primates live in dominance hierarchies. A social structure with "higher" and "lower" beings is part of our genes. We can always point to other animals as lower beings, but sometimes a higher being requires something unobvious.

- With the origin of symbolic thinking (which language requires), the abstract higher beings could be thought of in more specific terms.

- With language, gods could be talked about. From there, religion developed via cultural evolution.

- Fear of the unknown gives further reason for believing in gods. Dealings with a god, such as sacrificial offering or intercessory prayer, allow one the impression of some influence over events that are beyond one's control.

2. Religions themselves evolve (Cullen 1998; Gottsch 2001).

3. The claim is an argument from incredulity (see CA100). Even if no explanation is known, that does not mean there is none.

Further Reading: Boyer, P. 2001. *Religion Explained*; Burkert, W. 1996. *Creation of the Sacred* (It concentrates on early cultural aspects of religion but still has some relevance to religion's origin.); Dunbar, R. 2003. Evolution: Five big questions; Konner, M. 2002. *The Tangled Wing.*

BOTANY

CB501: Dendrochronology is suspect because two or more rings can grow per year.

Two or more growth periods frequently occur during a year, so dates derived from tree rings (dendrochronology) are suspect. (H. M. Morris 1985, 193)

1. For some trees, including bristlecone pine, ponderosa pine, and douglass fir, double rings are rare and easy to spot with a little practice. A bigger problem is missing rings; a bristlecone pine can have up to 5 percent of its rings missing. Thus, dates derived from dendrochronology, if they are suspect at all, should indicate ages too young.

For most of the dendrochronological record, dates are determined from more than one source, so errors can be spotted and corrected.

2. Dendrochronology is in rough agreement with C-14 dating, so even if it is off, it is not off by much—certainly not by orders of magnitude, as young-earth claims would require.

Further Reading: Matson, D. E. 1994. How good are those young-earth arguments? http://www.talkorigins.org/faqs/hovind/howgood-yea2.html#proof27.

CB510: Evolution cannot explain photosynthesis.

Evolution cannot explain the origin of photosynthesis. (C. Nelson 2001)

1. This is yet another argument from incredulity (see CA100). Not knowing how something evolves is a limit of ourselves, not of evolution.

We would not expect the evolution of photosynthesis to be easy to unravel. Photosynthesis has been evolving for more than three billion years, originating even before eukaryotes (Awramik 1992). Its early history involved gene transfer among several phyla of bacteria, making the trail harder to trace genetically (Raymond et al. 2003). Different components of photosynthesis have independent evolutionary pathways. However, much progress has been made in determining how photosynthesis evolved (Baymann et al. 2001; Blankenship 1992; Blankenship and Hartman 1998; Xiong and Bauer 2002).

Further Reading: Xiong, J., and C. E. Bauer. 2002. Complex evolution of photosynthesis.

ECOLOGY AND POPULATION BIOLOGY

CB601: The traditional peppered moth story is no longer supportable.

According to the traditional peppered moth story, cryptic coloration confers protection to the moths from predators, and as the habitat changed due to industrial pollution, natural selection caused the frequencies of different color varieties of the moth to change; as the trees became darker, the lighter moths stood out more, so the darker ones became more plentiful, and vice versa as the pollution cleared. That story is no longer supportable because of flaws found in the experiments, such as where the moths rested (see CB601.1), and the occurrence of contrary data, such as unaccountable frequencies of uncamouflaged moths in areas. (see CB601.2; J. Wells 1999; 2000, 137–157)

1. Although the experiments were not perfect, they were not fatally flawed. Even though Kettlewell released his moths in daylight when a night release would have been more true to nature, he used the same procedure in areas that differed only in the amount of industrial pollution, showing conclusively that industrial pollution was a factor responsible for the difference in predation between color varieties. Similar arguments can be made for all other experiments. Although no experiment is perfect (nor can be), even imperfect experiments can give supporting or disconfirming evidence. In the case of peppered moths, many experiments have been done, and they all support the traditional story (B. S. Grant 1999).

2. Even without the experiments, the peppered moth story would be well established. Peppered moth melanism has both risen and fallen with pollution levels, and they have done so in many sites on two continents (L. M. Cook 2003; B. S. Grant 1999).

3. The peppered moth story is consistent with many other experiments and observations of crypsis and coloration in other species. For example, bird predation maintains the colorations of *Heliconius cydno*, which has different coloration in different regions, in both regions mimicking a noxious *Heliconius* species (Kapan 2001). Natural selection acting on the peppered moth would be the parsimonious hypothesis even if there were no evidence to support it.

4. The peppered moth story is not simple. The full story as it is known today fills thousands of pages of journal articles. Familiarity with the literature and with the moths in the field is needed to evaluate all the articles. But the research and the debates over its implications have all been done in the open. Charges of fraud and misconduct stem from neglect and misrepresentation of the research by the people making the charges (B. S. Grant 2000). Of those familiar with the literature, none doubt that bird predation is of primary importance in the changing frequencies of melanism in peppered moths (Majerus 1999).

In teaching any subject to beginners, simplifying complex topics is proper. The peppered moth story is a valuable tool for helping students understand how nature really works. Teachers would be right to omit the complexities from the story if they judged that their students were not yet ready for that higher level of learning (Rudge 2000).

Further Reading: Gishlick, A. D. n.d. Icons of evolution? Peppered moths. http://www.ncseweb.org/icons/icon6moths.html; Majerus, M. E. N. 1998. *Melanism*; Rudge, D. W. 1999. Taking the peppered moth with a grain of salt; Rudge, D. W. 2000. Does being wrong make Kettlewell wrong for science teaching?; Tamzek, N. 2002. Icon of obfuscation. http://www.talkorigins.org/faqs/wells/iconob.html#moths.

CB601.1: Peppered moths do not rest on tree trunks, and pictures of them there were faked.

Peppered moths do not normally rest on tree trunks. In decades of field work, only one peppered moth was found resting on a tree trunk in the wild. Kettlewell released his moths near the ground in the morning, which would have caused the moths to land on the trunks unnaturally. Pictures showing moths on trunks were staged. This invalidates the research that was based on the assumption that they normally rest on trunks. (J. Wells 1999)

1. Peppered moths do not rest exclusively on tree trunks, but they do rest there. Of the forty-seven moths one researcher found in the wild, twelve were on trunks and twenty were on trunk or branch joints. (The other fifteen were on branches.) The numbers and proportion on trunks near light traps were even higher (Majerus 1998, 123). Wells's claim that the moths do not naturally land on trunks is simply a falsehood.

2. Branches provide a background similar to trunks. Photos showing moths on trunks were staged but only for purposes of illustration. The photographs depict what is found in the wild, whether trunk or branch. Furthermore, the photos played no part in the scientific research or its conclusions.

Further Reading: Frack, D. 1999. Peppered moths, round 2. http://www.calvin.edu/archive/evolution/199904/0100.html, . . . /0103.html, . . . /0200.html, . . . /0201.html; Grant, B. S. 1999. Fine tuning the peppered moth paradigm. http://mason.gmu.edu~jlawrey/biol471/melanism.pdf.

CB601.2: Peppered moths frequencies are inconsistent with selection.

*Various aspects of the change in frequencies of the varieties of peppered moth (*Biston betularia*) and their geographical distribution are inconsistent with the proposed explanation in terms of natural selection by birds reducing the less camouflaged varieties. (J. Wells 1999, 2000) For example:*

- Melanic (dark) moths did not replace lighter moths in heavily polluted Manchester, contrary to expectations based on their selective advantage.
- Melanic moths were more common than expected in East Anglia and southern Wales.
- There was no correlation between melanism and air pollution in southern England.
- Melanic moths increased in southern Britain after pollution controls began.
- Light moths increased in the Wirral Peninsula before the reappearance of lichens, which provide them camouflage.

1. Although differential predation by birds has long been considered an important factor—probably the most important—affecting relative frequencies of moth varieties,

it was never considered the only factor. Other factors affecting their relative frequencies are

- dispersal. Peppered moths can fly 2 km or more per night and probably disperse about that far, on average, per generation (Cook 2003).

- nonvisual selection. Bird predation aside, homozygous melanic moths (with two alleles for melanism) are more hardy than hybrids with nonmelanic moths (Creed et al. 1980).

- frequency-dependent selection. Predators tend to overlook rare forms and go after more common insects instead.

When these factors are included in models along with predation, the models match the observations very well (Mani 1982, 1990). In particular, these factors explain the commonness of melanic moths in East Anglia and southern Wales and the presence of light moths in Manchester.

2. Wells's claim that melanism showed no correlation with sulfur dioxide in southern England (J. Wells 2000, 146) is simply false. He apparently misinterpreted a statement that south of 52 degrees north, melanism was significantly correlated with sulfur dioxide but less so than with east–west location (Steward 1977). A strong east-west correlation is consistent with the distribution of England's heavy industry.

3. The increase of melanic moths in southwest England (not southern England, as Wells states) after pollution control was introduced was quite small. It remains unexplained, but it is plausible that the melanic moths had not yet reached equilibrium in that area when the pollution control started. Thus, while the pollution decreased slightly, selection still favored the melanic moths.

4. Light moths appearing before lichens may be explained by any of several factors:

- Surveys of the reappearing lichens were unsystematic and not entirely relevant to the biology of moth predation. For example, crustose lichens that recolonize trees when pollution lessens may grow first on new branches in the canopy, but lichen surveys looked mostly at the trunks (Majerus et al. 2000).

- Even though the lichens had not reappeared, the trees were lighter, giving the lighter moths more of a selective advantage than they had before.

- Light moths would migrate from surrounding areas where lichens had reappeared.

Further Reading: Grant, B.S. 1999. Fine tuning the peppered moth paradigm. http://mason.gmu.edu/~jlawrey/biol471/melanism.pdf.

CB601.3: Direct mutagenesis better explains peppered moth variation.

The differing frequencies of the different color varieties of peppered moths can better be explained by the direct effects of mutagenic pollutants, since the usual explanation of differential bird predation lacks support. (J. Wells 1999)

1. The explanation of differential bird predation affecting frequencies of color varieties does not lack support. On the contrary, support for the basic story is overwhelming

(B.S. Grant 1999; Majerus 1999). Wells and others simply ignore the bulk of the research and pick out the few dissenters and the disagreements about minor details.

2. An explanation in terms of the direct effect of pollutants fails. The early experiments (by Helsop Harrison) that claimed to show such an effect lacked appropriate controls, could not be replicated, and reported mutagenesis levels that were more than enough to induce complete sterility (Majerus 1999).

Further Reading: Majerus, M.E.N. 1998. *Melanism.*

CB601.4: An increased recapture rate suggests fraud in Kettlewell's data.

In one of Bernard Kettlewell's peppered moth studies, his moth recapture rate increased greatly beginning on July 1. July 1 was also the date on which E. B. Ford sent Kettlewell a letter commiserating with him for the low recapture rates. This suggests that Kettlewell cheated to increase his recapture rates. (J. Hooper 2002)

1. Kettlewell would not have received Ford's letter before the increase in his recapture rate had already begun. The collection of July 1 was completed by the early morning. Since Ford's letter would have arrived after, it could not have been a factor.

2. Kettlewell recaptured more moths after July 1 because he was releasing more moths then. The number of moths he collected is not significantly different from the collections one would expect on the basis of the number of moths released in the two prior days (M. Young 2004). He probably released more moths because the moths he was rearing reached adulthood then.

3. Much of the remaining variation in recapture rate might be explained by moonlight. The low recapture rates occurred when the moon was full, and for many species of moths, a full moon reduces the numbers caught in light traps.

4. Kettlewell's conclusions were based not only on his recapture experiment but also on three other investigations. The same conclusions were found by many other experiments on peppered and other moths by other researchers (B.S. Grant 1999).

Further Reading: Young, M. 2004. Moonshine. http://www.talkdesign.org/faqs/moonshine.htm.

CB610: The first individual of a new species would not find a mate.

The first individual of the new species would be very unlikely to find a mate. Hybrids are infertile, so a newly evolved individual would not be able to breed successfully with the original species. The mutation that caused that individual to be a new species would also have to occur in an individual of the opposite sex. (TCCOP 1998)

1. This objection falsely assumes that speciation must happen suddenly when one individual gives rise to an individual of another species. In fact, populations, not individuals, evolve, and most speciation occurs gradually. In one common mode of speciation

(allopatric speciation), two populations of the same species are split apart geographically. Small changes accumulate in both populations, causing them to be more and more different from each other. Eventually, the differences are great enough that the two populations cannot interbreed when they do get together (Otte and Endler 1989).

It is also possible for speciation to occur without the geographical separation (sympatric speciation; Diekmann and Doebeli 1999; Kondrashov and Kondrashov 1999; Otte and Endler 1989), but the process is still gradual.

2. Sometimes new species can form suddenly, but this occurs with species that are asexual or hermaphroditic and do not need to find mates.

Further Reading: Schilthuizen, M. 2001. *Frogs, Flies, and Dandelions.*

CB620: Human population growth indicates a young earth.

A reasonable assumption of population growth rate (0.5 percent) fits with a population that began with two people about 4,000 years ago, not with a human history of millions of years. (H. M. Morris 1985, 167–169)

1. This claim assumes that the population growth rate was always constant, which is a false assumption. Wars and plagues would have caused populations to drop from time to time. In particular, population sizes before agriculture would have been severely limited and would have had an average population growth of zero for any number of years.

2. There is no particular reason to choose a population growth rate of 0.5 percent for the calculation. The population growth from 1900 to 2000 has been closer to 0.132 percent per year (Encyclopaedia Britannica 1984; G. Martin 1999). At that rate, the population would have grown to its present size from the eight Flood survivors in 15,500 years. And recent population growth has been historically high.

3. The population growth rate proposed by the claim would imply unreasonable populations early in history. We will be more generous in our calculations and start with eight people in 2350 B.C.E. (a traditional date for the Flood). Then, assuming a growth rate of 0.5 percent per year, the population after N years is given by

$$P(N) = 8 \times (1.005)^N$$

The Pyramids of Giza were constructed before 2490 B.C.E., even before the proposed Flood date. Even if we assume they were built 100 years after the flood, then the world population for their construction was 13 people. In 1446 B.C.E., when Moses was said to be leading 600,000 men (plus women and children) on the Exodus, this model of population growth gives 726 people in the world. In 481 B.C.E., Xerxes gathered an army of 2,641,000 (according to Herodotus), when the world population, according to the model, was 89,425. Even allowing for exaggerated numbers, the population model makes no sense.

Further Reading: Elsberry, W. R. 1998. Population size and time of creation or Flood. http://www.rtis.com/nat/user/elsberry/evobio/evc/argresp/populate.html.

CB621: Humanity was traced back to an African Eve.

Scientists have traced humanity back to an African Eve. (W. Brown 2001, FAQ13)

1. The "mitochondrial Eve," to which this claim refers, is the most recent common female ancestor, not the original female ancestor. There were humans living before her and at the same time as her.

2. The results assume no paternal inheritance of mitochondrial DNA, but that assumption has been called into question. Male mtDNA resides in the tail of the sperm; the tail usually does not enter the egg that the sperm fertilizes, but rarely a little bit does. It is also possible that there is some recombination of mtDNA between lineages, which would also affect the results (Awadalla et al. 1999; Eyre-Walker et al. 1999).

3. The same principles find that the most recent human male common ancestor lived 84,000 years after the "mitochondrial Eve" (Hawkes 2000).

CB630: Mutually dependent species could not have evolved.

Many pairs of species are mutually and completely dependent upon each other. For example, fig trees require fig gall wasps to pollinate them, and the wasps require the figs to live. Another such example, among many, is the yucca and Pronuba *yucca moth. If one species evolved first, it could not have survived on its own; the mutualistic species must have come into existence at essentially the same time.*

1. Obligate mutualism can evolve gradually from nonobligate associations. For example, a yucca can be pollinated by many insects and gradually specialize to attract just one moth, and a moth may live off many species of yucca and gradually specialize on just one. In fact, some yucca species are still pollinated by more than one moth, and the yucca moth *Tegeticula yuccasella* pollinates several species of yucca (Powell 1992).

Further Reading: Dawkins, R. 1996. *Climbing Mount Improbable.* Chaps. 8 and 10; Pellmyr, O., et al. 1996. Evolution of pollination and mutualism in the yucca moth lineage.

DEVELOPMENTAL BIOLOGY

CB701: Haeckel falsified his embryo pictures.

Haeckel faked his pictures of embryos to make them look more alike than they are. (J. Wells 2000, 81–109)

1. Haeckel's pictures are irrelevant to the question of whether the embryos are similar. What matters are the embryos themselves. Within a group, early embryos do show many similarities. For example, all vertebrates develop a notochord, body segments, pharyngeal gill pouches, and a post-anal tail. These fundamental similarities indicate a common evolutionary history. Other embryological similarities are found in other lineages, such as mollusks, arthropods, and annelids. These similarities have been long known.

Professor Agassiz in 1849, for example, said, "We find, too, that the young bat, or bird, or the young serpent, in certain periods of their growth, resemble one another so much that he would defy any one to tell one from the other—or distinguish between a bat and a snake" (Scientific American 1849).

2. The embryos also show some differences, which Haeckel glossed over. However, differences should also be expected, since the animals are not all equally related. It is the pattern of both similarities and differences that displays patterns of descent. Organisms that are less closely related are expected to look less similar.

3. When Haeckel's inaccuracies were exposed, authors started using corrected versions. Science tends to be self-correcting.

Further Reading: Myers, P. Z. 2003. Wells and Haeckel's embryos. http://www.talkorigins.org/faqs/wells/haeckel.html; Richardson, M. K., and G. Keuck. 2002. Haeckel's ABC of evolution and development; Richardson, M. K., et al. 1998. Haeckel, embryos, and evolution.

CB701.1: Recapitulation theory is not supported.

The biogenetic law that ontogeny recapitulates phylogeny (that is, that the embryological stages of a developing organism follow the organism's evolutionary history) is false, yet embryological stages are still claimed as evidence for evolution. (H. M. Morris 1985, 76–77)

1. Haeckel's biogenetic law was never part of Darwin's theory and was challenged even in his own lifetime. Haeckel himself did not necessarily advocate the strict form of recapitulation commonly attributed to him (Richardson and Keuck 2002).

2. Irrespective of biogenetic law, embryological characters are still useful as evidence for evolution (in constructing phylogenies, for example), just as adult characters are. Furthermore, there is some degree of parallelism between ontogeny and phylogeny, especially when applied only to individual characters (Richardson and Keuck 2002). Various causes for this have been proposed. For example, there is selective pressure to retain embryonic structures that are needed for the development of other organs.

Further Reading: Chase, S. 1999. Is Haeckel's law of recapitulation a problem? http://www.talkorigins.org/origins/postmonth/feb99.html; Gould, S. J. 1977c. *Ontogeny and Phylogeny*; Richardson, M. K., and G. Keuck. 2002. Haeckel's ABC of evolution and development; Wilkins, J. 1996. Darwin's precursors and influences. http://www.talkorigins.org/faqs/precursors/precurstrans.html.

CB704: Human embryos do not have gill slits.

Human embryos do not have gill slits; they have pharyngeal pouches. In fish, these develop into gills, but in reptiles, mammals, and birds, they develop into other structures and are never even rudimentary gills. Calling them gill slits is reading Darwinian theory into the evidence. There is no way gill slits can serve as evidence for evolution. (J. Wells 2000, 105–107)

1. The pharyngeal pouches that appear in embryos technically are not gill slits, but that is irrelevant. The reason they are evidence for evolution is that the same structure, whatever you call it, appears in all vertebrate embryos. Agassiz (not a Darwinist himself) said, "The higher Vertebrates, including man himself, breathe through gill-like organs in

the early part of their life. These gills disappear and give place to lungs only in a later phase of their existence" (Agassiz 1874).

Darwinian evolution predicts, among other things, similar (not identical) structures in related organisms. That pharyngeal pouches in humans are similar to pharyngeal pouches (or whatever you call them) in fish is one piece of evidence that humans and fish share a common ancestor.

Further Reading: Gilbert, S. F. 1988. *Developmental Biology.*

CB710: Genes with major effects on development are conserved across phyla.

Evolution predicts that different kinds of organisms should have different genetic programs, but the genes responsible for major effects early in development are very similar across diverse phyla. For example, the gene distal-less *gives rise to limbs in several different phyla, but the limbs themselves are not structurally or evolutionarily homologous. (J. Wells 2000, 74–76)*

1. Wells's observation is quite in accord with what evolution predicts. Genes responsible for early developmental effects should not be expected to change greatly because small changes early in development produce large changes later, and those large changes are likely to be damaging. It makes more sense that the genes to change would be the ones that modify and regulate the early developmental genes. Thus, the genes that cause a limb to grow would be conserved, but many of the genes that regulate its form would not.

2. There are significant differences between phyla, too. Although any given HOX gene (the genes that determine much of an animal's basic body plan) may be very similar between phyla, different phyla differ in which HOX genes they carry and in how many copies of each they have (Carroll 1997).

Further Reading: Carroll, R. L. 1997. *Patterns and Processes of Vertebrate Evolution.* See especially chap. 10.

SYSTEMATICS

CB801: Science cannot define "species."

Complaints about creationists not defining "kind" are unfair since evolutionists cannot define "species" consistently.

1. Species are expected often to have fuzzy and imprecise boundaries because evolution is ongoing. Some species are in the process of forming; others are recently formed and still difficult to interpret. The complexities of biology add further complications. Many pairs of species remain distinct despite a small amount of hybridization between

them. Some groups are asexual or frequently produce asexual strains, so how many species to split them into becomes problematic.

Creation, defining things as kinds that were created once and for all, implies that all species should be clearly demarcated and that there should be a clear and universal definition of kind or species. Since there is not, creationism, not evolutionary theory, has something to explain.

2. Different definitions of species serve different purposes. Species concepts are used both as taxonomic units, for identification and classification, and as theoretical concepts, for modeling and explaining. There is a great deal of overlap between the two purposes, but a definition that serves one is not necessarily the best for the other. Furthermore, there are practical considerations that call for different species criteria as well. Species definitions applied to fossils, for example, cannot be based on genetics or behavior because those traits do not fossilize.

Further Reading: Claridge, M., et al. (eds.). 1997. *Species: The Units of Biodiversity*, pp. 357–424 (technical); Cracraft, J. 1987. Species concepts and the ontology of evolution; Schilthuizen, M. 2001. *Frogs, Flies, and Dandelions*. See especially chap. 1; Wilkins, J. S. 2003. How to be a chaste species pluralist–realist.

with John Wilkins

CB805: Evolution predicts a continuum of organisms, not discrete kinds.

Since evolution says organisms came from a common ancestor and since they lived in a continuity of environments, we should see a continuum of organisms. There should be a continuous series of animals between cats and dogs, so that one could not tell where cats left off and dogs began. (H. M. Morris 1985, 70–71)

1. The claim might be true if there were no such thing as extinction. But since species do become extinct, intermediates that once existed do not exist today. Since extinction is a one-way street, species can only become less connected over time. This is clear if we look at the fossil record, in which early members of separate groups are much harder to tell apart.

2. Environments (and ecological niches) are not really as continuous as the claim pretends. Dogs bring down their prey through long chases, and cats ambush their prey; dogs are made for long-distance running, and cats are made for short sprints with high acceleration from a standing start. These requirements are quite different, and it is hard to achieve both in a single body. Compromises between the two have disadvantages in competition with specialists for either type, and thus natural selection culls them. Intermediates are competitive only so long as specialists are absent; so when specialists evolve, the intermediates are likely to become extinct.

3. In part, distinctness is an illusion caused by our choice of which groups to give names to. Groups with unclear boundaries tend not to get separate names, or groups in which intermediate forms exist are chopped in half arbitrarily (especially obvious if fossil forms are considered; e.g., the line between dinosaurs and birds [see CC214] is arbitrary, increasingly so as new fossils are discovered).

4. There are indeed several cases of continua in nature. In many groups, such as some grasses and leafhoppers, different species are very hard to tell apart. At least 10 percent of bird species are similar enough to another species to produce fertile hybrids (Weiner 1994, 198–199). The most obvious continua are called ring species, because in the classic case (the herring gull complex) they form a ring around the North Pole. If we start in Western Europe and move west, similar populations, capable of interbreeding, succeed each other geographically. When we have traveled all the way around the world and reach Western Europe again, the final population is different enough that we call it a separate species, and it is incapable of interbreeding with herring gulls, even though they are connected by a continuous chain of interbreeding populations. This is a big problem for creationists. We expect kinds to be easily determined if they were created separately, but there are no such obvious divisions:

> They are mistaken, who repeat that the greater part of our species are clearly limited, and that the doubtful species are in a feeble minority. This seemed to be true, so long as a genus was imperfectly known, and its species were founded upon a few specimens, that is to say, were provisional. Just as we come to know them better, intermediate forms flow in, and doubts as to specific limits augment (de Condolle, quoted in Darwin, 1872, chap. 2).

Further Reading: Darwin, C. 1872. *The Origin of Species.* Chap. 4.

CB810: Homology cannot be evidence of ancestry if it is defined thus.

Homology is defined as similarity due to common ancestry. The claim then that it is evidence for common ancestry is a circular argument. (J. Wells 2000, 63–65)

1. Homology is not defined as similarity due to common ancestry and then used as evidence for common ancestry. Rather, the evidence for common ancestry comes from the patterns of similarity of many traits. These similarities show that organisms group naturally into a nested hierarchy. For example, that ladybugs and scarabs are both types of beetle is based on various common traits, such as hardened front wings; beetles, flies, and grasshoppers are types of insect; insects, scorpions, and centipedes are types of arthropod. Such grouping does not depend on any assumptions about origins and in fact was first codified by Linnaeus, a creationist. A grouping suggested by many common traits is evidence of common ancestry. This is true no matter what you choose to call the traits. The homology label gets added after the evidence for common ancestry is already in.

CB821: Phylogenetic analyses are inconsistent.

Modern versions of the phylogenetic "tree of life" are based on DNA and other molecular analyses. Inconsistent and bizarre results based on different molecular analyses "have now plunged molecular phylogeny into a crisis." (J. Wells 2000, 49–54; quote on 51)

1. A few inconsistencies are to be expected, because biology is messy. Genes need not always evolve at the same rate in different lineages. Some molecules may converge as a result of selection or chance. Horizontal gene transfer occasionally occurs. Such exceptions will be rare, but there will be a few of them among the vast body of consistent re-

sults. Most inconsistencies can be resolved by basing an analysis on multiple genes (Rokas et al. 2003).

Other inconsistencies will occur as a result of methodological and interpretive mistakes (Sanderson and Shaffer 2002). Phylogenetic analysis is a very complex subject; people who do not understand it well cannot be expected to get it right all the time. Publishing one's methods and results allows others to catch mistakes. Creationists looking for inconsistencies can dishonestly pick out the few there are while disregarding the vast body of consistent results and the reasons for the inconsistencies.

2. Some claimed inconsistencies are really consistent. Wells, for example, cited a study that "placed sea urchins among the chordates" (J. Wells 2000, 51), but sea urchins (and echinoderms in general) do group with chordates as a sister group. Wells also cited another study that "put cows closer to whales than to horses" (51), which is also entirely consistent with genetic, morphological, and fossil evidence.

Further Reading: Tamzek, N. 2002. Icon of obfuscation. http://www.talkorigins.org/faqs/wells/iconob.html#Treeoflife.

CB822: Evolution from a single ancestor is discredited.

According to Darwin, life evolved from one common ancestor. Recent scientific work contradicts this expectation. Molecular data indicate that the tree of life should be uprooted and discredit the homology concept. (Yahya 2003c)

1. The claim refers to results that indicate that horizontal gene transfer was common in the very earliest life. In other words, genetic information was not inherited only from one's immediate ancestor; some was obtained from entirely different organisms, too. As a result, the tree of life does not stem from a single trunk but from a reticulated collection of stems (Woese 2000). This does not invalidate the theory of evolution, though. It says only that another mechanism of heredity was once more common.

2. Horizontal gene transfer does not invalidate phylogenetics. Horizontal gene transfer is not a major factor affecting modern life, including all macroscopic life: "Although HGT does occur with important evolutionary consequences, classical Darwinian lineages seem to be the dominant mode of evolution for modern organisms" (Daubin et al. 2003; quote from Kurland et al. 2003, 9658). And it is still possible to compute phylogenies while taking horizontal gene transfer into account (Kim and Salisbury 2001).

Further Reading: Doolittle, W. F. 2000. Uprooting the tree of life; Tamzek, N. 2002. Icon of obfuscation. http://www.talkorigins.org/faqs/wells/iconob.html#Treeoflife.

EVOLUTION

CB901: Macroevolution has never been observed.

No case of macroevolution has ever been documented. (W. Brown 1995, 6)

1. We would not expect to observe large changes directly. Evolution consists mainly of the accumulation of small changes over large periods of time. If we saw something like a fish turning into a frog in just a couple generations, we would have good evidence against evolution.

2. The evidence for evolution does not depend, even a little, on observing macroevolution directly. There is a very great deal of other evidence (Theobald 2004; see also CA202).

3. As biologists use the term, macroevolution means evolution at or above the species level. Speciation has been observed and documented (see CB910).

4. Microevolution has been observed and is taken for granted even by creationists. And because there is no known barrier to large change (see CB902.1) and because we can expect small changes to accumulate into large changes (see CB902.2), microevolution implies macroevolution. Small changes to developmental genes or their regulation can cause relatively large changes in the adult organism (Shapiro et al. 2004).

5. There are many transitional forms that show that macroevolution has occurred (see CC200).

CB901.1: Range of variation is limited within kinds.
Species may undergo minor changes, but the range of variation is limited to variation within kinds. (H. M. Morris 1985, 51–52, 87–88; Watchtower 1985, 109)

1. What is a "kind"? Creationists have identified kinds with everything from species to entire kingdoms. By the narrower definitions, variation to new kinds has occurred. By the broader definitions, we would not expect to see it in historical time.

2. *Helacyton gartleri* shows one example of change that would be hard to call anything other than a change in kind. It is an amoeba-like life form that came from a human (Van Valen and Maoirana 1991; evolved from a carcinoma, it spreads by taking over other laboratory cell cultures).

3. Creationists have never hinted at, much less shown, any mechanism that would limit variation. Without such a mechanism, we would expect to see kinds vary over time, becoming more and more different from what they were at a given time in the past.

CB901.2: No new phyla, classes, or orders have appeared.
No new phyla, orders, or classes have been observed appearing. Macroevolution remains unobserved.

1. Evolution works almost exclusively by gradual changes. It has taken hundreds of millions of years of evolutionary divergence to produce the existing phyla, and probably hundreds of thousands of years at least for classes to develop. For a new phylum, order, or class to arise suddenly would be creationism, not evolution.

2. Macroevolution is evolution at or above the species level, which has been observed (see CB901).

3. Evidence is not limited to seeing something happen before our eyes. Evidence for macroevolution includes the pattern of homology between organisms, the fossil sequence (including abundant transitional fossils), biogeography, and other evidence. Furthermore, there are no plausible mechanisms that would prevent macroevolution, given the variation that we observe. Indeed, plausible mechanisms leading to diversity do exist (Lee et al. 2003).

Further Reading: Theobald, D. 2004. 29+ evidences for macroevolution. http://www.talk origins.org/faqs/comdesc.

CB901.3: Darwin's finches show only microevolution.

Darwin's finches show only microevolution. In a long-term study, the changes were small and oscillated back and forth. They show no evidence for macroevolution. (Yahya 2003d)

1. The extensive work on Darwin's finches done by the Grants shows in some detail how microevolution works, including details of transmutation and the power of natural selection (Weiner 1994). In the years that the Grants have been studying the finches, we would not expect to see macroevolution.

2. Darwin's finches show a pattern of morphological differences that indicate that they all derived from a common ancestor. The difference between the woodpecker finch and the large ground finch are about as great as those within the whole finch family. Darwin's finches do not show macroevolution occurring, but they are evidence that it has occurred.

Further Reading: Grant, B. R., and P. R. Grant 2003. What Darwin's finches can teach us about the evolutionary origin and regulation of biodiversity. Weiner, J. 1994. *The Beak of the Finch.*

CB902: Microevolution is distinct from macroevolution.

Microevolution is distinct from macroevolution. (T. Wallace 2002)

1. Microevolution and macroevolution are different things, but they involve mostly the same processes. Microevolution is defined as the change of allele frequencies (that is, genetic variation due to processes such as selection, mutation, genetic drift, or even migration) within a population. There is no argument that microevolution happens (although some creationists, such as Wallace, deny that mutations happen). Macroevolution is defined as evolutionary change at the species level or higher, that is, the formation of new species, new genera, and so forth. Speciation has also been observed (see CB910).

Creationists have created another category for which they use the word "macroevolution." They have no technical definition of it, but in practice they use it to mean evolution to an extent great enough that it has not been observed yet. (Some creationists talk about macroevolution being the emergence of new features, but it is not clear what they mean by this. Taking it literally, gradually changing a feature from fish fin to tetrapod limb to bird wing would not be macroevolution, but a mole on your skin which neither of your parents have would be.) I will call this category supermacroevolution to avoid confusing it with real macroevolution.

Speciation is distinct from microevolution in that speciation usually requires an isolating factor to keep the new species distinct. The isolating factor need not be biological; a new mountain range or the changed course of a river can qualify. Other than that, speciation requires no processes other than microevolution. Some processes, such as diruptive selection (natural selection that drives two states of the same feature further apart) and polyploidy (a mutation that creates copies of the entire genome), may be involved more often in speciation, but they are not substantively different from microevolution.

Supermacroevolution is harder to observe directly. However, there is not the slightest bit of evidence that it requires anything but microevolution. Sudden large changes probably do occur rarely, but they are not the only source of large change. There is no reason to think that small changes over time cannot add up to large changes, and every reason to believe they can. Creationists claim that microevolution and supermacroevolution are distinct, but they have never provided an iota of evidence to support their claim.

2. There is evidence for supermacroevolution in the form of progressive changes in the fossil record and in the pattern of similarities among living things showing an absence of distinct "kinds." This evidence caused evolution in some form to be accepted even before Darwin proposed his theory.

Further Reading: Wilkins, J. 1997e. Macroevolution. http://www.talkorigins.org/faqs/macro evolution.html.

CB902.1: There are barriers to large change.
Evolutionists claim that biological evolution is extrapolated from the minor variations we observe. They ignore that there are barriers to large change. (J. Morris 2003)

1. What barriers? The only barrier that anyone has ever proposed is time, and the hundreds of millions of years available for evolution show that time is not a real barrier.

2. Evolution is not just extrapolated from observed microevolution; it is also interpolated from observed changes in the fossil record and from the pattern of observed similarities and differences between present species.

CB902.2: Small changes do not imply large changes.
Creationists recognize that small microevolutionary changes occur, but small changes do not imply large changes, so the theory of macroevolution is unjustified.

1. This claim falsely assumes that the conclusion of macroevolution is based solely on the observation of microevolution. In fact, microevolution is just one piece of the evidence that demonstrates evolution as a whole. Other evidence includes the fossil record, patterns of similarities and differences between living species, and genetic comparisons (see CA202).

2. Small changes do imply large changes under some common circumstances. If there is some selective pressure for the changes to go in one direction, the changes will add up. Such a condition can happen, for example, under a gradual climate change or in evolutionary arms races. Even if there is no selective pressure at all, the changes will tend to diverge further and further from the starting point. Small changes will not lead to large changes only

- if there is stabilizing selection for organisms to remain as they are, or
- if there is too little time for much to happen, or
- if there are genetic mechanisms limiting change.

Stabilizing selection occurs sometimes but is far from universal. We know that the earth, and life on it, is very old (see CH210). And there is no hint of a mechanism to limit variation (see CB901.1). Therefore, we expect large changes based on basic principles.

Further Reading: Darwin, C. 1872. *The Origin of Species*. http://www.talkorigins.org/faqs/origin.html; Theobald, D. 2004. 29+ evidences for macroevolution. http://www.talkorigins.org/faqs/comdesc.

CB904: No entirely new features have evolved.
No entirely new features or biological functions have evolved.

1. Most, if not all, "entirely new features" are modifications of previously existing features. Bird wings, for example, are modified tetrapod forelimbs, which are modified sarcopterygian pectoral fins. A complex, entirely new feature, appearing out of nowhere, would be evidence for creationism.

2. New features have evolved from older different features. There are several examples of microorganisms evolving the ability to degrade or metabolize novel manmade compounds:

- arsenobetaine degradation (Jenkins et al. 2003)
- naphthalene and related compound degradation (Annweiler et al. 2002)
- chlorocatechol degradation (Moiseeva et al. 2002)
- 2,4–dinitrotolule degradation (G.R. Johnson et al. 2002)

Also, a unicellular organism has been evolved to form mulicellular colonies (Boraas et al. 1998; see also CB101.2).

3. An arbitrary genetic sequence can evolve to acquire functionality (Hayashi et al. 2003).

Further Reading: Harris, A.N. 2000. An observed example of morphological evolution. http://www.talkorigins.org/origins/postmonth/jul00.html; Thomas, D. n.d. Evolution and information. http://www.nmsr.org/nylon.htm.

CB910: No new species have been observed.
No new species have been observed. (H.M. Morris 1986)

1. New species have arisen in historical times. For example:

- A new species of mosquito, *Culex molestus*, isolated in London's Underground, has speciated from *Culex pipiens* (Byrne and Nichols 1999; Nuttall 1998).

- *Helacyton gartleri* is the HeLa cell culture, which evolved from a human cervical carcinoma in 1951. The culture grows indefinitely and has become widespread (Van Valen and Maiorana 1991).

- Several new species of plants have arisen via polyploidy (when the chromosome count multiplies by two or more; de Wet 1971). One example is *Primula kewensis* (Newton and Pellew 1929).

2. Incipient speciation, where two subspecies interbreed rarely or with only little success, is common. Here are just a few examples:

- *Rhagoletis pomonella*, the apple maggot fly, is undergoing sympatric speciation. Its native host in North America is Hawthorn (*Crataegus* spp.), but in the mid-1800s, a new population formed on introduced domestic apples (*Malus pumila*). The two races are kept partially isolated by natural selection (Filchak et al. 2000).

- The mosquito *Anopheles gambiae* shows incipient speciation between its populations in northwestern and southeastern Africa (Lehmann et al. 2003).

- Silverside fish show incipient speciation between marine and estuarine populations (Beheregaray and Sunnucks 2001).

3. Ring species show the process of speciation in action. In ring species, the species is distributed more or less in a line, such as around the base of a mountain range. Each population is able to breed with its neighboring population, but the populations at the two ends are not able to interbreed. (In a true ring species, those two end populations are adjacent to each other, completing the ring.) Examples of ring species are

- the salamander *Ensatina*, with seven different subspecies on the west coast of the United States. They form a ring around California's central valley. At the south end, adjacent subspecies *klauberi* and *eschscholtzi* do not interbreed (C. W. Brown n.d.; Wake 1997).

- greenish warblers (*Phylloscopus trochiloides*), around the Himalayas. Their behavioral and genetic characteristics change gradually, starting from central Siberia, extending around the Himalayas, and back again, so two forms of the songbird coexist but do not interbreed in that part of their range (Irwin et al. 2001; Whitehouse 2001).

- the deer mouse (*Peromyces maniculatus*), with over fifty subspecies in North America.

- many species of birds, including *Parus major* and *P. minor*, *Halcyon chloris*, *Zosterops*, *Lalage*, *Pernis*, the *Larus argentatus* group, and *Phylloscopus trochiloides* (Mayr 1942, 182–183).

- the American bee, *Hoplitis (Alcidamea) producta* (Mayr 1963, 510).

- the subterranean mole rat, *Spalax ehrenbergi* (Nevo 1999).

4. Evidence of speciation occurs in the form of organisms that exist only in environments that did not exist a few hundreds or thousands of years ago. For example:

- In several Canadian lakes, which originated in the last 10,000 years following the last ice age, stickleback fish have diversified into separate species for shallow and deep water (Schilthuizen 2001, 146–151).

- Cichlids in Lake Malawi and Lake Victoria have diversified into hundreds of species. Lake Malawi in particular originated in the nineteenth century and has about 200 cichlid species (Schilthuizen 2001, 166–176).

- A *Mimulus* species adapted for soils high in copper exists only on the tailings of a copper mine that did not exist before 1859 (Macnair 1989).

5. Some young-earth creationists claim that speciation is essential to explain Noah's ark. The ark was not roomy enough to carry and care for all species, so speciation is invoked to explain how the much fewer "kinds" aboard the ark became the diversity we see today. Also, some species have special needs that could not have been met during the flood (e.g., fish requiring fresh water). Creationists assume that they evolved from other, more tolerant organisms since the Flood (Woodmorappe 1996).

Further Reading: Callaghan, C. A. 1987. Instances of observed speciation; Kimball, J. W. 2003. Speciation. http://users.rcn.com/jkimball.ma.ultranet/BiologyPages/S/Speciation.html; Otte, D., and J. A. Endler. 1989. *Speciation and Its Consequences* (technical); Schilthuizen, M. 2001. *Frogs, Flies, and Dandelions.* See especially chap. 1; Stassen, C., et al. 1997. Some more observed speciation events. http://www.talkorigins.org/faqs/speciation.html.

CB910.1: Fruit fly experiments produce only fruit flies.

Fruit flies have been mutated and bred in laboratories for generations, but they are still fruit flies. (Watchtower 1985, 104)

1. Biological classification is hierarchical; when a new species evolves, it branches at the very lowermost level, and it remains part of all groups it is already in. Anything that evolves from a fruit fly, no matter how much it diverges, would still be classified as a fruit fly, a dipteran, an insect, an arthropod, an animal, and so forth.

2. There are about 3,000 described species of fruit flies (family Drosophilidae; Wheeler 1987). "Still fruit flies" covers a wide range.

3. Fruit flies do not remain the same species of fruit flies. *Drosophila melanogaster* populations evolved reproductive isolation as a result of contrasting microenvironments within a canyon (Korol et al. 2000). We would not expect to see much greater divergence in historical times.

CB910.2: Peppered moths remained the same species.

Peppered moths are commonly claimed as evidence of evolution, but falsely so. The moths only changed color; they remain peppered moths. (Watchtower 1985, 105–107)

1. Peppered moths were never claimed as evidence of a new species. They show evolutionary change as a result of selection. Creationists themselves admit such microevolution is a reality. Peppered moths show microevolution and one of its causes.

Further Reading: Tamzek, N. 2002. Icon of obfuscation http://www.talkorigins.org/faqs/wells/iconob.html#moths.

CB920: No new body parts have evolved.

The evolution of new body parts has never been observed. (Kepler n.d.)

1. We would not expect to directly observe the evolution of new body parts. Major changes occur gradually over long periods of time. Finding a new body part one day where there was none the day before, or even a generation before, would be better evidence for creationism than for evolution.

2. What exactly is a new body part? Most evolutionary changes are changes to existing structures, not additions de novo. We have transitional sequences showing the evolutionary transition of fins to legs, plus some understanding of the genetic changes involved (Zimmer 1998, 57–85). Do legs qualify as a new body part?

We also sometimes see duplication of body parts. It is not uncommon for cats to have extra toes, for example. Should not these qualify as new body parts?

Further Reading: Zimmer, C. 1998. *At the Water's Edge.*

CB921: New structures would be useless until fully developed.

New structures or organs would not develop incrementally because they would not function until fully developed. For example, what use is half an eye? (H. M. Morris 1985, 53)

1. The assumption made by the claim is false. Structures and organs function quite well when they are not fully developed. Six-year-olds may not have the strength and agility of adults, but their arms, legs, and so forth function well enough to do a great deal.

2. "Fully developed" is not even well defined. Human eyes do not have the acuity of hawks, the dark sight ability of owls, the color discrimination of some fish, or the bee's ability to see in ultraviolet (see CB921.1). With so much more potential possible for the human eye, how can one claim that our own eyes are fully developed?

Further Reading: Gould, S. J. 1977b. The problem of perfection, or how can a clam mount a fish on its rear end? http://www.indiana.edu~ensiweb/lessons/contriv.pdf.

CB921.1: What use is half an eye?

What use is half an eye? (R. Paley 2000)

1. Half an eye is useful for vision. Many organisms have eyes that lack some features of human eyes. Examples include the following:

- Dinoflagellates are single cells, but they have eyespots that allow them to orient toward light sources (Kreimer 1999).
- Starfish and flatworms have eyecups; clustering light-sensitive cells in a depression allows animals to more accurately detect the direction from which the light is coming.
- Some monkeys have only two kinds of color photoreceptors, allowing less color discrimination than most humans have. Some deep-sea fish can see only black and white.

2. Humans themselves have far from perfect vision:

- Humans see in only three colors. Some fish see five. (A very few women are tetra-chromats; they have four types of color receptors; Zorpette 2000.)

- Humans cannot see into the ultraviolet, like bees.

- Humans cannot see infrared, like pit vipers and some fish.

- Humans cannot easily detect the polarization of light, like ants and bees.

- Humans can see only in front of themselves. Many other animals have far greater fields of view; examples are sandpipers and dragonflies.

- Human vision is poor in the dark; the vision of owls is 50 to 100 times more sensitive in darkness. Some deep-sea shrimp can detect light hundreds of times fainter still (Zimmer 1996).

- The range of distances on which one may focus is measured in diopters. A human's range is about fourteen diopters as a child, dropping to about one diopter in old age. Some diving birds have a fifty-diopter range.

- The resolution of human vision is not as good as that of hawks. A hawk's vision is about 20/5; they can see an object from about four times the distance of a human with 20/20 vision.

- Humans have a blind spot caused by the wiring of their retinas; octopuses do not.

- The four-eyed fish (*Anableps microlepis*) has eyes divided in half horizontally, each eye with two separate optical systems for seeing in and out of the water simultaneously. Whirligig beetles (family Gyrinidae) also have divided compound eyes, so one pair of eyes sees underwater and a separate pair sees above.

- The vision of most humans is poor underwater. The penguin has a flat cornea, allowing it to see clearly underwater. Interestingly, the Moken (sea gypsies) from Southeast Asia have better underwater vision than other people (Gislén et al. 2003).

- Humans close their eyes to blink, unlike some snakes.

- Chameleons and seahorses can move their eyes independent of the other.

If you want to know what use is half an eye, ask yourself how you survive with much less than half of what eyes are capable of.

Further Reading: Bahar, S. 2002. Evolution of the eye. http://www.aps.org/units/dbp/news letter/jun02.pdf.

CB921.2: What use is half a wing?

What use is half a wing? A leg evolving into a wing would be a bad leg long before it was a good wing. (W. Brown 1995, 7)

1. Half a wing can have any of several uses:

- In insects, half a wing is useful for skimming rapidly across the surface of water (Marden and Kramer 1995).

- In larger animals, half a wing is useful for gliding. Airfoils for gliding appear in several different forms in many different animals, including
  - skin between legs on flying squirrels, flying phalangers, flying lemurs, some lizards (e.g., *Saurus soarus*), and some frogs (e.g., *Rana dermoptera*)

- flattened body of the flying snake (*Chrysopelea*)
- large webbed feet on gliding tree frogs (*Rhacophorus* and *Polypedates*)
- fins on flying fish (Exocoetidae) and flying squid (*Onychoteuthis*)
- expanded lateral membranes supported by elongated flexible ribs on gliding lizards (e.g., *Draco*)
- expanded lateral membranes supported by elongated jointed ribs on the Kuehneosauridae from the late Triassic
- lateral membrane supported by bones separate from the rest of the skeleton on *Coelurosauravus jaekeli*, an Upper Permian flying reptile (Frey et al. 1997).

- In immature chickens, wing-flapping enhances hindlimb traction, allowing the chickens to ascend steeper inclines. This function could be an intermediate to the original flight of birds (Dial 2003).

- In some flightless birds (e.g., penguins), wings are used for swimming.

- In some flightless birds, wings are probably used for startling potential predators.

- Partial wings may have other useful functions that nobody has thought of yet.

Further Reading: Brodsky, A. K. 1996. *The Evolution of Insect Flight.*

CB922: No two-celled life exists intermediate between one- and multicelled.

There are no two-celled life forms intermediate between unicellular and multicellular life, demonstrating that the intermediate stage is not viable. (W. Brown 1995, 9)

1. The intermediate stage between one-celled and multicelled life need not have been two-celled. The first requirement is for signals between cells, which is necessary if cells are to cooperate in division of labor to break down a food source. Many bacteria utilize a variety of different signals. The evolution of a signal for cooperative swarming has been observed in one bacterium (Velicer and Yu 2003).

The transition to multicellularity has been studied in experiments with *Pseudomonas fluorescens*, which showed that "transitions to higher orders of complexity are readily achievable" (P. Rainey and Rainey 2003, 72).

2. The bacterium *Neisseria* tends to form two-celled arrangements, although, as noted above, this is not relevant to the evolution of multicellularity.

Further Reading: Bonner, J. T. 2000. *First Signals.*

CB925: We do not see creatures in various stages of completion.

We do not see creatures in various stages of completion.

1. Evolution does not predict incomplete creatures. In fact if we ever saw such a thing it would pretty much disprove evolution. In order to survive, all creatures must be sufficiently adapted to their environment; thus, they must be complete in some sense.

The basic false assumption here is twofold: first, that intermediates are necessarily incomplete; and second, that once variation beyond the "type" is allowed, any and all variation is allowed (this latter is typological or essentialist thinking).

2. We see many creatures in transitional stages. These may be considered incomplete in that they do not have all the same features and abilities of similar or related creatures:

- Various gliding animals, such as the flying squirrel, which may be on their way to becoming more batlike
- The euglena, which is halfway to plant
- Aquatic snakes
- Reptiles with a "third eye" that only gets infrared
- Various fish that can live out of water for long periods, use their fins as legs, and breathe air
- The various jaw bones of *Probainognathus* that were in the process of migrating toward the middle ear (see CC215)
- Various Eocene whales, which had hooved forelimbs and hindlimbs (see CC216.1)

with Stanley Friesen

CB926: Preadaptation implies that organs evolved before they were needed.

Evolutionists invoke preadaptation as an explanation for how some features arose. Preadaptation says that organs and other features evolved before they were needed. But an unneeded feature would never be selected, so the whole concept contradicts the theory of evolution.

1. Preadaptation implies that features evolved before they were needed for the function they eventually served, but not before they were needed at all. Many organs and features originally evolved for use in one manner, which incidentally predisposed them for use in another manner. Once an organ or feature used for one purpose also becomes usable in a new and more advantageous manner, natural selection will adapt it for the new use. The following are some potential examples:

- Hard dermal scales in shark skin originally selected for abrasion resistance being predisposed to become teeth
- Feathers on dinosaur/reptile skin originally selected for thermal protection and/or sexual displays being predisposed to become flight feathers
- Lobe fins on fish originally selected for the ability to "walk" through dense underwater vegetation being predisposed to become legs

In each case the organs and features evolved for use in ways that were different from how they eventually ended up being used. At no point did they evolve prior to being useful in some manner, only prior to being useful in the manner they would eventually be used.

Some scientists prefer to use the term "exaption" to explain this phenomenon, specifically to avoid the common misinterpretation of the term "preadaptation" that leads to the above claim.

2. Some preadaptation comes from adapting to niches that are similar to an ultimate niche. For example, living within talus preadapts ground beetles and other organisms for living in caves.

Further Reading: Gould, S.J. 1977b. The problem of perfection, or how can a clam mount a fish on its rear end? http://www.indiana.edu~ensiweb/lessons/contriv.pdf.

with Derek Mathias

CB928: Why are beneficial traits not evolved more often?

Why are beneficial traits not evolved more often? If wings were beneficial for protobirds, for example, why have they not evolved on gazelles and apes?

1. Different organisms make their living in different ways, so a trait that is beneficial for one organism may not be beneficial for another. For example, if the ability to eat a certain kind of hard seed is beneficial for one bird, it may not be beneficial to another for the simple reason that the first bird has a monopoly on those seeds already.

Beneficial traits have drawbacks, too. They usually cost extra energy to grow and use, and often they have other costs. If a trait's advantages do not outweigh its disadvantages, it will not evolve. The existence of an organism that already has the trait often means it is not worth it for another organism to evolve it.

2. Evolution can work only (or almost only; there may be rare exceptions) by making slight modifications to existing features. Most of the modifications must be adaptive. If the raw materials for a trait do not exist, the trait will not evolve even if it is beneficial.

CB928.1: If intelligence is adaptive, other apes should have evolved it.

If human intelligence is beneficial, why have other apes not evolved anything equivalent?

1. A trait that is beneficial to one organism is not necessarily beneficial to another (see CB928). Other life shows that intelligence is not necessary to survive and thrive. Human intelligence in particular is energetically costly. Other apes may simply be better off without it.

CB929: Evolution does not explain our using one tenth of our brain.

According to evolution, we do not evolve organs that are not adaptive. Therefore, evolution does not explain our using only one-tenth of our brains.

1. The belief that we use only 10 percent of our brain is an urban legend and is false (Radford 2000). We use all of it.

Further Reading: Radford, B. 2000. Ten-percent myth. http://66.165.133.65/science/stats/10percnt.htm.

CB930: Some fossil species are still living.

Some species, such as the tuatara, horseshoe crab, cockroach, ginkgo, and coelacanth (see CB930.1), are "fossil species." They have not evolved for millions of years. (Whitcomb and Morris 1961, 176–180)

1. The theory of evolution does not say that organisms must evolve morphologically. In fact, in an unchanging environment, stabilizing selection would tend to keep an organism largely unchanged. Many environments around today are not greatly different from environments of millions of years ago.

2. Some so-called fossil species have evolved significantly. Cockroaches, for example, include over 4,000 species of various shapes and sizes. Species may also evolve in ways that are not obvious. For example, the immune system of horseshoe crabs today is probably quite different from that of horseshoe crabs of millions of years ago.

CB930.1: The coelacanth, thought extinct for ages, is still living.

The coelacanth, thought to have been extinct for seventy million years and used as an example of a fish–tetrapod transition, is found still alive, unchanged in form, today. (H. M. Morris 1985, 82–83, 89)

1. The modern coelacanth is *Latimeria chalumnae*, in the family Latimeriidae. Fossil coelacanths are in other families, mostly Coelacanthidae, and are significantly different in that they are smaller and lack certain internal structures. *Latimeria* has no fossil record, so it cannot be a "living fossil."

2. Even if the modern coelacanth and fossil coelacanths were the same, it would not be a serious problem for evolution. The theory of evolution does not say that all organisms must evolve. In an unchanging environment, natural selection would tend to keep things largely unchanged morphologically.

3. Coelacanths have primitive features relative to most other fish, so at one time they were one of the closest known specimens to the fish–tetrapod transition. We now know several other fossils that show the fish–tetrapod transition quite well (see CC212).

Further Reading: Forey, P. L. 1998. *History of the Coelacanth Fishes*; Lindsay, D. 2000c. Living fossils like the coelacanth. http://www.don-lindsay-archive.org/creation/coelacanth.html.

CB930.2: A plesiosaur was found by a Japanese trawler.

In 1977, a plesiosaur carcass was netted by a Japanese trawler. (Swanson 1978)

1. The carcass was a partly decomposed basking shark, according to tissue samples and descriptions and sketches by witnesses. Reports that it was a plesiosaur were based on a superficial resemblance caused by the pattern in which sharks decay and were spread and exaggerated by news media hype (Kuban 1997).

2. Even if the carcass were a plesiosaur, the find would not have been a challenge to evolution. The theory of evolution does not demand that species go extinct after a certain amount of time.

Further Reading: Kuban, G. J. 1997. Sea monster or shark?

CB930.3: Dinosaurs may still be alive in the Congo.
An apatosaurus-like dinosaur may still be alive in swamps in the Congo. (Doolan 1993; Hovind 1998)

1. The reputed "dinosaur," Mokele-mbembe, is folklore. O'Hanlon (1997, 373) reported the answer upon asking a native if he had seen Mokele-mbembe:

> "What a stupid question," said Doubla, looking genuinely surprised. . . . "Mokele-mbembe is not an animal like a gorilla or python. And Mokele-mbembe is not a sacred animal. It doesn't appear to people. It is an animal of mystery. It exists because we imagine it. But to see it—never. You don't *see* it."

Other reports, though some treat Mokele-mbembe as real, are also folkloric; they make good stories, but there is no tangible evidence. It is nigh impossible that a population of very large, very distinctive land animals could have eluded human exploration so completely.

2. A living dinosaur would not be a problem for evolution even if it existed (see CB930).

CB932: Some modern species are apparently degenerate, not higher forms.
Evolution is supposed to advance species, but some modern species, such as many parasites and species living in caves, appear to be degenerate forms of other species.

1. Evolution adapts species to their environment. Sometimes that means becoming more complex, but not always. If losing a feature allows a species to survive better, the species likely will gradually lose the feature. Losing a feature is advantageous, for example, if the feature is energetically costly to grow and maintain (which is almost always the case) and if it serves no other function (which often happens when a population moves to a new environment, such as a cave).

CB940: Pure chance cannot create new structures.
Complex structures could not have arisen by chance. (H. M. Morris 1985, 59–69)

1. Evolutionists the world over are, and always have been, unanimous in their agreement that complex structures did not arise by chance. The theory of evolution does not say they did, and to say otherwise is to display a profound absence of understanding of evolution. The novel aspect that Darwin proposed is natural selection. Selection is the very opposite of chance.

Selection of randomly introduced variation is known to be able to produce complex formations, including functional circuits (C. Davidson 1997; T. Thompson 1996) and robots (Lipson and Pollack 2000). Creationists have never proposed a reason to explain why the same processes would not produce the same results in nature.

2. The principles by which evolution works, including random variation and recombination and natural selection, have proven successful and useful for designing new drugs (Coghlan 1998), for designing better enzymes for detergents (Pollack 2000), and, as genetic algorithms, for many other applications.

Further Reading: Adami, C., et al. 2000. Evolution of biological complexity; Darwin, C. 1872. *The Origin of Species*. http://www.talkorigins.org/faqs/origin.html; Dawkins, R. 1986. *The Blind Watchmaker*.

CB940.1: Odds of many successive beneficial mutations are minuscule.

Successful production of a 200-component functioning organism requires at least 200 beneficial mutations. The odds of getting that many successive beneficial mutations is r^{200}, where r is the rate of beneficial mutations. Even if r is 0.5 (and it is really much smaller), that makes the odds worse than 1 in 10^{60}, which is impossibly small. (H. M. Morris 2003b)

1. Morris's calculation assumes that all the beneficial mutations must occur consecutively with no other mutations occurring in the meantime. When one allows harmful mutations that get selected out along the way, 200 beneficial mutations would accumulate fairly quickly—in 200/r generations using the assumptions of Morris's model. (The real world is quite a bit more complicated yet. In particular, large populations and genetic recombination via sex can allow beneficial mutations to accumulate at a greater rate.)

CB941: How do things know how to evolve?

How do things know how to evolve? For example, how do plants know what flavor of berries to evolve so the birds will eat and disperse them? How does a coconut tree know there is an ocean nearby?

1. The organism does not know anything. It is all just the "blind watchmaker" at work. All sorts of mutations happen all the time. Most of them are useless or even detrimental in the environment they find themselves in, and those are mostly eliminated. The occasional mutation that increases its bearer's reproductive success is preserved and spreads throughout the population in future generations. For example, different individual plants produce (over time) different flavors of seeds. The flavors that birds like more get dispersed more and thrive more; those that birds do not like fall by the wayside. The end result is that berries evolve to become more flavorful to birds.

2. Individual organisms do not evolve; populations evolve. The individual's role is to survive and reproduce, or not. Those with beneficial variation are more likely to survive and reproduce.

Further Reading: Dawkins, R. 1986. *The Blind Watchmaker*.

CB950: Overspecialization with no adaptive value sometimes occurs.

Evolution says that complex structures would not be formed without a selective advantage for them. But overspecialization with no adaptive value sometimes occurs. An example is the enormous antlers of the Irish elk. Despite the seeming disadvantages, animals with such extreme features seem to do as well as other animals. (Macbeth 1971, 70–73)

1. There are no established cases of overspecialization that did not have adaptive value. In some cases, the adaptive value may not be obvious, but there are none where its absence is plausible. In particular, sexual selection can produce exaggerated and seemingly detrimental features, such as the stalks of stalk-eyed flies and the huge antlers of Irish elk.

2. Adaptation can lead to overspecialization. If a flower becomes dependant on one pollinator, for example, the loss of the pollinator can cause the extinction of the flower.

CB990: Proposed evolution scenarios are just-so stories.

Evolutionist explanations are just-so stories. They are entirely speculative and do not qualify as evidence. (Dembski 2002b, chap. 6)

1. It is indeed wrong to offer a just-so story as evidence that something happened a certain way. However, such stories still serve a purpose as hypotheses. They present a model that can be tested by further research and either rejected or qualified as more probable. For example, the just-so story that horns on horned lizards evolved as defense has now been supported with experiments (K. V. Young et al. 2004). Science makes little progress without hypotheses to test.

2. Such stories also function to rebut claims that something could not have happened. If a plausible story is presented, the claim of impossibility is shown to be false. This is true whether or not the story is speculative.

3. Creationists have almost nothing but just-so stories to back up their models (such as they are). For example, every detail creationists give about the Flood is a just-so story, due to a lack of basis for anything more than the broad outline given in Genesis. And no research is ever done to test their stories.

Further Reading: Lindsay, D. 1998b. Scenarios and "just so" stories. http://www.don-lindsay-archive.org/creation/stories.html.

PALEONTOLOGY

PHYSICAL ANTHROPOLOGY

CC001: Piltdown man was a hoax.

In 1912, Charles Dawson and Arthur Smith Woodward announced the discovery of a mandible and part of a skull from a gravel pit near Piltdown, England. The mandible was apelike except for humanlike wear on the teeth; the skull was like a modern human. These bones became the basis for Eanthropus dawsoni, commonly known as Piltdown Man, interpreted as a 500,000-year-old British ape–man. But in the early 1950s, it was found that the jawbone was stained and filed down to give its appearance and that the skull was a recent human fossil. In short, Piltdown Man was a fraud. British scientists believed it because they wanted to. The failure to expose it sooner shows that scientists tend to be guided by their preconceptions. (Gish 1985, 188–190)

1. Piltdown man was exposed by scientists. The fact that it took forty years is certainly no shining example of science in action, but it does show that science corrects errors.

Preconceptions are an unavoidable problem in just about any investigation, but they are less so in science because first, different scientists often have different preconceptions, and second, the physical evidence must always be accounted for (see CA230.1). Many scientists from America and Europe did not accept Piltdown Man uncritically, and the hoax unraveled when the fossils could not be reconciled with other hominid fossil finds.

2. One hoax cannot indicate the inferiority of conventional archeology, because creationists have several of their own, including Paluxy footprints (see CC101), the Calaveras skull (see CC112), Moab and Malachite Man (see CC110 and CC111), and others. More telling is how people deal with these hoaxes. When Piltdown was exposed, it stopped being used as evidence. The creationist hoaxes, however, can still be found cited

as if they were real. Piltdown has been over and done with for decades, but the dishonesty of creationist hoaxes continues.

Further Reading: Harter, R. 1996. Piltdown Man: The bogus bones caper. http://www.talkorigins.org/faqs/piltdown.html.

CC002: Nebraska man was a hoax.

Nebraska Man (Hesperopithecus haroldcookii) *was described on the basis of a single tooth that turned out to come from a peccary. This tooth was used to construct an entire species, complete with illustrations of the primitive man and his family. (Gish 1985, 187–188)*

1. The tooth was never held in high regard by scientists. Osborn, who described it, was unsure whether it came from a hominid or from another kind of ape, and others were skeptical that it even belonged to a primate. The illustration was done for a popular publication and was clearly labeled as highly imaginative.

Nebraska Man is an example of science working well. An intriguing discovery was made that could have important implications. The discoverer announced the discovery and sent casts of it to several other experts. Scientists were initially skeptical. More evidence was gathered, ultimately showing that the initial interpretation was wrong. Finally, a retraction was prominently published.

Further Reading: Foley, J. 2001. Creationist arguments: Nebraska Man. http://www.talkorigins.org/faqs/homs/a_nebraska.html; Gould, S. J. 1991. An essay on a pig roast; Mellett, J. S., and J. Wolf. 1985. The role of "Nebraska man" in the creation–evolution debate. http://www.talkorigins.org/faqs/homs/wolfmellett.html.

CC003: Lucy's knee was found far from the rest of the skeleton.

The knee of the "Lucy" fossil (the most complete Australopithecus afarensis *fossil) was found over a mile away from the rest of the skeleton, so it cannot be used as evidence that Lucy walked upright. That evolutionists have never admitted this fact in print shows their dishonesty. (J. D. Morris 1989)*

1. The claim is false. The skeleton called Lucy does not have an intact knee. A different, isolated knee fossil was found two to three kilometers away (Johanson and Edey 1981). Confusion over the two fossils apparently led to the false claim.

2. Far from indicating evolutionist dishonesty, this claim shows how creationists fail to check their claims (Lippard 1999).

Further Reading: Lippard, J. 1999. Lucy's knee joint. http://www.talkorigins.org/faqs/knee-joint.html.

CC030: All human fossils would fit on a billiard table.

All known fossils of ancient humans would fit on a billiard table. (or in a coffin; H. M. Morris 1985, 202; Watchtower 1985, 86)

1. That may have been true at one time, but there are thousands of hominid fossils now. Lubenow (1992) found that there were fossils from almost 4,000 hominid individuals catalogued as of 1976. As of 1999, there were fossils of about 150 *Homo erectus* individuals, 90 *Australopithecus robustus*, 150 *Australopithecus afarensis*, 500 Neanderthals, and more (Handprint 1999). Foley (2004b) listed some of the more prominent fossils.

2. It takes only a handful of fossils to show that hominid forms have changed over time.

Further Reading: Foley, J. 2004b. Prominent hominid fossils. http://www.talkorigins.org/faqs/homs/specimen.html; Handprint Media. 1999. Human evolution. http://www.handprint.com/LS/ANC/evol.html chart; Lubenow, M. 1992. *Bones of Contention.* Chap. 3.

CC040: Anthropologists disagree.
Anthropologists disagree about what the human family tree looks like. Every new discovery seems to give reason to redraw the tree, whereas we would expect the tree to become clearer as discoveries accumulate. (Watchtower 1985, 88)

1. Pointing to the disagreements is a ruse to distract from the areas where there is agreement. There is no significant disagreement among professionals that modern humans evolved from an African australopithecine or that other hominids sometimes coexisted with the lineage that led to humans.

2. Much of the disagreement is hype. When someone discovers yet another *Homo erectus* fossil from the same region and era as other *Homo erectus* fossils, newspapers do not trumpet the headline, "Another Fossil Supports Hominid Lineage." Such fossils are not news except to the paleoanthropologists who work on them. The headlines go to the truly novel finds.

Disagreement is also hyped because it makes a better emotional story. Anthropologists would be glad to make a discovery that overturns conventional understanding, and news reporters favor such stories as well, so the significance of small disagreements tends to get magnified.

3. Disagreement and uncertainty are routine in areas opened by new scientific discoveries. Paleoanthropology is a field in which new discoveries are not uncommon, so there will be uncertainty at first around those discoveries. However, paleoanthropology is also a mature science at its core; the uncertainty and disagreement there is at a minimum.

4. Disagreements get resolved. This is an important feature of science never found in creationism. As more data are discovered, the data answer the questions we have. For example, it was once unknown whether Neanderthals and modern humans were separate species. Molecular evidence now strongly indicates that they were (Krings et al. 2000). The record may be insufficient to answer some of our questions, such as when language began, but by and large, our questions can and do get answered.

CC041: *Homo habilis* is an invalid taxon.

Homo habilis *is an invalid taxon. It is a taxonomic waste bin in which several different types have been placed.* (Lubenow 1992, 157–166)

1. At least two different species have been assigned to *Homo habilis*, but that does not make the taxon invalid. OH7, its type specimen (which defines the taxon), clearly differs from other named species. As more fossils are found, we will get a better idea of the ranges of variation and thus of what constitutes different species.

Further Reading: Foley, J. 2003. Homo habilis: is it an invalid taxon? http://www.talkorigins .org/faqs/homs/invalidtaxon.html.

CC050: All hominid fossils are fully human or fully ape.

All hominid fossils are fully human or fully ape.

1. There is a fine transition between modern humans and australopithecines and other hominids. The transition is gradual enough that it is not clear where to draw the line between human and not (see Figure CC1).
Intermediate fossils include

- *Australopithecus afarensis*, from 3.9 to 3.0 million years ago (Ma). Its skull is similar to a chimpanzee's, but with more humanlike teeth. Most (possibly all) creationists would call this an ape, but it was bipedal.

- *Australopithecus africanus* (3 to 2 Ma); its brain size, 420–500 cc, was slightly larger than A. *afarensis*, and its teeth yet more humanlike.

- *Homo habilis* (2.4 to 1.5 Ma), which is similar to australopithecines, but which used tools and had a larger brain (650-cc average) and less projecting face.

- *Homo erectus* (1.8 to 0.3 Ma); brain size averaged about 900 cc in early *H. erectus* and 1,100 cc in later ones. (Modern human brains average 1,350 cc.)

- A Pleistocene *Homo sapiens* which was "morphologically and chronologically intermediate between archaic African fossils and later anatomically modern Late Pleistocene humans" (T. D. White et al. 2003, 742).

- A hominid combining features of, and possibly ancestral to, Neanderthals and modern humans (Bermúdez de Castro et al. 1997).

And there are fossils intermediate between these (Foley 1996–2004).

2. Creationists themselves disagree about which intermediate hominids are human and which are ape (Foley 2002).

3. There is abundant genetic evidence for the relatedness between humans and other apes:

- Humans have twenty-three chromosome pairs; apes have twenty-four. Twenty-two of the pairs are similar between humans and apes. The remaining two ape chromosomes appear to have joined; they are similar to each half of the remaining human chromosome (chromosome 2; Yunis and Prakash 1982; see Fig. CB1).

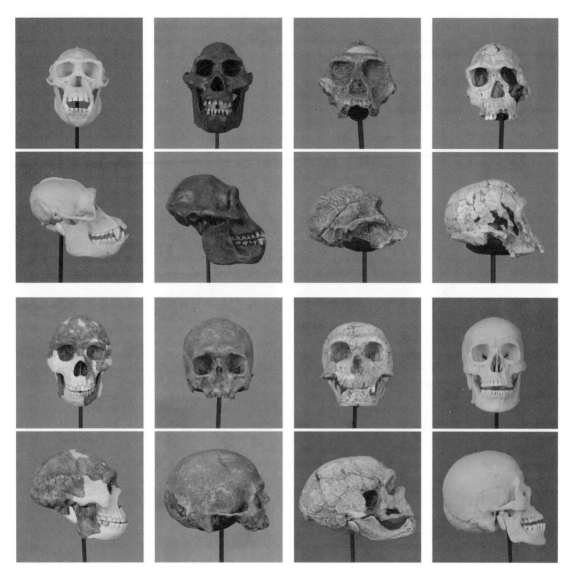

Figure CC1
Hominid fossil skulls and chimpanzee skull. Left to right (top two rows): Modern chimpanzee (*Pan troglodytes*), *Australopithecus afarensis*, *Australopithecus africanus*, *Homo habilis*; (bottom two rows) *Homo erectus*, Cro Magnon *Homo sapiens*, *Homo neanderthalensis* (an extinct side branch), modern human (*Homo sapiens*). Photos by Wesley Elsberry.

- The ends of chromosomes have repetitious telomeric sequences and a distinctive pretelomeric region. Such sequences are found in the middle of human chromosome 2, just as one would expect if two chromosomes joined (IJdo et al. 1991).

- A centromere-like region of human chromosome 2 corresponds with the centromere of the ape chromosome (Avarello et al. 1992).

- Humans and chimpanzees have innumerable sequence similarities, including shared pseudogenes such as genetic material from ERVs (endogenous retroviruses; D. M. Taylor 2003; Max 2003).

Further Reading: Drews, C. 2002. Transitional fossils of hominid skulls. http://www.theistic-evolution.com/transitional.html; Foley, J. 1996–2004. Fossil hominids. http://www.talkorigins.org/faqs/homs; Johanson, D. C., and B. Edgar. 1996. *From Lucy to Language*; Tattersall, I. 1995. *The Fossil Trail.*

CC051: Neanderthal was based on a disfigured human.
Neanderthal man was reconstructed from a fossil skeleton badly deformed by disease. Neanderthals really are not much different from modern humans. (Watchtower 1985, 95)

1. An early stereotype of Neanderthals was that they were stooped, very hairy, and had divergent big toes. Straus and Cave (1957) showed that they were fully human in posture. However, Neanderthals do have distinctive features that distinguish them from modern humans (Straus and Cave 1957). Some of these features—powerful bones and muscles, in particular—cannot plausibly be attributed to pathology or injury.

2. Neanderthals are known from many specimens. It is extremely unlikely that all of them would be suffering from exactly the same illness.

Further Reading: Foley, J. 2002. Creationist arguments: Neandertals. http://www.talkorigins.org/faqs/homs/a_neands.html.

CC051.1: Neanderthals were humans with rickets.
Neanderthals were modern humans with rickets. (Lubenow 1992, 149–156)

1. The signs of rickets differ from Neanderthal fossils in several respects, including the following:

a. People with rickets are undernourished and calcium-poor; their bones are weak. Neanderthal bones are 50 percent thicker than the average human's.

b. Evidence of rickets is easily detectable, especially on the ends of the long bones of the body. This evidence is not found in Neanderthals.

c. Rickets causes a sideways curvature of the femur. Neanderthal femurs bend backward.

Virchow, who first reported the possibility of rickets in a Neanderthal, did not cite it alone. He said the fossil had rickets in early childhood, head injuries in middle age, and arthritis in old age. It is doubtful that an entire population suffered these same afflictions.

2. Lubenow attributes rickets to a post-Flood ice age, with heavy cloud cover, shelter, and clothing, and a lack of vitamin D. But the greatest differences from modern humans, seen in *Homo erectus*, are found mostly in tropical areas.

Further Reading: Foley, J. 2002. Creationist arguments: Neandertals. http://www.talkorigins.org/faqs/homs/a_neands.html.

CC052: Laetoli footprints were human.

The Laetoli footprints, dated 3.7 million years old, appear to be those of modern humans. (Gish 1985, 174–176)

1. How similar the Laetoli footprints look to australopithecine feet is a matter of debate. Tuttle (1990) thought that they were too humanlike for *Australopithecus afarensis* and may have belonged to another species of australopithecine or to an early *Homo* species. Other anthropologists think they are significantly different from *Homo* and could be *A. afarensis* (reviewed by Foley 2004a). Creationists tend to cite only Tuttle because he best supports their view.

Further Reading: Foley, J. 2004a. Creationist arguments: Anomalous fossils. http://www.talk origins.org/faqs/homs/a_anomaly.html.

CC054: KP 271 (a fossil humerus) was human.

The fossil humerus KP 271 is an apparently human fossil from four million years ago, which, according to the standard evolutionary model, is well before the appearance of modern humans. (Lubenow 1992, 54–58)

1. Even a humerus from a chimpanzee looks similar to a human humerus; it should not be surprising that the humerus from a closer relative would look even more similar. However, the anatomical evidence strongly indicates that the specimen is not human and is a good match with *Australopithecus anamensis* (Lague and Jungers 1996).

Further Reading: Foley, J. 2004a. Creationist arguments: Anomalous fossils. http://www.talk origins.org/faqs/homs/a_anomaly.html#kp271.

CC061: French scientists called Peking Man "monkey-like."

French anthropologist Marcellin Boule said that the skulls of Peking Man (Sinanthropus, now classified with Homo erectus) *are "monkey-like." Others have pointed out that this was based on a poor translation, but the monkey quote does not misrepresent Boule's text.* (Gish 1979, 139–140; 1997)

1. Without question, calling the skulls monkeylike is a major misrepresentation of Boule. In the pages immediately following the quote in question, Boule emphasized that he was not dismissing Peking Man as a monkey.

Some creationists have rationalized that the French word *singe* used by Boule can mean either monkey or ape, but it is clear from context that the ape meaning was intended.

2. Events surrounding use of this quote show poor behavior by several creationists:

- O'Connell (1969) misrepresented Boule's ideas, apparently deliberately, by mistranslating him.
- Gish (1979) failed to reference O'Connell when repeating his mistranslation.
- Gish and others show abysmally poor scholarship in the first place in thinking that calling Peking Man monkeylike is even a viable scientific view.

- When corrected about the mistranslation, Gish and Answers in Genesis attacked their correctors and refused to fully correct their errors (Foley 2003).

Further Reading: Foley, J. 2003. Creationist arguments: The monkey quote. http://www.talk origins.org/faqs/homs/monkeyquote.html.

CC080: *Australopithecus* was fully ape, closer to chimp.

Australopithecus *was fully ape, closer to chimp. World-renowned anatomists and evolutionists Solly Zuckerman and Charles Oxnard have shown that* Australopithecus *did not walk upright in a human manner.* (Watchtower 1985, 93–94)

1. *Australopithecus africanus* and *robustus* are far more humanlike than apelike. *A. afarensis* falls somewhere between human and ape, perhaps more on the side of ape. *A. ramidus* is even more apelike.

The claims that *africanus* and *robustus* are fully ape are based on old and discredited accounts from just a couple of papers. Oxnard himself considered *Australopithecus* to be a human ancestor (Groves 1999). The vastly greater evidence that those species are fully bipedal and had other humanlike traits is either ignored, distorted, or baselessly dismissed by creationists (Foley 1997).

Further Reading: Foley, J. 1997. Creationist arguments: Australopithecines. http://www.talk origins.org/faqs/homs/a_piths.html.

CC101: Human footprints have been found with dinosaur tracks at Paluxy.

Human and dinosaur footprints have been found together in the Glen Rose formation at Paluxy River, Texas. (H. M. Morris 1985, 122)

1. The footprints reputed to be of human origin are not. For example:

- Some of the footprints are dinosaur footprints. Processes such as erosion, infilling, and mud collapse obscure the dinosaurian features of some footprints, making them look like giant human footprints, but careful cleaning reveals the three-toed tracks of dinosaurs (Hastings 1987; Kuban 1989).

- Some of the reputed prints are erosional features or other irregularities. They show no clear human features without selective highlighting.

- Some of the prints show evidence of deliberate alteration (Godfrey 1985).

2. The Paluxy tracks are illustrative of creationists' wishful thinking and of their unwillingness to face evidence. Although some creationists have repudiated the Paluxy claim, many others still cling to it (Schadewald 1986).

Further Reading: Cole, J. R., and L. R. Godfrey (eds.). 1985. The Paluxy River footprint mystery—solved. (special issue devoted to the topic); Gillette, D. D., and M. G. Lockley (eds.). 1989. *Dinosaur Tracks and Traces*; Hastings, R. J. 1988. Rise and fall of the Paluxy man tracks; Kuban, G. J. 1996b. The Texas dinosaur/"man track" controversy. http://www.talkorigins.org/faqs/ paluxy.html; Matson, D. E. 1994. How good are those young-earth arguments? http://www.talk origins.org/faqs/hovind/howgood-gc.html#G4d.

CC102: Sandal footprints have been found associated with trilobites.

Apparent human shoe prints were found in a slab with obvious trilobite fossils in 1968, by William J. Meister, at Antelope Springs, forty-three miles northwest of Delta, Utah. The heel of one print was worn just as shoe heels wear today. (W. Brown 1995, 25)

1. The trilobites are real (*Elrathia kingii*), but the "sandal print" is a spall pattern. The heel of the Meister print is not worn down but is caused by a long crack running across the rock. It lacks the diagnostic features that a real sandal print has. There are many other weathering features in the area identical in character to the so-called sandal prints but in a variety of shapes. They do not occur in a trail but as isolated prints (Conrad 1981).

2. Geochemical processes, such as solution penetrations, spalling, and other weathering, have been well documented to produce such features on the shales of the Wheeler Formation, where the prints were found (Stokes 1986).

Further Reading: Kuban, G.J. 1998b. The "Meister Print." http://www.talkorigins.org/faqs/paluxy/meister.html; Strahler, A.N. 1987. *Science and Earth History*. Chap. 48, pp. 459–472.

CC110: Moab man was found in Cretaceous sandstone.

Two greenish human skeletons were excavated from Jurassic sediments in the Big Indian Copper Mine near Moab, Utah. (F.A. Barnes 1975)

1. The bones were found fifteen feet deep in loose sand, not in a rock matrix. Their postures were similar to known Indian burials. The bones were unfossilized and partly decayed, and dating them yielded an age of 210 plus or minus 70 years. In short, they were a fairly recent burial (Kuban 1998a).

Further Reading: Kuban, G.J. 1998a. The life and death of Malachite Man. http://members.aol.com/gkuban/moab.htm.

CC111: Malachite man was found in Cretaceous sandstone.

Ten modern human skeletons were excavated from fifty-eight feet deep in the Lower Cretaceous Dakota Sandstone, which is dated as 140 million years old and is known for the same dinosaurs as in Dinosaur National Monument. (Patton n.d.b)

1. The skeletons are the same bones as the discredited Moab man bones (see CC110), apparently with skeletons from eight nearby Indian burials added (Kuban 1998a).

2. All details given in the account are apparently false. The bones were found fifteen feet deep in soft, unconsolidated sand. They were clearly intrusive (i.e., buried there long after the sediments were laid down). The Dakota Formation is approximately 90–115 million years old, straddling the Early and Late Cretaceous. Dinosaur National Monument is in the Morrison Formation, which is Jurassic (Kuban 1998a).

3. The people making claims about Malachite Man have not been cooperative in supplying information that might be used to verify their claim. This would be surprising if they thought their claims could actually be verified.

Further Reading: Kuban, G.J. 1998a. The life and death of Malachite Man. http://members .aol.com/gkuban/moab.htm.

CC112: Castenedolo, Olmo, and Calaveras skulls were found in Pliocene strata.

The Castenedolo and Olmo skulls from Italy and the Calaveras skull from California were modern skulls, but all were found in undisturbed Pliocene strata. (H.M. Morris 1985, 177)

1. The Castendolo bones belong to skeletons of several men, women, and children. They are a recent burial in Pliocene sediments, evidenced by the fact that other fossils but not the human bones were impregnated with salt.

The Olmo skull is from upper Pleistocene gravel, placing it in the Upper Paleolithic (Stone Age) period. It is not out of place.

The Calaveras skull was a deliberate hoax and was immediately recognized as such by scientists (Conrad 1982).

Further Reading: Matson, D.E. 1994. How good are those young-earth arguments? http:// www.talkorigins.org/faqs/hovind/howgood-gc.html#G4d.

CC120: Baugh found a fossilized finger from the Cretaceous.

A fossilized human finger has been found from the Cretaceous. (Lines 1995)

1. The finger looks remarkably similar in size and shape to the cylindrical sandstone infillings of *Ophiomorpha* or *Thalassinoides* shrimp burrows commonly found in Cretaceous rocks. Although its general shape is fingerlike, it has none of the fine structure one would expect from a finger.

2. The fossil was not found in situ, so it cannot be conclusively associated with Cretaceous formations (Kuban 1996a). Even if it were a real fossil finger, it would be of no value as evidence against evolution.

Further Reading: Kuban, G.J. 1996a. A review of NBC's "The Mysterious Origins of Man." http://www.talkorigins.org/faqs/paluxy/nbc.html.

CC130: A petrified hammer was found in Cretaceous rocks.

An iron hammer with wooden handle was found embedded in rock in Cretaceous sediments (or Ordovician, by some accounts) near London, Texas. The enclosing rock contains Lower Cretaceous fossils. (Patton n.d.a)

1. The hammer is encrusted with calcium carbonate, which can happen quickly. The fossils are in nearby rocks, not part of the material encrusting the hammer. There is no evidence that the hammer is more than a few decades old.

Further Reading: Cole, J. R. 1985. If I had a hammer; Kuban, G. J. 1999. The London hammer. http://paleo.cc/paluxy/hammer.htm; Matson, D. E. 1994. How good are those young-earth arguments? http://www.talkorigins.org/faqs/hovind/howgood-gc.html#G4d.

CC150: If we are descended from apes, why are there still apes around?

If we are descended from apes, why are there still apes around? (Robinson 2003, no. 11)

1. Humans and other apes are descended from a common ancestor whose population split to become two (and more) lineages. The question is rather like asking, "If many Americans and Australians are descended from Europeans, why are there still Europeans around?" Creationists themselves recognize the invalidity of this claim (AIG n.d.a).

Further Reading: Darwin, C. 1872. *The Origin of Species*. Chap. 4; Foley, J. 2002. Fossil hominids: Frequently asked questions. http://www.talkorigins.org/faqs/homs/faqs.html#apes.

TRANSITIONAL FOSSILS

CC200: Transitional fossils are lacking.

There are no transitional fossils between one major kind and another. There are systematic gaps in the fossil record, and new forms appear suddenly. (H. M. Morris 1985, 78–90; Watchtower 1985, 57–59)

1. There are many transitional fossils. The only way that the claim of their absence may be remotely justified, aside from ignoring the evidence completely, is to redefine "transitional" as referring to a fossil that is a direct ancestor of one organism and a direct descendant of another. However, direct lineages are not required; they could not be verified even if found. What a transitional fossil is, in keeping with what the theory of evolution predicts, is a fossil that shows a mosaic of features from an older and more recent organism.

2. Transitional fossils may coexist with gaps. We do not expect to find finely detailed sequences of fossils lasting for millions of years. Nevertheless, we do find several fine gradations of fossils between species and genera, and we find many other sequences between higher taxa that are still very well filled out.

The following are fossil transitions between species and genera:

a. Human ancestry (see CC050). There are many fossils of human ancestors, and the differences between species are so gradual that it is not always clear where to draw the lines between them.

b. The horns of titanotheres (extinct Cenozoic mammals) appear in progressively larger sizes, from nothing to prominence (see Figure CC2). Other head and neck features also evolved. These features are adaptations for head-on ramming analogous to sheep behavior (Stanley 1974).

c. A gradual transitional fossil sequence connects the foraminifera *Globigerinoides trilobus* and *Orbulina universa* (Pearson et al. 1997). *O. universa*, the later fossil, features a spherical test surrounding a "Globigerinoides-like" shell, showing that a feature was added, not lost. The evidence is seen in all major tropical ocean basins. Several intermediate morphospecies connect the two species, as may be seen in the figure included in Lindsay (1997c).

d. The fossil record shows transitions between species of *Phacops* (a trilobite; *Phacops rana* is the Pennsylvania state fossil; Eldredge 1972, 1974).

e. Planktonic forminifera (Malmgren et al. 1984). This is an example of punctuated gradualism. A ten-million-year foraminifera fossil record shows long periods of stasis and other periods of relatively rapid but still gradual morphologic change.

f. Fossils of the diatom *Rhizosolenia* are very common (they are mined as diatomaceous earth), and they show a continuous record of almost two million years, which includes a record of a speciation event (K.R. Miller 1999, 44–45).

g. Lake Turkana mollusk species (Lewin 1981).

h. Cenozoic marine ostracodes (T.M. Cronin 1985).

i. The Eocene primate genus *Cantius* (Gingerich 1976, 1980, 1983).

j. Scallops of the genus *Chesapecten* show gradual change in one "ear" of their hinge over about thirteen million years. The ribs also change (Pojeta and Springer 2001; Ward and Blackwelder 1975).

k. *Gryphaea* (coiled oysters) become larger and broader but thinner and flatter during the Early Jurassic (Hallam 1968; see Figure CC3).

The following are fossil transitionals between families, orders, and classes:

a. Human ancestry. *Australopithecus*, though its leg and pelvis bones show it walked upright, had a bony ridge on the forearm, probably vestigial, indicative of knuckle walking (Richmond and Strait 2000).

b. Dinosaur–bird transitions (see CC214).

c. *Haasiophis terrasanctus* is a primitive marine snake with well-developed hind limbs. Although other limbless snakes might be more ancestral, this fossil shows a relationship of snakes with limbed ancestors (Tchernov et al. 2000). *Pachyrhachis* is another snake with legs that is related to *Haasiophis* (Caldwell and Lee 1997).

d. The jaws of mososaurs are also intermediate between snakes and lizards. Like the snake's stretchable jaws, they have highly flexible lower jaws, but unlike snakes, they do not have highly flexible upper jaws. Some other skull features of mososaurs are intermediate between snakes and primitive lizards (Caldwell and Lee 1997; M.S.T. Lee et al. 1999; Tchernov et al. 2000).

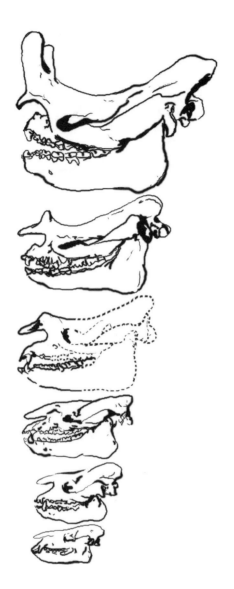

Figure CC2
Skulls representing stages in titanothere evolution. *Bottom to top*: *Eotitanops borealis*, Lower Eocene (Wind River Formation); *Limnohyops priscus*, Middle Eocene (upper Bridger Formation); *Manteoceras manteoceras*, Middle Eocene (upper Bridger Formation); *Protitanotherium emargination*, Upper Eocene (upper Uinta Formation); *Brontotherium leidyi*, Lower Oligocene (lower Chadron Formation); *Brontotherium gigas*, Lower Oligocene (upper Chadron Formation). From H. F. Osborn, 1929, The titanotheres of ancient Wyoming, Dakota, and Nebraska. U.S.G.S. Monograph 55.

   e. Transitions between mesonychids and whales (see CC216.1).

   f. Transitions between fish and tetrapods (see CC212).

   g. Transitions from condylarths (a kind of land mammal) to fully aquatic modern manatees. In particular, *Pezosiren portelli* is clearly a sirenian, but its hind limbs and pelvis are unreduced (Domning 2001a, 2001b).

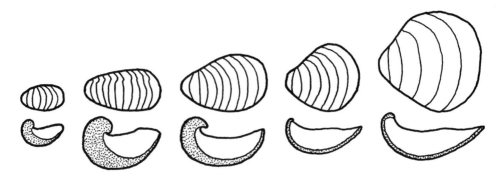

Figure CC3
A lineage of *Gryphaea* (coiled oysters) during the Early Jurassic, from Angulata zone (*left*; approximately 197 million years ago) to Spinatum zone (approximately 183 Ma). *Left to right*: G. *arcuata obliquata*, G. *arcuata incurva*, G. *mccullochii*, two varieties of G. *gigantea*. Lower views show lateral cross-section, with stippled area showing thickness of the shell. After A. Hallam, 1968, "Morphology, palaeoecology and evolution of the genus *Gryphaea* in the British Lias," *Philosophical Transactions of the Royal Society of London B* 254: 124.

The following are fossil transitionals between kingdoms and phyla:

a. The Cambrian fossils *Halkiera* and *Wiwaxia* have features that connect them with each other and with the modern phyla of Mollusca, Brachiopoda, and Annelida. In particular, one species of halkieriid has brachiopod-like shells on the dorsal side at each end. This is seen also in an immature stage of the living brachiopod species *Neocrania*. It has setae identical in structure to polychaetes, a group of annelids. *Wiwaxia* and *Halkiera* have the same basic arrangement of hollow sclerites, an arrangement that is similar to the chaetae arrangement of polychaetes. The undersurface of *Wiwaxia* has a soft sole like a mollusk's foot, and its jaw looks like a mollusk's mouth. Aplacophorans, which are a group of primitive mollusks, have a soft body covered with spicules similar to the sclerites of *Wiwaxia* (Conway Morris 1998, 185–195).

b. Cambrian and Precambrain fossils *Anomalocaris* and *Opabinia* are transitional between arthropods and lobopods (see CC220).

c. An ancestral echinoderm has been found that is intermediate between modern echinoderms and other deuterostomes (Shu et al. 2004).

Further Reading: Cohn, M. J., and C. Tickle. 1999. Developmental basis of limblessness and axial patterning in snakes; Cuffey, C. A. 2000. The fossil record: Evolution or "scientific creation." http://www.nogs.org/cuffeyart.html; Elsberry, W. R. 1995. Transitional fossil challenge. http://www.rtis.com/nat/user/elsberry/evobio/evc/argresp/tranform.html; Godfrey, L. R. (ed.). 1983. *Scientists Confront Creationism*, p. 206; Hunt, K. 1997. Transitional vertebrate fossils FAQ. http://www.talkorigins.org/faqs/faq-transitional.html; Miller, K. B. n.d. Taxonomy, transitional forms, and the fossil record. http://www.asa3.org/ASA/resources/Miller.html; Morton, G. R. 2000b. Phylum level evolution. http://home.entouch.net/dmd/cambevol.htm; Patterson, B. 2002. Transitional fossil species and modes of speciation. http://www.origins.tv/darwin/transitionals.htm; Pojeta, J., Jr., and D. A. Springer. 2001. *Evolution and the fossil record*. http://www.agiweb.org/news/evolution.pdf, p. 2; Strahler, A. N. 1987. *Science and Earth History*, pp. 398–400; Thompson, T. 1999. On creation science and transitional fossils. http://www.tim-thompson.com/trans-fossils.html; Zimmer, C. 2000a. In search of vertebrate origins: Beyond brain and bone.

CC200.1: There should be billions of transitional fossils.

Given all the species that exist and have existed, there should be billions of transitional fossils in the fossil record; we should have found tens of thousands at least. (Gish 1994)

1. Some important factors prevent the formation of fossils from being common:

   a. Fossilization itself is not a particularly common event. It requires conditions that preserve the fossil before it becomes scavenged or decayed. Such conditions are common only in a very few habitats, such as river deltas, peat bogs, and tar pits. Organisms that do not live in or near these habitats will be preserved only rarely.

   b. Many types of animals are fragile and do not preserve well.

   c. Many species have small ranges. Their chance of fossilization will be proportionally small.

   d. The evolution of new species probably is fairly rapid in geological terms, so the transitions between species will be uncommon.

Passenger pigeons, once numbered in the billions, went extinct less than 200 years ago. How many passenger pigeon fossils can you find? If they are hard to find, why should we expect to find fossils that are likely from smaller populations and have been subject to millions of years of potential erosion?

2. Other processes destroy fossils. Erosion (and/or lack of deposition in the first place) often destroys hundreds of millions of years or more of the geological record, so the geological record at any place usually has long gaps. Fossils can also be destroyed by heat or pressure when buried deep underground.

3. As rare as fossils are, fossil discovery is still rarer. For the most part, we find only fossils that have been exposed by erosion and only if the exposure is recent enough that the fossils themselves do not erode.

As climates change, species will move, so we cannot expect a transition to occur all at one spot. Fossils often must be collected from all over a continent to find the transitions.

Only Europe and North America have been well explored for fossils because that is where most of the paleontologists lived. Furthermore, regional politics interfere with collecting fossils. Some fabulous fossils have been found in China only recently because before then the politics prevented most paleontology there.

4. The shortage is not just in fossils but in paleontologists and taxonomists. Preparing and analyzing the material for just one lineage can take a decade of work. There are likely hundreds of transitional fossils sitting in museum drawers, unknown because nobody knowledgeable has examined them.

5. Description of fossils is often limited to professional literature and does not get popularized. This is especially true of marine microfossils, which have the best record.

6. If fossilization were so prevalent, we should find indications in the fossil record of animals migrating from the Ark to other continents.

Further Reading: Hunt, K. 1997. Transitional vertebrate fossils FAQ. Part 1A. http://www.talk origins.org/faqs/faq-transitional/part1a.html#gaps; Kidwell, S.M., and S.M. Holland. 2002. The quality of the fossil record.

CC201: We should see smooth change through the fossil record, not gaps.

If evolution proceeds via the accumulation of small steps, we should see a smooth continuum of creatures across the fossil record. Instead, we see long periods in which species do not change, and there are gaps between the changes. (P. E. Johnson 1990; H. M. Morris 1985, 78)

1. The idea that gradual change should appear throughout the fossil record is called *phyletic gradualism*. It is based on the following tenets:

a. New species arise by the transformation of an ancestral population into its modified descendants.

b. The transformation is even and slow.

c. The transformation involves most or all of the ancestral population.

d. The transformation occurs over most or all of the ancestral species' geographic range.

However, all but the first of these is false far more often that not. Studies of modern populations and incipient species show that new species arise mostly from the splitting of a small part of the original species into a new geographical area. The population genetics of small populations allow this new species to evolve relatively quickly. Its evolution may allow it to spread into new geographical areas. Since the actual transitions occur relatively quickly and in a relatively small area, the transitions do not often show up in the fossil record. Sudden appearance in the fossil record often simply reflects that an existing species moved into a new region.

Once species are well adapted to an environment, selective pressures tend to keep them that way. A change in the environment that alters the selective pressure would then end the "stasis" (or lead to extinction).

It should be noted that even Darwin did not expect the rate of evolutionary change to be constant; for example:

> It is a more important consideration . . . that the period during which each species underwent modification, though long as measured by years, was probably short in comparison with that during which it remained without undergoing any change. (Darwin 1872b, 428)

2. The imperfection of the fossil record (due to erosion and periods unfavorable to fossil preservation) also causes gaps, although it probably cannot account for all of them.

3. Some transitional sequences exist, which, despite an uneven rate of change, still show a gradual continuum of forms (see CC200).

4. The fossil record still shows a great deal of change over time. The creationists who make note of the many gaps almost never admit the logical conclusion: If they are due to creation, then there have been hundreds, perhaps even millions, of separate creation events scattered through time.

Further Reading: Elsberry, W. R. 1996. Punctuated equilibria. http://www.talkorigins.org/faqs/punc-eq.html.

CC201.1: Punctuated equilibrium was ad hoc to justify gaps.

The theory of punctuated equilibrium was proposed ad hoc to explain away the embarrassing gaps in the fossil record. (Yahya 2003b)

1. The theory of punctuated equilibrium is based on positive evidence, including extensive studies of living and extinct species groups (Eldredge and Gould 1972).

2. The idea of phyletic gradualism, which is invoked to justify a lack of gaps, fails to fit the evidence of population biology (see CC201).

3. There is nothing wrong with proposing theories to fit the data.

Further Reading: Elsberry, W.R. 1996. Punctuated equilibria. http://www.talkorigins.org/faqs/punc-eq.html; Gould, S.J. 1983. Evolution as fact and theory.

CC211: There are gaps between invertebrates and vertebrates.

No fossils have been found transitional between invertebrates and vertebrates. (H.M. Morris 1985, 82)

1. There are Cambrian fossils transitional between vertebrate and invertebrate:

a. *Pikaia*, an early invertebrate chordate. It was at first interpreted as a segmented worm until a reanalysis showed it had a notochord.

b. *Yunnanozoon*, an early chordate.

c. *Haikouella*, a chordate similar to *Yunnanozoon*, but with additional traits, such as a heart and a relatively larger brain (Chen et al. 1999).

d. Conodont animals had bony teeth, but the rest of their body was soft. They also had a notochord (Briggs et al. 1983; Sansom et al. 1992).

e. *Cathaymyrus diadexus*, the oldest known chordate (535 million years old; Shu et al. 1996).

f. *Myllokunmingia* and *Haikouichthys*, two early vertebrates that still lack a clear head and bony skeletons and teeth. They differ from earlier invertebrate chordates in having a zigzag arrangement of segmented muscles, and their gill arrangement is more complex than a simple slit (Monastersky 1999).

2. There are living invertebrate chordates (Branchiostoma [*Amphioxus*], urochordates [tunicates]) and living basal near-vertebrates (hagfish, lampreys) that show plausible intermediate forms.

Further Reading: Monastersky, R. 1999. Waking up to the dawn of vertebrates. http://www.sciencenews.org/sn_arc99/11_6_99/fob1.htm; Speer, B.R. 2000. Introduction to the Deuterostomia. http://www.ucmp.berkeley.edu/phyla/deuterostomia.html; Waggoner, B. 1996. Introduction to the Cephalochordata. http://www.ucmp.berkeley.edu/chordata/cephalo.html.

CC212: There are gaps between fish and amphibians.

There are no transitional fossils between fish and tetrapods. (H.M. Morris 1985, 82–83; Watchtower 1985, 72)

1. There are several good transitional fossils (see Figure CC4):

a. A fossil shows eight bony fingers in the front fin of a lobed fish, offering evidence that fingers developed before land-going tetrapods (Daeschler and Shubin 1998).

b. A Devonian humerus has features showing that it belonged to an aquatic tetrapod that could push itself up with its forelimbs but could not move it limbs back and forth to walk (Shubin et al. 2004).

c. *Acanthostega*, a Devonian fossil, about 60 cm long, probably lived in rivers (Coates 1996). It had polydactyl limbs with no wrists or ankles (Coates and Clack 1990). It was predominantly, if not exclusively, aquatic: It had fishlike internal gills (Coates and Clack 1991), and its limbs and spine could not support much weight. It also had a stapes and a lateral sensory system like a fish.

d. *Ichthyostega*, a tetrapod from Devonian streams, was about 1.5 m long and probably amphibious. It had seven digits on its rear legs (its hands are unknown). Its limbs and spine were more robust than those of *Acanthostega*, and its rib cage was massive. It had fishlike spines on its tail, but these were fewer and smaller than *Acanthostega*'s. Its skull had several primitive fishlike features, but it probably did not have internal gills (Murphy 2002).

e. *Tulerpeton*, from estuarine deposits roughly the same age as *Acanthostega* and *Ichthyostega*, had six digits on its front limbs and seven on its rear limbs. Its shoulders were more robust than *Acanthostega*, suggesting it was somewhat less aquatic, and its skull appears to be closer to later Carboniferous amphibians than to *Acanthostega* or *Ichthyostega*.

Further Reading: Clack, J. A. 2002. *Gaining Ground*; Morton, G. R. 1997a. Fish to amphibian transition. http://home.entouch.net/dmd/transit.htm; Murphy, D. C. 2002. Devonian times. http://www.mdgekko.com/devonian; Pojeta, J., Jr., and Springer, D. A. 2001. *Evolution and the fossil record.* http://www.agiweb.org/news/evolution.pdf; Zimmer, C. 1998. *At the Water's Edge.* Chaps. 1–4.

CC213: There are gaps between amphibians and reptiles.

No unambiguous transitional fossils have been found between amphibians and reptiles. Distinguishing transitionals between these two groups is problematic because their bone structures are similar. Just because it is hard to tell whether a fossil is reptile or amphibian does not mean it is transitional between the two. (H. M. Morris 1985, 83; Watchtower 1985, 73)

1. The main character that separates amphibians (primitive tetrapods) from reptiles (amniotes) is possession of an amnion, which does not fossilize. We have a lot of Permian creatures, some of which are early amniotes and some of which likely are not. There are no unambiguous intermediates between the two groups like *Acanthostega* between fish and tetrapods (see CC212) or *Morganucodon* between reptiles and mammals (see CC215). However, the same uncertainty means there is no clear gap between the amphibians and reptiles, either.

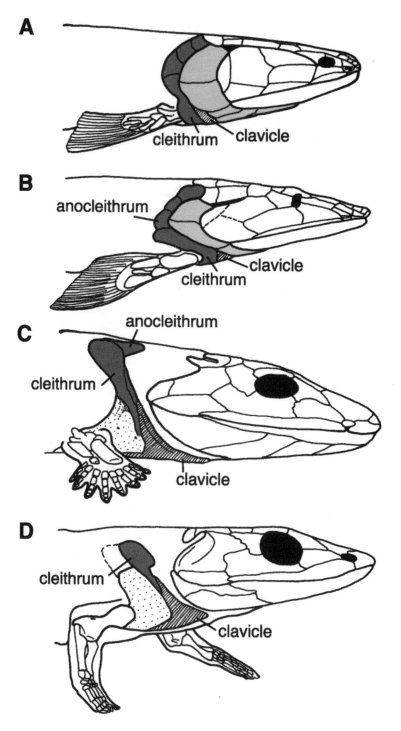

Figure CC4
Skulls and shoulder girdles of (A) *Eusthenopteron*, (B) *Panderichthys*, (C) *Acanthostega*, and (D) *Dendrerpeton*, to show changes to the opercular bones and to the dermal and endoskeletal parts of the girdle. Dermal parts of the girdle are dark-shaded and hatched, the opercular series is light-shaded, and the scapulocoracoid is stippled. From J. A. Clack, 2002, *Gaining Ground*, Indiana University Press.

CC214: There are gaps between reptiles and birds.

There are no transitional fossils between reptiles and birds. (Watchtower 1985, 75)

1. Many new bird fossils have been discovered in the last couple of decades, revealing several intermediates between theropod dinosaurs (such as *Allosaurus*) and modern birds:

- *Sinosauropteryx prima*. A dinosaur covered with primitive feathers, but structurally similar to unfeathered dinosaurs *Ornitholestes* and *Compsognathus* (P. Chen et al. 1998; Currie and Chen 2001).

- Ornithomimosaurs, therizinosaurs, and oviraptorosaurs. The oviraptorosaur *Caudipteryx* had a body covering of tufted feathers and had feathers with a central rachis on its wings and tail (Ji et al. 1998). Feathers are also known from the therizinosaur *Beipiaosaurus* (Xu et al. 1999a). Several other birdlike characters appear in these dinosaurs, including unserrated teeth, highly pneumatized skulls and vertebrae, and elongated wings. Oviraptorids also had birdlike eggs and brooding habits (Clark et al. 1999).

- Deinonychosaurs (troodontids and dromaeosaurs). These are the closest known dinosaurs to birds. *Sinovenator*, the most primitive troodontid, is especially similar to *Archaeopteryx* (Xu et al. 2002). *Byronosaurus*, another troodontid, had teeth nearly identical to primitive birds (Makovicky et al. 2003). *Microraptor*, the most primitive dromaeosaur, is also the most birdlike; specimens have been found with undisputed feathers on their wings, legs, and tail (Hwang et al. 2002; Xu et al. 2003). *Sinornithosaurus* also was covered with a variety of feathers and had a skull more birdlike than later dromaeosaurs (Xu, Wang, and Wu 1999; Xu and Wu 2001; Xu et al. 2001).

- *Protarchaeopteryx*, alvarezsaurids, *Yixianosaurus*, and *Avimimus*. These are birdlike dinosaurs of uncertain placement, each potentially closer to birds than deinonychosaurs are. *Protarchaeopteryx* has tail feathers, uncompressed teeth, and an elongated manus (hand/wing; Ji et al. 1998). *Yixianosaurus* has an indistinctly preserved feathery covering and hand/wing proportions close to birds (Xu and Wang 2003). Alvarezsaurids (Chiappe et al. 2002) and *Avimimus* (Vickers-Rich et al. 2002) have other birdlike features.

- *Archaeopteryx* (see CC214.1.1). This famous fossil is defined as a bird, but it is actually less birdlike in some ways than some genera mentioned above (Maryanska et al. 2002; Paul 2002).

- *Shenzhouraptor* (Zhou and Zhang 2002), *Rahonavis* (Forster et al. 1998), *Yandangornis*, and *Jixiangornis*. All of these birds were slightly more advanced than *Archaeopteryx*, especially in characters of the vertebrae, sternum, and wing bones.

- *Sapeornis* (Zhou and Zhang 2003), *Omnivoropteryx*, and confuciusornithids (e.g., *Confuciusornis* and *Changchengornis*; Chiappe et al. 1999). These were the first birds to possess large pygostyles (bone formed from fused tail vertebrae). Other new birdlike characters include seven sacral vertebrae, a sternum with a keel (some species), and a reversed hallux (hind toe).

- Enantiornithines, including at least nineteen species of primitive birds, such as *Sinornis* (Sereno and Rao 1992), *Gobipteryx* (Chiappe et al. 2001), and the primitive *Protopteryx* (Zhang and Zhou 2000). Several birdlike features appeared in enantiornithines, including twelve or fewer dorsal vertebrae, a narrow V-shaped furcula (wishbone), and reduction in wing digit bones.

- *Patagopteryx*, *Apsaravis*, and *yanornithids* (Chiappe 2002b; Clarke and Norell 2002). More birdlike features appeared in this group, including changes to vertebrae and development of the sternal keel.

- *Hesperornis*, *Ichthyornis*, *Gansus*, and *Limenavis*. These birds are almost as advanced as modern species. New features included the loss of most teeth and changes to leg bones.

- Modern birds.

Further Reading: Chiappe, L. M., and G. J. Dyke. 2002. The Mesozoic radiation of birds; Dingus, L., and T. Rowe. 1997. *The Mistaken Extinction*; Paul, G. S. 2002. *Dinosaurs of the Air*. Pojeta, J., Jr., and D. A. Springer. 2001. *Evolution and the fossil record*. http://www.agiweb.org/news/evolution.pdf; Wang, J. 1998. Scientists flock to explore China's "site of the century."

CC214.1: *Archaeopteryx* was probably not an ancestor of modern birds.

Modern birds were probably not descended from Archaeopteryx, *so it is wrong to claim* Archaeopteryx *as a missing link between dinosaurs and birds.* (J. Wells 2000, 115–116)

1. "Transitional" does not mean "ancestral." It means that the transitional fossil shows a mosaic of features from organisms before and after. It is wrong to say that *Archaeopteryx* was ancestral to modern birds. But it is also wrong to say that it is not transitional. It is indisputable that *Archaeopteryx* is intermediate between dinosaurs and modern birds. That makes it transitional and gives evidence of the relatedness between dinosaurs and birds.

2. Several other recently discovered dinosaur, bird, and intermediate dinosaur–bird fossils are starting to fill in the gaps and are providing further evidence that this interpretation of *Archaeopteryx* is correct (see CC214).

Further Reading: Nedin, C. 1999. All about *Archaeopteryx*. http://www.talkorigins.org/faqs/archaeopteryx/info.html.

CC214.1.1: *Archaeopteryx* is fully bird.

Archaeopteryx *was fully bird. It had fully formed wings and feathers.* (H. M. Morris 1985, 85; Watchtower 1985, 79–80)

1. *Archaeopteryx* is defined to be a bird (technically, an avialan). However, it had many more dinosaurian traits than bird traits. Its main bird traits are

- long external nostrils
- quadrate and quadratojugal (two jaw bones) not sutured together
- palatine bones that have three extensions
- all teeth lacking serrations
- large lateral furrows in top rear body of the vertebrae

Other birdlike traits of *Archaeopteryx* are found also on several non-avian dinosaurs. These traits include feathers, a furcula (wishbone) fused at the midline, and a pubis elongated and directed backward. The birdlike hallux (toe) attributed to *Archaeopteryx* may be an error due to poor preservation (Middleton 2002).

Dinosaurian traits include the following:

- no bill
- teeth on premaxilla and maxilla bones
- nasal opening far forward, separated from the eye by a large preorbital fenestra (hole)
- neck attached to skull from the rear
- center of cervical vertebrae that have simple concave articular facets
- long bony tail; no pygostyle
- ribs slender, without joints or uncinate processes, and not articulated with the sternum
- sacrum that occupies six vertebrae
- small thoracic girdle
- metacarpals free (except third metacarpal), wrist hand joint flexible
- claws on three unfused digits
- pelvic girdle and femur joint shaped like those of archosaurs in many details
- bones of pelvis unfused

and over 100 other differences from birds (Chiappe 2002a; Norell and Clarke 2001).

In addition, *Archaeopteryx* was intermediate between dinosaurs and modern birds in the shape of the coracoid and humerus bones and the brain (Elzanowski 2002; Nedin 1999).

Further Reading: Nedin, C. 1999. All about *Archaeopteryx*. http://www.talkorigins.org/faqs/archaeopteryx/info.html.

CC215: There are gaps between reptiles and mammals.

There is a great gulf between reptiles and mammals, with no transitional fossils between them. (Watchtower 1985, 80–81)

1. The transition from reptile to mammal has an excellent record. The following fossils are just a sampling. In particular, these fossils document the transition of one type of jaw joint into another. Reptiles have one bone in the middle ear and several bones in the lower jaw. Mammals have three bones in the middle ear and only one bone in the lower jaw. These species show transitional jaw–ear arrangements (Hunt 1997; T. White 2002b). The sequence shows transitional stages in other features, too, such as skull, vertebrae, ribs, and toes.

　　a. *Sphenacodon* (late Pennsylvanian to early Permian, about 270 Ma). Lower jaw is made of multiple bones; the jaw hinge is fully reptilian. No eardrum.

　　b. *Biarmosuchia* (late Permian). One of the earliest therapsids. Jaw hinge is more mammalian. Upper jaw is fixed. Hindlimbs are more upright.

　　c. *Procynosuchus* (latest Permian). A primitive cynodont, a group of mammal-like therapsids. Most of the lower jaw bones are grouped in a small complex near the jaw hinge.

d. *Thrinaxodon* (early Triassic). A more advanced cynodont. An eardrum has developed in the lower jaw, allowing it to hear airborne sound. Its quadrate and articular jaw bones could vibrate freely, allowing them to function for sound transmission while still functioning as jaw bones. All four legs are fully upright.

e. *Probainognathus* (mid-Triassic, about 235 Ma). It has two jaw joints: mammalian and reptilian (T. White 2002a).

f. *Diarthrognathus* (early Jurassic, 209 Ma). An advanced cynodont. It still has a double jaw joint, but the reptilian joint functions almost entirely for hearing.

g. *Morganucodon* (early Jurassic, about 220 Ma). It still has a remnant of the reptilian jaw joint (Kermack et al. 1981).

h. *Hadrocodium* (early Jurassic). Its middle ear bones have moved from the jaw to the cranium (Luo et al. 2001; T. White 2002b).

Further Reading: Benton, M. J. 1991. *The Rise of the Mammals*; Flank, L. 1995b. The therapsid–mammal transitional series. http://www.geocities.com/CapeCanaveral/Hangar/2437/therapsd.htm; Hunt, K. 1997. Transitional vertebrate fossils FAQ, Part 1B. http://www.talkorigins.org/faqs/faq-transitional/part1b.html#syn2mamm; Theobald, D. 2004. 29+ evidences, part 1: The unique universal phylogenetic tree. http://www.talkorigins.org/faqs/comdesc/section1.html#morphological_intermediates_ex2.

CC216.1: There are gaps between land mammals and whales.

There are gaps between land mammals and whales. (Gish 1994)

1. The transitional sequence from mesonychids (an extinct wolflike mammal) to whales is quite robust. See http://fp.bio.utk.edu/darwin/1997/whale.html for pictures of some of these.

a. *Pakicetus inachus:* latest Early Eocene (Gingerich et al. 1983; Thewissen and Hussain 1993).

b. *Ambulocetus natans:* Early to Middle Eocene, above *Pakicetus*. It had short front limbs and hind legs adapted for swimming; undulating its spine up and down helped its swimming. It apparently could walk on land as well as swim (Thewissen et al. 1994).

c. *Indocetus ramani:* earliest Middle Eocene.

d. *Dorudon:* the dominant cetacean of the late Eocene. Their tiny hind limbs were not involved in locomotion.

e. *Basilosaurus:* middle Eocene and younger. A fully aquatic whale with structurally complete legs (Gingerich et al. 1990).

f. an early baleen whale with its blowhole far forward and some structural features found in land animals but not later whales (Stricherz 1998).

Further Reading: Babinski, E. T. 2003. Cetacean evolution (whales, dolphins, porpoises). http://www.edwardtbabinski.us/babinski/whale_evolution.html; Gould, S. J. 1995. Hooking leviathan by its past; Sutera, R. 2001. The origin of whales and the power of independent evidence. http://www.talkorigins.org/features/whales; Thewissen, J.G.M. (ed.). 1998. *The Emergence of Whales*. Thewissen, J.G.M., et al. 1998. Whale ankles and evolutionary relationships; Thewissen, J.G.M., and E. M. Williams. 2002. The early radiations of *Cetacea* (Mammalia); Zimmer, C. 1995. Back to the sea; Zimmer, C. 1998. *At the Water's Edge*. Chaps. 6–10.

CC216.2: Horse fossils do not show evolution.

The fossil record does not show a gradual development from a small animal to the large modern horse. The horse family tree is not simple and direct; some scientists say Eohippus was not an ancestor of the modern horse; and the different types of fossils show stability, not gradual change. (Watchtower 1985, 66–67)

1. The fossil record does not show a gradual, linear progression from *Hyracotherium* (*Eohippus*) to *Equus*. Nor is there any reason to think it should. The fossil record of equids shows that various lineages split into several branches. Evolution was not smooth and gradual; traits evolved at different rates and occasionally reversed. Some species arose gradually, others suddenly. All of this is in accord with the messiness we expect from evolution and from biology in general.

2. Some creationists consider all the species in the horse family to be the same "kind." They accept "microevolution" from *Hyracotherium* at the time of the Flood, to modern horses and donkeys first recorded less than four hundred years later (Wood and Cavanaugh 2003). This rate of change is far greater than biologists accept.

Further Reading: FLMNH. n.d. Fossil horses in hyperspace. http://www.flmnh.ufl.edu/natsci/vertpaleo/fhc/fhc.htm; Gould, S. J. 1991. Life's little joke; Hunt, K. 1995. Horse evolution. http://www.talkorigins.org/faqs/horses/horse_evol.html.

CC220: Arthropods arose suddenly.

Arthropods arose suddenly in the fossil record. There are no transitionals leading up to them.

1. *Anomalocaris*, from the Cambrian and Precambrian, has a pair of segmented appendages, indicating arthropod affinities, but it also seems to have lobopod legs. The related *Opabinia* also apparently had lobopod legs. Other rare Cambrian and Precambrian fossils show some promise of shedding more light on relationships. For example, *Spriggina*, another Precambrian animal, has a head shield similar to trilobites (Conway Morris 1998, 184–185).

Further Reading: Morton, G. R. 2000b. Phylum level evolution. http://home.entouch.net/dmd/cambevol.htm.

CC220.1: There are no fossil ancestors of insects.

Insect fossils appear in abundance, but the record of insect origins is completely blank. (W. Brown 1995, 10; H. M. Morris 1985, 86)

1. Insect fossils before the major diversification of insects (in the Carboniferous) are far from abundant. Insects are believed, from genomic data, to have originated near the beginning of the Silurian (434.2–421.1 Ma; Gaunt and Miles 2002), but the first two hexapod fossils are from Rhynie chert, about 396–407 Ma (Engel and Grimaldi 2004; Whalley and Jarzembowski 1981). As of 2004, only two other insect fossils were known from the Devonian (Labandeira et al. 1988). Two of these fossils consist only of

mandibles, and another is a crushed head. In short, the first eighty-five million years of the history of insects is preserved in only four fossils, three of them quite fragmentary. With such a scarcity of fossils, the lack of fossils showing the origins of insects is unremarkable.

CC250: There are no fossil ancestors of plants.

The fossil record does not show the origin of any group of modern plants from its beginning. (H.M. Morris 1985, 86–87)

1. The fossil record shows the origins of several groups of modern plants. The groups listed here are some of the most prominent:

- Land plants. Several fossils exist showing their origin (Bateman et al. 1998; Kenrick and Crane 1997). Molecular data combined with the fossil evidence shows that the first land plants were liverworts (Qiu et al. 1998). A fossil Ordovician fungus (about 460 Ma) has the same form as modern arbuscular mycorrhizal fungi, indicating that the earliest land plants had this kind of symbiotic relationship with fungi (Redecker et al. 2000).

- Seed plants. The first trees were also among the first free-sporing plants sharing characteristics with seed plants (Meyer-Berthaud et al. 1999).

- Angiosperms. Jurassic fossils of *Archaefructus* show some of the earliest and most primitive angiosperms (Sun et al. 2002). Ren (1998) described some Jurassic flies adapted for pollination, suggesting that angiosperms may have originated by then. Dilcher (2000) briefly reviews angiosperm paleobotany. Major events in their evolution were the appearance of closed carpels, bilaterally symmetrical flowers, and large fruits.

- Monocotyledons. The early fossils of this group are meager, but some fossils exist (Gandolfo et al. 1998).

Further Reading: Bateman, R.M., et al. 1998. Early evolution of land plants; McCourt, R.M., et al. 1996. Green plants. http://tolweb.org/tree?group=Green_plants&contgroup=Eukaryotes. (This is part of the Tree of Life Web Project. This and the pages it links to have abundant additional references.)

CC251: Progymnosperms are imaginary evolutionary ancestors.

Evolutionists claim that conifers are descended from a group of plants called progymnosperms. But there is no evidence that progymnosperms ever existed. (A. Williams 2002)

1. Progymnosperms are not hypothetical. They are an important group of fossil plants, now extinct, which include the earliest trees (Meyer-Berthaud et al. 1999). Creationists can choose to believe that progymnosperms are not transitional forms, but they certainly existed.

Further Reading: Arens, N.C. 1998. Progymnosperms. http://www.ucmp.berkeley.edu/IB181/VPL/Osp/Osp2.html; Scheckler, S.E. 1999. Progymnosperms; Speer, B.R., and Arens, N.C. 1996.

Introduction to the progymnosperms. http://www.ucmp.berkeley.edu/seedplants/progymnosperms .html.

with Mike Hopkins

FOSSIL RECORD

CC300: The Cambrian explosion shows all kinds of life appearing suddenly.
Complex life forms appear suddenly in the Cambrian explosion, with no ancestral fossils. (H. M. Morris 1985, 80–81; Watchtower 1985, 60–62)

1. The Cambrian explosion was the seemingly sudden appearance of a variety of complex animals about 540 million years ago (Ma), but it was not the origin of complex life. Evidence of multicellular life from about 590 and 560 Ma appears in the Doushantuo Formation in China (Chen et al. 2000, 2004), and diverse fossil forms occurred before 555 Ma (Martin et al. 2000; the Cambrian began 543 Ma, and the Cambrian explosion is considered by many to start with the first trilobites, about 530 Ma). Testate amoebae are known from about 750 Ma (Porter and Knoll 2000). There is evidence in the form of wormlike burrows in the Chorhat Sandstone in India of animals over a billion years ago (Seilacher et al. 1998) and tracelike fossils more than 1,200 Ma in the Stirling Range Formation of Australia (Rasmussen et al. 2002). Eukaryotes (which have relatively complex cells) may have arisen 2,700 Ma, according to fossil chemical evidence (Brocks et al. 1999). Apparent fossil microorganisms have been found from 3,465 Ma (Schopf 1993). There is isotopic evidence of sulfur-reducing bacteria from 3.47 billion years ago (Shen et al. 2001).

2. There are transitional fossils within the Cambrian explosion fossils. For example, there are lobopods (basically worms with legs) which are intermediate between arthropods and worms (Conway Morris 1998).

3. Only some phyla appear in the Cambrian explosion. In particular, all plants postdate the Cambrian, and flowering plants, by far the dominant form of land life today, only appeared about 140 Ma (K. S. Brown 1999).

Even among animals, not all types appear in the Cambrian. Cnidarians, sponges, and probably other phyla appeared before the Cambrian. Molecular evidence shows that at least six animal phyla are Precambrian (Wang et al. 1999). Bryozoans appear first in the Ordovician. Many other soft-bodied phyla do not appear in the fossil record until much later. Although many new animal forms appeared during the Cambrian, not all did. According to one reference (A. G. Collins 1994), eleven of thirty-two metazoan phyla appear during the Cambrian, one appears Precambrian, eight after the Cambrian, and twelve have no fossil record.

And that just considers phyla. Almost none of the animal groups that people think of as groups, such as mammals, reptiles, birds, insects, and spiders, appeared in the Cambrian. The fish that appeared in the Cambrian were unlike any fish alive today.

4. The length of the Cambrian explosion is ambiguous and uncertain, but five to ten million years is a reasonable estimate; some say the explosion spans forty million years

or more, starting about 553 million years ago. Even the shortest estimate of five million years is hardly sudden.

5. There are some plausible explanations for why diversification may have been relatively sudden:

- The evolution of active predators in the late Precambrian likely spurred the co-evolution of hard parts on other animals. These hard parts fossilize much more easily than the previous soft-bodied animals, leading to many more fossils but not necessarily more animals.

- Early complex animals may have been nearly microscopic. Apparent fossil animals smaller than 0.2 mm have been found in the Doushantuo Formation, China, forty to fifty-five million years before the Cambrian (Chen et al. 2004). Much of the early evolution could have simply been too small to see.

- The earth was just coming out of a global ice age at the beginning of the Cambrian (Hoffman 1998; Kerr 2000). A "snowball earth" before the Cambrian explosion may have hindered development of complexity or kept populations down so that fossils would be too rare to expect to find today. The more favorable environment after the snowball earth would have opened new niches for life to evolve into.

- Hox genes, which control much of an animal's basic body plan, were likely first evolving around that time. Development of these genes might have just then allowed the raw materials for body plans to diversify (Carroll 1997).

- Atmospheric oxygen may have increased at the start of the Cambrian (Canfield and Teske 1996; Logan et al. 1995; Thomas 1997).

- Planktonic grazers began producing fecal pellets that fell to the bottom of the ocean rapidly, profoundly changing the ocean state, especially its oxygenation (Logan et al. 1995).

- Unusual amounts of phosphate were deposited in shallow seas at the start of the Cambrian (Cook and Shergold 1986; Lipps and Signor 1992).

6. Cambrian life was still unlike almost everything alive today. Using number of cell types as a measure of complexity, we see that complexity has been increasing more or less constantly since the beginning of the Cambrian (Valentine et al. 1994).

7. Major radiations of life forms have occurred at other times, too. One of the most extensive diversifications of life occurred in the Ordovician, for example (A. I. Miller 1997).

Further Reading: Conway Morris, S. 1998. *The Crucible of Creation*; Conway Morris, S. 2000. The Cambrian "explosion"; Schopf, J. W. 2000. Solution to Darwin's dilemma.

with John Harshman

CC301: Cambrian explosion contradicts evolutionary "tree" pattern.

In the Cambrian explosion, all major animal groups appear together in the fossil record fully formed instead of branching from a common ancestor, thus contradicting the evolutionary tree of life. (J. Wells 2000, 40–45)

1. The Cambrian explosion does not show all groups appearing together fully formed. It shows some animal groups (and no plant, fungus, or microbe groups) appearing over many millions of years in forms very different, for the most part, from the forms that are seen today (see CC300).

2. During the Cambrian, there was the first appearance of hard parts, such as shells and teeth, in animals. The lack of readily fossilizable parts before then ensures that the fossil record would be very incomplete in the Precambrian. The old age of the Precambrian era contributes to a scarcity of fossils.

3. The Precambrian fossils that have been found are consistent with a branching pattern and inconsistent with a sudden Cambrian origin. For example, bacteria appear well before multicellular organisms, and there are fossils giving evidence of transitionals leading to halkierids (see CC200) and arthropods (see CC220).

4. Genetic evidence also shows a branching pattern in the Precambrian, indicating, for example, that plants diverged from a common ancestor before fungi diverged from animals.

CC310: Fossils are dated from strata; strata are dated from fossils.

Fossils are used to determine the order and dates of the strata in which they are found. But the fossil order itself is based on the order of strata and the assumption of evolution. Therefore, using fossil progression as evidence for evolution is circular reasoning. (H.M. Morris 1985, 95–96,136)

1. Many strata are not dated from fossils. Relative dates of strata (whether layers are older or younger than others) are determined mainly by which strata are above others. Some strata are dated absolutely via radiometric dating. These methods are sufficient to determine a great deal of stratigraphy.

Some fossils are seen to occur only in certain strata. Such fossils can be used as index fossils. When these fossils exist, they can be used to determine the age of the strata, because the fossils show that the strata correspond to strata that have already been dated by other means.

2. The geological column, including the relative ages of the strata and dominant fossils within various strata, was determined before the theory of evolution.

Further Reading: MacRae, A. 1998. Radiometric dating and the geological time scale. http://www.talkorigins.org/faqs/dating.html.

CC331: Polystrate fossils indicate massive sudden deposition.

Polystrate fossil trees show tree trunks passing through many layers and several meters of sediments. Obviously, the sediments must have been laid down suddenly, not at the gradual rates proposed by uniformitarian geology. (H.M. Morris 1985, 107–108)

Figure CC5
Fossil tree stumps ("polystrate" trees) near Joggins, Nova Scotia, Canada. A: Panoramic view of Joggins section (Upper Carboniferous) from Coal Mine Point. B: Detail of cliff, with five fossil trees visible (see arrows), the rightmost one rooted at a higher stratigraphic level than the others. The section also includes two narrow coal seams. C–E: Close-ups of (C) leftmost stump, (D) stump at beach level, (E) rightmost stump cast. Note the roots. (The hammer in the lower right of D is 20cm long.) F: A modern tree in the process of being buried in a salt marsh, Lusby's Marsh, near Amherst, Nova Scotia, in 1999. Old aerial photos show that dead trees have been standing at this site since the 1960s at least and have not decayed away. At spring tides, this location is covered by water. The tree roots are now buried 30–40cm deep in the mud (a 15cm-long pen is stuck in the side of the hole for scale). Photos by Andrew MacRae.

1. Sudden deposition is not a problem for uniformitarian geology. Single floods can deposit sediments up to several feet thick. Furthermore, trees buried in such sediments do not die and decay immediately; the trunks can remain there for years or even decades (see Figure CC5 F).

2. Sudden deposition does not explain other features found with buried trees, such as trees fossilized in place at multiple stratigraphic levels (see Figure CC5 A–E; see also CC332).

Further Reading: Birkeland, B. 2004. Fossil soils (paleosols) at Joggins. http://www.evcforum. net/ubb/Forum7/HTML/000116.html#7; MacRae, A. 1994b. "Polystrate" tree fossils. http://www

.talkorigins.org/faqs/polystrate/trees.html; Matson, D. E. 1994. How good are those young-earth arguments? http://www.talkorigins.org/faqs/hovind/howgood-gc.html#G4a.

CC332: Yellowstone's Specimen Creek fossil forests were evidently transported.

The Specimen Creek fossil forests in Yellowstone National Park show up to fifty layers of fossil forests with upright trees. The conventional explanation is that each new forest grew atop the previous ones as the previous forests were buried in volcanic ash. This explanation fails; instead, the fossil forests result from the deposition of trees uprooted from elsewhere (Sarfati 1999). The prostrate fossil trees show similar orientation. This is consistent with the trees having been transported (as by a flood), not with the trees having grown in place over many years. (J. D. Morris 1995)

1. Specimen Creek and the surrounding area show evidence of some trees transported and others buried in place. The rock layers include conglomerates from meandering and braided streams, conglomerates from mud flows, and tuffaceous sandstones composed mostly of water-transported volcanic ash. In the Eocene, the area would have been between two volcanic chains. Occasional mud flows from the slopes of the mountains would have uprooted and transported trees from the slopes and buried lowland trees in place (Fritz 1980, 1984).

Evidence that some of the trees, especially those on Specimen Ridge, were buried in place, includes the following (Retallack 1981; Yuretich 1984a, 1984b):

- There are tree stumps that are rooted in fine-grained tuffaceous sandstone but buried in conglomerates.
- Upper parts of some stumps and logs, surrounded by conglomerates, were severely abraded, but the lower parts in sandstone have good root systems.
- Flow structures in some conglomerates show they buried in-place trees.
- Thin sections show evidence of soil around the roots.
- There are clear soil horizons around some root systems.

Most of the horizontal logs and some stumps (up to 15 percent of the upright stumps, according to Fritz 1984) were transported, but most of the upright Specimen Ridge fossil trees were buried in place.

2. Upright stumps and trees on Specimen Ridge cannot be explained as floating stumps settling out of standing water, as proposed by Coffin (1983) based on observations of trees washed into Spirit Lake.

- Most tree fossils in Yellowstone occur in sediments from high-energy flows, not a low-energy lake environment (Fritz 1983).
- The percentage of erect trees in Yellowstone is more than 50 percent on Specimen Ridge, but only about 10 to 20 percent of trees stay upright in flows such as in Spirit Lake (Fritz 1983).
- Many of the Specimen Ridge trees are rooted in soils, as noted above.

3. Arct (1991) tried to show that the trees lived all at one time, but even his data indicate that the trees did not all die in the same year (Arct 1991, 38), thus ruling out the possibility that they were uprooted by a common event. (However, Arct's data are insufficient to show overlap in ages; MacRae 1994c.)

Further Reading: MacRae, A. 1994c. Yellowstone National Park (U.S.) fossil forests. http://www.talkorigins.org/faqs/polystrate/yellowstone.html.

CC335: A fossil whale was found vertically through several strata.

Near Lompoc, California, an eighty-foot whale fossil was found in a diatomaceous earth quarry. It was oriented vertically (standing on its tail), passing through millions of years of strata. Only a cataclysmic deposition could account for this. (Ackerman 1986, 81–83)

1. The fossil was not vertical. It was 40 to 50 degrees off horizontal, and the fossil was oriented parallel to the strata. In other words, the whale was horizontal when buried. The strata were later uplifted and folded into their present orientation.

2. There is no evidence for catastrophic deposition. The strata show laminations such as occur from slow accumulation onto an anoxic bottom. A partially buried whale skeleton has been observed off the coast of California; it exemplifies how such fossilization could occur.

Further Reading: South, D. 1995. A whale of a tale. http://www.talkorigins.org/faqs/polystrate/whale.html.

CC340: Many fossils are out of place.

There are over 200 published occurrences of anomalously occurring fossils, or fossils that show up in strata of ages much different than expected. (Woodmorappe 1982)

1. Few if any of the "anomalous" fossils are truly anomalous. It is fairly common for fossils to erode out of an old formation and be redeposited in a younger formation. (It is usually easy to recognize such reworked fossils by the extra wear they show.) Pollen, spores, and other very small fossils can also be blown or washed into tiny cracks to appear in older formations (see CC341). The few anomalies that remain might be explained by genuine range extensions (see below), misidentification of the fossil, or uncertain attribution of where the fossil came from.

2. For most species, the fossil record is quite spotty. The earliest known fossil of a species is likely to be quite a bit later than the earliest appearance of the species; likewise, the latest known fossil is earlier than the species' extinction. There are plenty of opportunities for the discovery of new fossils to extend the known range of a species. It is inappropriate to refer to such new discoveries as anomalies.

3. Even 200 anomalies is an insignificant amount compared with the estimated 250 million fossils that have been catalogued and the much larger number that have been discovered.

CC341: Recent pollen has been found in old rocks.

Pollen has been found in Cambrian and Precambrian rocks, particular the Hakatai Shales of the Grand Canyon. By standard evolutionary models, these rocks pre-date the evolution of pollen-bearing plants. (AIG 1990b; Burdick 1966, 1972)

1. Most of the palynology work was done by Clifford Burdick, who had very little knowledge of geological techniques. Creationists themselves admit that his results come from contamination of old rocks by recent pollen (Chadwick 1973, 1981; Flank 1995a).

2. Intrusion of pollen in older rocks is very common. Pollen is ubiquitous, and its small size allows it to be carried into even small cracks by water seepage. To verify that pollen is fossil pollen rather than a contamination, one must look at several factors:

- What color is the pollen? Pollen darkens as it ages. If it is yellow or clear, it is recent.
- Have the rocks been cooked? Vulcanism around the rocks would burn up the pollen.
- Are the pollen grains flattened? Fossil pollens would be flattened as they are buried and compressed.

There is no indication that the out-of-place pollen passes any of these tests. In particular, the Hakatai Shales have lava intrusions, so we would expect any fossil pollen in them to have burned up.

Further Reading: Morton, G.R. 1997b. Precambrian pollen. http://www.asa3.org/archive/asa/199709/0101.html.

CC352: *Archaeoraptor* was a fake.
Archaeoraptor was touted by scientists as the dinosaur–bird transition (Sloan 1999), *but it was revealed as a fake, a composite of an avian body and a non-avian dinosaur's tail.* (Austin 2000)

1. *Archaeoraptor* was not a scientific fraud. It was put together by the Chinese fossil hunter who discovered it. The pieces were assembled to make the fossil more marketable to collectors, not to researchers. This worker may or may not have known that the tail came from a separate fossil (Simons 2000).

2. *Archaeoraptor* was published in the popular press, not in peer-reviewed journals. The main author of the article about it was *National Geographic's* art editor, not a scientist. *Nature* and *Science* both rejected papers describing it, citing suspicions that it was doctored and illegally smuggled (Dalton 2000; Simons 2000). Normal scientific procedures worked to uphold high standards.

3. The two halves of *Archaeoraptor* (*Yanornis martini*, the body, and *Microraptor zhaoianus*, the tail) are valuable fossils in their own right (Rowe et al. 2001; Zhou et al. 2002).

Further Reading: Simons, L.M. 2000. Archaeoraptor fossil trail.

CC360: No new fossils are being formed.
No new fossils are being formed. Fossils must have been produced by radically different conditions (a global flood) in the past. (Whitcomb and Morris 1961, 128–129)

1. The mechanisms by which fossils form are still occurring. We do not often observe them because they are generally out of sight or rare, but they still happen.

- The La Brea Tar Pits have trapped and preserved animals (and at least one person) in recent times.

- Pompeii preserved many bodies in 79 C.E.

- Sediments from rivers still cover corpses in their deltas.

- Insects and plants are being covered by geyser deposits.

CC361: Fossils can form quickly.

Fossils can form rapidly, so fossils are not a problem for a young earth.

1. Most fossils, by themselves, are not a problem for a young earth. The problems come from geological context, including the following:

- Independent dating of sediments via any number of techniques.

- Multiple layers of fossils. Sometimes each layer preserves an entire ecosystem, which would have taken decades to establish.

- Large number of fossils, beyond what the earth could support at once, showing multiple generations were necessary.

- In-place marine fossils on mountains, showing that the mountain must have risen since the fossil was deposited.

- Reworked fossils, showing that a mountain must have risen and eroded since the fossil was deposited.

2. Many fossils occur in amber, and the formation of amber cannot happen rapidly. First, plant resin polymerizes to produce copal, which takes thousands of years. Then the volatile oils must evaporate, which can take millions of years more.

CC361.1: Coal and oil can form quickly.

Oil and coal can form rapidly. Their formation is more a matter of heat and pressure than of time. Millions of years are not necessary to account for them. (H.M. Morris 1985, 109–110)

1. Coal deposits show evidence of a history. Most coals are found in sedimentary rocks deposited in flood plains. They often contain stream channels, roots, and soil horizons. Long time may not be necessary to form the coal itself, but it is necessary to account for the context where coal is found.

Further Reading: MacRae, A. 1994a. Could coal deposits be explained by a global flood? http://www.talkorigins.org/faqs/polystrate/coal.html.

CC361.2: Mammoths that were quickly frozen have been found.

The bodies of mammoths that apparently froze suddenly have been found. Their flesh was well preserved, and they still had food in their mouths. This shows that they were quick-frozen in some sort of catastrophe. (Watchtower 1985, 203)

1. The reports of frozen mammoths with well-preserved flesh are greatly exaggerated. Parts of cadavers have been well preserved, but in all cases, the internal organs were rotted, or the body was partly eaten by scavengers, or both, before the animal became frozen. The Berezovka mammoth, perhaps the most famous example, showed evidence of very slow decay and was putrefied to the point that the excavators found its stench unbearable (Weber 1980). The best preserved mammoth, Dima, was an infant; its small size and starved condition permitted quicker freezing, and even it had a little decomposition (Guthrie 1990, 7).

There are probably several different causes of the deaths of frozen mammoths and other animals, including the following:

- Fall in a landslide, as a thawed riverbank gives way under the animal's weight. The landslide and subsequent soil creep can bury and preserve the animal (Kurtén 1986, chap. 9).
- Sinking in muddy silt (Guthrie 1990, 7–24).
- Drowning/burial in flash floods carrying a heavy load of silt.
- Predation, followed by winter freezing, followed by burial in silt carried by snowmelt (Guthrie 1990, 81–113).

The food found with the mammoths were arctic species. Some mammoth deaths would have been sudden, but there is no evidence of sudden climate change.

2. Frozen mammoths are not common. As of 1961, only thirty-nine have been found with some flesh preserved, and only four of those were more or less intact (Farrand 1961).

Further Reading: Guthrie, R. D. 1990. *Frozen Fauna of the Mammoth Steppe*; Kurtén, B. 1986. *How to Deep-Freeze a Mammoth*; Weber, C. G. 1980. Common creationist attacks on geology.

CC361.3: Contorted positions of fossil animals indicate rapid burial.

The contorted positions of fossil animals indicate that they were rapidly buried alive (W. Brown 1995, 9).

1. As carcasses dry, ligaments contract and distort the body (Weber 1980). Also, dead animals are often disturbed by scavengers and/or water currents before their remains become buried. This can account for the contorted positions.

2. Some fossils do form by rapid burial, but these indicate only local catastrophes, such as landslides of a river bank.

CC362: Large collections of fossils indicate catastrophism.

There are many places where fossils occur in great numbers. These vast fossil beds indicate catastrophic rapid burial, not gradualistic conditions. (H. M. Morris 1985, 97–100)

1. Great numbers of fossils in one area indicate great numbers of animals dying in that area (or, in some cases, their bodies being transported there). Usually, this argues against rapid burial, because that many animals are not found together at once in life. A simpler explanation is that animals have died in the area over many years. For example, one mass burial is at the La Brea Tar Pits, which have been trapping animals for thousands of years.

2. In fact, vast fossil beds are evidence against catastrophic rapid burial. One formation alone (the Karroo Formation in Africa) is estimated to contain 800 billion vertebrate fossils. If that is just 1 percent of the world's fossils, there must be 2,100 vertebrate animals per acre, far more than we see today (Schadewald 1982). Fossil plant remains, such as coal, are almost 100 times more massive than living plant biomass (Poldervaart 1955; Ricklefs 1993).

3. Mass kills can occur through normal processes. Every year, hundreds of wildebeests drown during river crossings on their annual migration. Their bodies wash up on river banks. Collapse of the stream banks could bury many. Other local catastrophes can also kill many animals at once.

CC363: Fossilization requires sudden burial.

Fossilization requires rapid burial, or the organism will decay. This suggests that a catastrophe is responsible for fossils. (Whitcomb and Morris 1961, 128–129)

1. Bones can survive for over a year before being buried. Shells can last decades or even centuries. In fact, some fossils that have been eroded or encrusted or bored by other animals have been found, showing that long times passed before they were buried, and discrediting catastrophic burial. Only soft tissues need to be preserved quickly.

2. Rapid burial is not necessary for rapid preservation. Fossils can also be preserved by falling in a peat bog or on an anoxic lake bottom, areas where decay is slow or nonexistent. Other fossils are preserved in tree sap, which can become amber over time.

3. Rapid burial is common as a result of processes that are local catastrophes or that can scarcely be considered catastrophes at all, such as

- burial in sediments in a river delta
- burial in sediments from a local river flood
- burial in a small landslide, as along an eroded stream bank
- burial in ash from a volcano
- burial in a blown sand dune

4. Patterns of fossilization are consistent with noncatastrophic processes such as those mentioned above. Fossilization occurs as a result of all those different processes, not as a result of a single catastrophe. And it occurs where we would expect on the basis of commonplace processes. Bison fossils, for example, are found in active floodplains, not in upland areas.

Further Reading: Littleton, K. 2002. Fish fossils. http://www.talkorigins.org/origins/postmonth/sep02.html.

CC364: Sea fossils have been found on mountaintops.

Seashells and other marine fossils have been found on mountaintops, even very tall ones. These indicate that the sea once covered the mountains, which is evidence for a global flood. (Watchtower 1985, 203)

1. Shells on mountains are easily explained by uplift of the land. Although this process is slow, it is observed happening today, and it accounts not only for the seashells on mountains but also for the other geological and paleontological features of those mountains. The sea once did cover the areas where the fossils are found, but they were not mountains at the time; they were shallow seas.

2. A flood cannot explain the presence of marine shells on mountains for the following reasons:

- Floods erode mountains and deposit their sediments in valleys.

- In many cases, the fossils are in the same positions as they grow in life, not scattered as if they were redeposited by a flood. This was noted as early as the sixteenth century by Leonardo da Vinci (S. J. Gould 1998b).

- Other evidence, such as fossilized tracks and burrows of marine organisms, show that the region was once under the sea. Seashells are not found in sediments that were not formerly covered by sea.

CC365: Footprints in the Coconino Sandstone appear to have been made underwater.

Footprints in the Coconino Sandstone are attributed to animals making tracks on damp sand dunes in a desert. However, they appear to have been made underwater instead. Leonard Brand compared the Coconino footprints with footprints made by actual reptiles under various conditions, and the Coconino footprints best matched the footprints made underwater. (Brand 1978; Brand and Tang 1991; Snelling and Austin 1992)

1. The evidence for footprints being made underwater comes from rather ambiguous statistical studies, but is contradicted by evidence (Lockley 1992; Lockley and Hunt 1995; Loope 1992), including the following:

- "One of the most common observations is that the tracks have bulges or sand crescents on one side, thereby proving that they were made on inclined surfaces" (Lockley and Hunt 1995).

- Tracks showing possible loping, running, and galloping gaits are found throughout the Coconino Sandstone. These can only have been made on dry land.

- Tracks of small arthropods, attributable to spiders, centipedes, millipedes, and scorpions, occur abundantly in the Coconino Sandstone (see Figure CC6). Some of these trackways can only be made on completely dry sand.

- Raindrop impressions also appear.

2. The Coconino Sandstone covers an area of 200,000 square miles. Snelling and Austin (1992) proposed that thousands of cubic miles of sand were transported from hundreds of miles north. Forces violent enough to transport the sand would have killed any

Figure CC6
Scorpion trackway from Coconino sandstone. Photo by C. Schur. See Schur 2000 for other such trackways.

animals that got in the way. There would have been nothing alive within a hundred miles of where the footprints were found.

3. Brand himself, in the conclusion to one of his papers, wrote that "The data do suggest that the Coconino Sandstone fossil trackways may have been produced in either subaqueous sand or subaerial damp sand" (1996). So Brand's own work, taken at face value, does not necessarily indicate that the footprints were made underwater.

4. There is abundant geological evidence that the Coconino Sandstone was eolian (see CC365.1).

with Keith Littleton

CC365.1: Coconino sandstone was deposited underwater.

The Coconino Sandstone, the origin of which is conventionally attributed to desert sand dunes, was deposited by water. Evidence for this (in addition to the character of fossil footprints therein; see CC365) includes cross-bedding angles of only 25 degrees, not the 30–34 degrees one expects from desert dunes. (Snelling and Austin 1992)

1. Eolian (wind-blown) and subaqueous dunes have superficial similarities, but they differ in particulars. There is a great deal of diverse evidence that the Coconino Sandstone originated as eolian desert dunes. As McKee (1979, 204) stated:

> The basis for considering the Coconino Sandstone to be of eolian origin involves numerous criteria, some of which are distinctive of an eolian environment and others merely compatible with but not diagnostic of it. No single type of evidence seems entirely conclusive, but, together, the various features

present very strong evidence. The principal criteria of dune deposition are as follows:

1. The extent and homogeneity of the sand body.

2. The tabular-planar and wedge-planar type and large scale of cross-stratification. The common high-angle deposits are interpreted as slipfaces on the lee sides of dunes, and the relatively rare low-angle cross-strata that dip toward the opposite quadrant apparently represent deposits of windward slopes.

3. Slump marks of several varieties preserved on the steeply dipping surfaces of lee-side deposits. These are distinctive of dry sand avalanching.

4. Ripple marks which are common on surfaces of high-angle crossbedding suggest eolian deposition both by their high indexes (above 15) and by their orientation with axes parallel to dip slope.

5. The local preservation of a distinctive type of rain pit. Such pits illustrate the cohesion of sand grains with added moisture and a reorientation of the crater axes with respect to bedding slopes.

6. Successions of miniature rises or steps ascending dip slopes of crossbeds.

7. The preservation in fine sand of reptile footprints and probable millipede trails with sharp definition and clear impression.

8. The consistent orientation of reptilian tracks up (not down) the steep foreset slopes.

Since McKee published, additional types of terrestrial trace fossils, paleosols, and other distinctive eolian sedimentary structures have been recognized in Coconino and related eolian strata.

If a person looks carefully at modern dunes—for example, the Great Sand Dunes, White Sands, and Nebraska Sand Hills—he or she will find an abundance of climbing translatent beds, with coarsening-up laminae and rare foreset laminae that form only by the migration and accretion of low-amplitude wind ripples in eolian environments. Such beds form only in terrestrial eolian environments and are completely absent from marine or lacustrine environments because the wind ripples that create them simply do not form under water and underwater analogues of these sedimentary. The fact that wind ripple and the distinctive bedding and laminations occur throughout the Coconino Sandstone and other similar strata—for example, the Navajo and Entrada—clearly refutes the marine hypothesis for their origin.

2. Sand waves deposited in water possess very low angle cross-beds, rarely steeper than 10 degrees. Cross-bedding in eolian dunes occurs at various angles. The general range in slope of the cross-beds is from 11 to 34 degrees. The average appears to be close to 25–28 degrees. The average slope of cross-bedding does not have to be equal to 30 to 34 degrees, which is the maximum slope of dry sand, to be from a sand dune. The maximum slope of cross-bedding within the Coconino Sandstone does get as steep as 30 to 34 degrees (McKee 1979; Reineck and Singh 1980). The 30–34 degree slope is produced from sand avalanching down the lee slip face of the dune. The beds and laminae produced by wind ripple migration can form cross-bedding and lamination that has slopes up to 20 degrees within a sand dune. Given that this cross-bedding is present everywhere in the Coconino Sandstone, it greatly decreases the average slope of the cross-bedding within the Coconino Sandstone.

In addition, grain-fall processes produce low, inclined lamination and beds with slopes that average between 20 and 30 degrees and range from 0 to 40 degrees. The presence of grain-fall bedding and lamination within the Coconino not only refutes the hypotheses concerning the underwater or marine origin of the Coconino Sandstone but also again greatly decreases the average slope of the cross-bedding found in the Coconino Sandstone. Thus, it is completely reasonable that the average slope of the cross-bedding in the Coconino Sandstone is less than the average slope of dry sand—that is, 30 to 34 degrees—because the cascading of sand down the lee side of the sand dune is not the only process producing cross-beds and laminations in dune sands (Hunter 1977).

Further Reading: Boggs, S. 1995. *Principles of Sedimentology and Stratigraphy.*

with Keith Littleton

CC371: Evidence of blood in a *Tyrannosaurus* bone indicates recent burial.

Schweitzer et al. (1997a) found evidence of hemoglobin and red blood cells in an unfossilized Tyrannosaurus rex *bone. This indicates that the dinosaur died rather recently, not millions of years ago, which in turn proves invalid the radiometric dating methods that indicate that dinosaurs (and the earth) are ancient. (Ham et al. 2000, 246–247; Wieland 1997)*

1. Schweitzer et al. did not find hemoglobin or red blood cells. Rather, they found evidence of degraded hemoglobin fragments and structures that might represent altered blood remnants. They emphasized repeatedly that even those results were tentative, that the chemicals and structures may be from geological processes and contamination (Schweitzer and Horner, 1999; Schweitzer and Staedter, 1997; Schweitzer et al. 1997a, 1997b). The bone is exceptionally well preserved, so much so that it may contain some organic material from the original dinosaur, but the preservation should not be exaggerated.

2. The bone that Schweitzer and her colleagues studied was fossilized, but it was not altered by "permineralization or other diagenetic effects" (Schweitzer et al. 1997b, 349). Permineralization is the filling of the bone's open parts with minerals; diagenetic effects include alterations like cracking. Schweitzer commented that the bone was "not completely fossilized" (Schweitzer and Staedter 1997, 35), but lack of permineralization does not mean unfossilized.

3. The ancient age of the bone is supported by the (nonradiometric) amino racemization dating technique.

Further Reading: Hurd, G. S. 2004. Dino-blood and the young earth. http://www.talkorigins .org/faqs/dinosaur/blood.html or http://home.austarnet.com.au/stear/YEC_and_dino_blood.htm.

METHODOLOGY

CC401: Paleontologists reconstruct an entire animal from a single bone.

Many of the fossils on which evolution is based are reconstructed from the flimsiest evidence, even from a single tooth or bone. The conclusions based on such fossils are mere speculation. (Lenner et al. 1995)

1. Evolution is not based on fragmentary fossils. The theory would still be extremely robust with no fossils at all, based on evidence from modern life. Furthermore, there are more than enough substantially complete skeletons to support evolution. The whale transitional sequence, for example, is based on several excellent skeletons (see CC216.1).

2. A single bone, even in isolation, can give a surprising amount of information. A tooth, for example, can show generally what kind of food an animal ate and give an idea of its size. These conclusions, in turn, tell how the animal fits into the ecology.

3. Bones are never considered in isolation; rather, they are compared with other bones from more complete skeletons. If you have a bone that looks like an *Iguanodon* femur but smaller, to give a simple example, the reconstruction would look a lot like a smaller *Iguanodon*. A complete reconstruction, however, is possible only if you can match the single bone to an animal for which there is a complete skeleton already.

The ability to deduce much about a fossil from a single tooth or bone was made famous by anatomist and paleontologist Georges Cuvier. In 1804, for example, he confidently announced that a French fossil was an opossum (then unknown from France) on the basis of only its teeth (Zimmer 1998, 135–137). Cuvier was a creationist.

GEOLOGY

GEOCHRONOLOGY

CD001: Radiometric dating falsely assumes rocks are closed systems.

Radiometric dating falsely assumes that the rocks being dated are closed systems. It inappropriately assumes that no parent or daughter isotopes were added or removed via other processes through the history of the sample. (H. M. Morris 1985, 139)

1. Absolutely closed systems do not exist even under the best laboratory conditions. Nevertheless, many rocks so closely approximate closed systems that multiple radiometric dating methods produce consistent results within 1 percent of each other.

2. Some rocks may be open to outside contamination, but not all of them are. Most ages are determined from multiple mineral and rock samples, which give a consistent date within 1 and 3 percent. It is extremely unlikely that contamination would affect all samples by the same amount.

3. Isochron methods can detect contamination and, to some extent, correct for it. Isochrons are determined from multiple samples, and contamination would have to affect all of the samples the same way in order to create an isochron that appeared okay but was wrong (see CD002).

With uranium–lead dating, closure of the system may be tested with a concordia diagram. This takes advantage of the fact that there are two isotopes of uranium (^{238}U and ^{235}U) that decay to different isotopes of lead (^{206}Pb and ^{207}Pb, respectively). If the system has remained closed, then a plot of ^{206}Pb / ^{238}U versus ^{207}Pb / ^{235}U will fall on a known line called the concordia. Even if samples are discordant, reliable dates can often be derived (Faure 1998, 287–290).

4. Geochronologists are well aware of the dangers of contamination, and they take pains to minimize it. For example, they do not use weathered samples.

Further Reading: Dalrymple, G. B. 1991. *The Age of the Earth*; Dickin, A. P. 1995. *Radiogenic Isotope Geology*. (A standard text); Faure, G. 1986. *Principles of Isotope Geology*. (A standard text); Jessey, D. n.d. Isotope geochemistry. http://geology.csupomona.edu/drjessey/class/gsc300/isotope1 .pdf; Stassen, C. 1998. Isochron dating. http://www.talkorigins.org/faqs/isochron-dating.html.

CD002: Radiometric dating falsely assumes initial conditions are known.

Radiometric dating falsely assumes that initial conditions are known, that none of the daughter components are in the mineral initially. (H. M. Morris 1985, 139)

1. Isochron methods do not assume that the initial parent or daughter concentrations are known. In basic radiometric dating, a parent isotope (call it P) decays to a daughter isotope (D) at a predictable rate. The age can be calculated from the ratio daughter isotope to parent isotope in a sample. However, this assumes that we know how much of the daughter isotope was in the sample initially. (It also assumes that neither isotope entered or left the sample; see CD001.)

With isochron dating, we also measure a different isotope of the same element as the daughter (call it D_2), and we take measurements of several different minerals that formed at the same time from the same pool of materials. Instead of assuming a known amount of daughter isotope, we only assume that D/D_2 is initially the same in all of the samples. Plotting P/D_2 on the *x* axis and D/D_2 on the *y* axis for several different samples gives a line that is initially horizontal. Over time, as P decays to D, the line remains straight, but its slope increases. The age of the sample can be calculated from the slope, and the initial concentration of the daughter element D is given by where the line meets the *y* axis (see Figure CD1). If D/D_2 is not initially the same in all samples, the data points tend to scatter on the isochron diagram, rather than falling on a straight line.

2. For some radiometric dating techniques, the assumed initial conditions are reasonable. For example:

- K-Ar dating assumes that minerals form with no argon in them. Since argon is an inert gas, it will usually be excluded from forming crystals. This assumption can be tested by looking for argon in low-potassium minerals (such as quartz), which would not produce argon as a daughter product. $^{40}Ar/^{39}Ar$ dating and K-Ar isochron dating can also identify the presence of initial excess argon.

- The concordia method is used on minerals, mostly zircon, that reject lead as they crystalize.

- Radiocarbon dating is based on the relative abundance of carbon-14 in the atmosphere when a plant or animal lived. This varies somewhat, but calibration with other techniques (such as dendrochronology) allows the variations to be corrected.

- Fission-track dating assumes that newly solidified minerals will not have fission tracks in them.

Further Reading: Dalrymple, G. B. 1991. *The Age of the Earth*; Stassen, C. 1998. Isochron dating. http://www.talkorigins.org/faqs/isochron-dating.html.

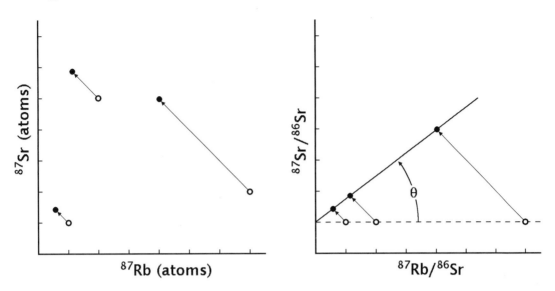

Figure CD1

Isochron method. *Left*: Plot of Rb-87 vs Sr-87 for three samples. Over time, Rb-87 decays to Sr-87, so initial values (open circles) move toward decreasing Rb-87 and increasing Sr-87, by an amount proportional to the Rb-87 content of the sample and the elapsed time. *Right*: The same data, but normalized to Sr-86. Because Sr-87 and Sr-86 are chemically almost identical, Sr-87/Sr-86 is constant when a rock forms. The age of the rock can be calculated from the slope of the line (θ) formed by the present values (closed circles). Initial concentrations need not be known. Adapted from Dalrymple 1991.

CD004: Cosmic rays and free neutrinos affect U and Ar decay rates.

Cosmic rays and free neutrinos, such as might be produced by nearby supernovas or the reversal of the earth's magnetic field, might affect the decay rates of radioactive elements, invalidating such radiometric dating methods as carbon-14, uranium-lead, and potassium-argon. (H. M. Morris 1985, 142–143,146)

1. Where is there the slightest bit of evidence that cosmic rays or neutrinos do affect decay rates? The following shows the contrary:

- Inside standard nuclear fission power generators, neutrino radiation is intense, but the uranium that is not fissioned decays at the usual rate.

- Some spacecraft are powered by nuclear decays. Some of them fly in very intense cosmic ray fields (like near Jupiter). If cosmic rays affected decay rates, the power generated would be different from expectations.

- To get unweathered rocks, rocks for radiometric dating are usually taken from some depth into an outcrop, where cosmic rays have insignificant effect.

2. Radiation high enough to affect nuclear decay rates by several orders of magnitude (a change great enough to allow young-earth timescales) would sterilize the planet.

3. Reversals of the earth's magnetic field do not produce cosmic rays or neutrinos. They may allow more cosmic rays to reach the earth's surface, but not much beyond that, and most rocks used for dating have been buried for most of their history.

4. C-14 dating is calibrated by independent clocks (see CD011.1).

Further Reading: Young, D. A. 1988. *Christianity and the Age of the Earth.*

CD010: Radiometric dating gives unreliable results.

Radiometric dating gives unreliable results. (W. Brown 1995, 24)

1. Independent measurements, using different and independent radiometric techniques, give consistent results (Dalrymple 2000; Lindsay 1999a; Meert 2000). Such results cannot be explained either by chance or by a systematic error in decay rate assumptions (see also CD020).

2. Radiometric dates are consistent with several nonradiometric dating methods. For example:

- The Hawaiian archipelago was formed by the Pacific ocean plate moving over a hot spot at a slow but observable rate. Radiometric dates of the islands are consistent with the order and rate of their being positioned over the hot spot (Rubin 2001).

- Radiometric dating is consistent with Milankovitch cycles, which depend only on astronomical factors such as precession of the earth's tilt and orbital eccentricity (Hilgen et al. 1997).

- Radiometric dating is consistent with the luminescence dating method (T. Thompson n.d.c; Thorne et al. 1999).

- Radiometric dating gives results consistent with relative dating methods such as "deeper is older" (Lindsay 2000a).

3. The creationist claim that radiometric dates are inconsistent rests on only a few examples. Creationists ignore the vast majority of radiometric dates showing consistent results (e.g., Harland et al. 1990).

Further Reading: Thompson, T. 2003. A radiometric dating resource list. http://www.timthompson.com/radiometric.html; Wiens, R. C. 2002. Radiometric dating: A Christian perspective. http://www.asa3.org/ASA/resources/Wiens.html.

CD011: Carbon dating gives inaccurate results.

Carbon-14 dating gives unreliable results. (R. E. Lee 1981)

1. Any tool will give bad results when misused. Radiocarbon dating has some known limitations. Any measurement that exceeds these limitations will probably be invalid. In particular, radiocarbon dating works to find ages as old as 50,000 years but not much older. Using it to date older items will give bad results. Samples can be contaminated with younger or older carbon, again invalidating the results. Because of excess carbon-12 released into the atmosphere from the Industrial Revolution and excess carbon-14

produced by atmospheric nuclear testing during the 1950s, materials less than 150 years old cannot be dated with radiocarbon (Faure 1998, 294).

In their claims of errors, creationists do not consider misuse of the technique. It is not uncommon for them to misuse radiocarbon dating by attempting to date samples that are millions of years old (see, for example, CD011.5) or that have been treated with organic substances. In such cases, the errors belong to the creationists, not the carbon-14 dating method.

2. Radiocarbon dating has been repeatedly tested, demonstrating its accuracy. It is calibrated by tree-ring data, which gives a nearly exact calendar for more than 11,000 years back. It has also been tested on items for which the age is known through historical records, such as parts of the Dead Sea scrolls and some wood from an Egyptian tomb (MNSU n.d.; Watson 2001). Multiple samples from a single object have been dated independently, yielding consistent results. Radiocarbon dating is also concordant with other dating techniques (e.g., Bard et al. 1990).

Further Reading: Higham, T. 1999. Radiocarbon WEB-Info. http://www.c14dating.com; Thompson, T. 2003. A radiometric dating resource list. http://www.tim-thompson.com/radiometric .html#reliability.

CD011.1: Variable C-14/C-12 ratio invalidates C-14 dating.

Carbon dating is based on the atmospheric C-14/C-12 ratio, but that ratio varies. Thus, the carbon dating method is not valid. (H. M. Morris 1985, 162–166)

1. The variability of the C-14/C-12 ratio, and the need for calibration, has been recognized since 1969 (Dickin 1995, 364–366). Calibration is possible by analyzing the C-14 content of items dated by independent methods. Dendrochronology (age dating by counting tree rings) has been used to calibrate C-14/C-12 ratios back more than 11,000 years before the present (B. Becker and Kromer 1993; Becker et al. 1991). C-14 dating has been calibrated back more than 30,000 years by using uranium–thorium (isochron) dating of corals (Bard et al. 1990; Edwards et al. 1993), and to 45,000 years ago by using U-Th dates of glacial lake varve sediments (Kitagawa and van der Plicht 1998; see Fig. CD2).

Further Reading: Dalrymple, G. B. 1991. *The Age of the Earth*; Matson, D. E. 1994. How good are those young-earth arguments? http://www.talkorigins.org/faqs/hovind/howgood-c14.html#R1.

CD011.2: Vollosovitch and Dima mammoths yielded inconsistent C-14 dates.

Widely different radiocarbon dates are obtained from the same frozen mammoths. Different parts of the Vollosovitch mammoth date to 29,500 and 44,000 radiocarbon years (RCY). One part of Dima, a frozen baby mammoth, was 40,000 RCY, another part 26,000 RCY, and wood immediately around it was 9,000–10,000 RCY. Two parts of the Fairbanks Creek mammoth date to 15,380 and 21,300 RCY. (W. Brown 2001; Hovind n.d.a.)

1. The dates come from different mammoths. The reference cited by Brown and cribbed by Hovind likely refers only to a Fairbanks mammoth, which Brown also men-

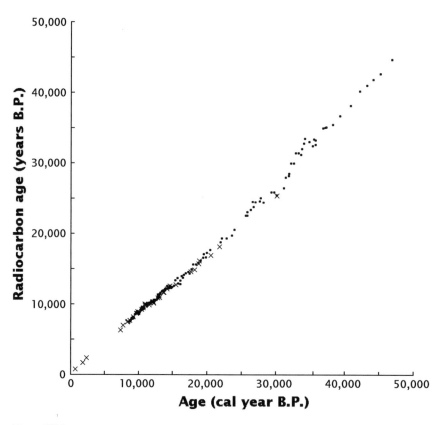

Figure CD2
Radiocarbon calibration to 45,000 years ago from varves of Lake Suigetsu, Japan (dots) and from U-Th dating of corals (xs). Adapted from Kitagawa and van der Plicht 1998.

tions (Péwé 1975, 30). The 15,380 and 21,300 RCY dates come from separate mammoths, and it is noted that the 21,300 date is invalid because it comes from a hide soaked in glycerin. It is uncertain what is Brown's source for the 29,500 and 44,000 dates.

Ukraintseva (1993) reviews the Kirgilyakh mammoth, also known as Dima, and cites three dates obtained for it. All are around 40,000 years before present. Dates for deposits surrounding the mammoth are consistent with dates for the mammoth.

Further Reading: Heinrich, P. 1996. The mysterious origins of man. http://www.talkorigins.org/faqs/mom/atlantis.html.

CD011.3: Living snails were C-14 dated at 2,300 and 27,000 years old.
Living snails were carbon-14 dated at 2,300 and 27,000 years old, showing that the dating method is invalid. (Hovind n.d.a.)

1. The source of the 2,300-years-old radiocarbon date (Keith and Anderson 1963, discussed by Strahler 1987, 156–157), has been abused and misused to discredit radiocarbon dating. The article discussed the potential errors that the presence of "dead carbon" would introduce into the dating of mollusks. For example, carbon dioxide in the water can partially come from Paleozoic limestone, which lacks carbon-14. As a result, the carbon dioxide in the water is deficient in carbon-14 relative to the atmosphere, and mollusks living in the water build shells that give apparent dates older than they really are. This is a type of "reservoir effect."

The 27,000-year-old date comes from Riggs (1984, 224), who wrote:

> Carbon-14 contents as low as 3.3 ± 0.2 percent modern (apparent age, 27,000 years) measured from the shells of snails *Melanoides tuberculatus* living in artesian springs in southern Nevada are attributed to fixation of dissolved HCO_3^- with which the shells are in carbon isotope equilibrium.

In other words, the apparent age of 27,000 years for these snail shells is another example of the reservoir effect. The springs, from which the snails came, were fed by carbonate aquifers. As this water percolated through the enclosing carbonates, it dissolved limestone and dolomite hundreds of millions of years old. The dissolution of limestone and dolomite introduced considerable quantities of "dead carbon" into the groundwater. As a result, the groundwater which fed the spring and in which the snails lived was significantly deficient in carbon-14 relative to what is found in the atmosphere. When the snails made their shells, they incorporated an excess amount of "dead carbon," relative to modern atmosphere, into their shells, which resulted in the excessively old apparent date.

Contrary to the complaints of creationists, conventional scientists are well aware of this problem. They test for it and take it into account when interpreting radiocarbon data. In cases where corrections for presence of dead carbon cannot be made, such dates are readily recognized as erroneous and can be safely disregarded. This is not the fatal flaw to radiometric dating that some creationists claim it to be. It just shows that dates from mollusks from streams and lakes need to be carefully evaluated as to their reliability. Other materials, such as wood, charcoal, bone, and hide, would remain unaffected by this type of reservoir effect. If found with shells in the same layer, these materials could be dated to determine if shells are locally affected by the reservoir effect and, if so, how much their radiocarbon dates have been skewed by it.

Further Reading: Aitken, M.J. 1990. *Science-Based Dating in Archaeology*; Bowman, S. 1990. *Radiocarbon Dating*; Faure, G. 1986. *Principles of Isotope Geology*; Matson, D.E. 1994. How good are those young-earth arguments? http://www.talkorigins.org/faqs/hovind/howgood-c14.html#R3; Taylor, R.E. 1987. *Radiocarbon Dating*.

by Keith Littleton

CD011.4: A freshly killed seal was C-14 dated at 1,300 years old.

A freshly killed seal was carbon-14 dated at 1,300 years old. (Hovind n.d.a.)

1. This claim derives from D. Wakefield (1971):

> Radiocarbon analysis of specimens obtained from mummified seals in southern Victoria Land has yielded ages ranging from 615 to 4,600 years. However, Antarctica sea water has significantly lower carbon-14 activity than that accepted as the world standard. Therefore, radiocarbon dating of marine organisms yields apparent ages that are older than true ages, but by an unknown and possibly variable amount. Therefore, the several radiocarbon ages determined for the mummified seal carcasses cannot be accepted as correct. For example, the apparent radiocarbon age of the Lake Bonney seal known to have been dead no more than a few weeks was determined to be 615 ± 100 years. A seal freshly killed at McMurdo had an apparent age of 1,300 years.

This is the well-known reservoir effect that occurs also with mollusks (see CD011.4) and other animals that live in the water. It happens when "old" carbon is introduced into the water. In the above case of the seal, old carbon dioxide is present within deep ocean bottom water that has been circulating through the ocean for thousands of years before upwelling along the Antarctic coast.

The seals feed off of animals that live in a nutrient-rich upwelling zone. The water that is upwelling has been traveling along the bottom for a few thousand years before surfacing. The carbon dioxide in it came from the atmosphere before the water sank. Thus, the carbon in the sea water is a couple of thousand years "old" from when it was in the atmosphere, and its radiocarbon content reflects this time. Plants incorporate this "old" carbon in them as they grow. Animals eat the plants; seals eat the animals, and the "old" carbon from the bottom waters is passed through the food chain. As a result, the radiocarbon content reflects a mixture of old radiocarbon, which is thousands of years old, and contemporaneous radiocarbon from the atmosphere. The result is an apparent age that differs from the true age of the seal.

The reservoir effect is well known by scientists, who work hard to understand the limitations of their tools. It is explained, for example, in Faure (1986) and Higham (n.d.). Contrary to creationist propaganda, limitations of a tool do not invalidate the tool.

> **Further Reading:** Aitken, M.J. 1990. *Science-Based Dating in Archaeology*; Bowman, S. 1990. *Radiocarbon Dating*; Taylor, R.E. 1987. *Radiocarbon Dating*.
>
> *with Keith Littleton*

CD011.5: Triassic wood from Australia was dated at 33,000 years old.

A piece of wood is fossilized in the Hawkesbury Sandstone, Australia, which most geologists date to the middle Triassic, about 225 to 230 million years ago. The wood was dated by Geochron (a commercial dating laboratory) using the carbon-14 method. Geochron determined its age to be only 33,720 ± 430 years before present. Contamination by recent microbes or fungi cannot explain the discrepant age. (Snelling 1999)

1. It is doubtful that the sample was even wood. Snelling was not even sure what the sample was. Nor could the staff at Geochron tell what the sample was (T. Walker 2000). If it was Triassic, it probably contained no original carbon (there are no known cases of any 225-Ma wood retaining any of its original carbon). Using carbon dating was pointless from the start since it would inevitably give meaningless results.

2. The sample was porous, making it likely that it would have absorbed organic carbon from the groundwater. It was probably this contaminating carbon that produced the date.

Another possibility is that some carbon-14 was created in situ by natural radioactivity in the surrounding rocks (Hunt 2002).

3. Furthermore, 33,720 years is still significantly older than the age that many creationists, Snelling included, ascribe to the earth, and there are no plausible sources of error to make the age younger than 33,000 years.

Further Reading: Meert, J. 2003a. Andrew Snelling and the iron concretion? http://gondwanaresearch.com/hp/crefaqs.htm#who.

CD011.6: Ancient coal and oil are C-14 dated as only 50,000 years old.

Coal and oil are supposedly millions of years old. Effectively all of the carbon-14 in a sample would have decayed in that time. But carbon-14 still exists in coal, implying an age of only about 50,000 years. (Baumgardner 2003)

1. New carbon-14 is formed from background radiation, such as radioactivity in the surrounding rocks. In some cases, carbon-14 from the atmosphere can contaminate a sample. Minute amounts of contamination from these sources can cause apparent ages around 50,000 years, which is near the limit of the maximum age that carbon dating can measure.

Further Reading: Hunt, K. 2002. Carbon-14 in coal deposits. http://www.talkorigins.org/faqs/c14.html.

CD012: U-Th dating gives inaccurate results for modern volcanic rocks.

U-Th dating of volcanic rocks formed in historic times gives dates vastly older than their true age. (Clementson 1970)

1. This claim is based on a single obscure reference (Cherdyntsev, V. V. et al., Geological Institute Academy of Sciences, USSR, Earth Science Section, 172, p. 178). Until we get hold of that reference, we cannot address it directly, but some general comments are possible.

First, dating techniques, like any tool, can be misused. This particular case could be an example, especially if xenoliths (older inclusions) are incorporated in the volcanic rocks (see CD013). The misuse could be accidental; it could also be deliberate so as to illustrate how not to do things; or it could be someone deliberately trying for a discrepant age. Other examples of discordant dates among these claims show that a date alone does not invalidate the method; one must also consider how the method was applied.

Second, there is a vast body of literature showing that the U-Th method does work.

It would take more than one published counterexample to discredit it. And if that counterexample were a serious challenge to the method, there would be plenty of publications about it.

CD013: K-Ar dating gives inaccurate results for modern volcanic rocks.

K-Ar dating of rocks from lava flows known to be modern gave ages millions to billions of years older. (H. M. Morris 1985, 146-147)

1. Argon may be incorporated with potassium at time of formation. This is a real problem, but it is easily overcome either by careful selection of the material being dated or by using $^{40}Ar/^{39}Ar$ dating instead of K-Ar dating.

In the case of the claim about recent lava yielding dates that are millions to billions of years old, H. M. Morris (1985) misstated the facts concerning these "anomalous" dates as published in Funkhouser and Naughton (1968). The main misstatements of fact by Morris are as follows:

- It was not the lava that was dated, but inclusions of olivine, called "xenoliths," present within the lava. These gave anomalously old age because they contained excess argon that the enclosing lava did not.

- Morris failed to mention that the lava matrix without the xenoliths was dated and found to be too young to date using potassium–argon. (Funkhouser and Naughton [1968, 4603], stated that the matrix rock "can be said to contain no measurable radiogenic argon within experimental error.") This is consistent with the recent age of lavas and the state of the art of K-Ar dating at that time. The presence of excess argon was only a problem for the xenoliths but not for the lava containing them.

Morris cited other examples of anomalous dates produced by excess argon and falsely claimed that it is a universal problem for K-Ar dating. The problem is not universal, as the majority of minerals and rocks dated by K-Ar do not contain the excess argon. Where excess argon is a problem, accurate, reliable dates typically can be obtained by using $^{40}Ar/^{39}Ar$ dating, as demonstrated by Dalrymple (1969) and Renne et al. (1997) and discussed by Dalyrmple (2000).

2. Morris's complaints are dated in that, for the most part, geologists no longer use the K-Ar dating technique as was practiced in 1974. Instead, K-Ar dating has been largely replaced by the related $^{40}Ar/^{39}Ar$ dating technique. This change also solved other problems that Morris complained about in his discussion of the K-Ar dating technique. These complaints were as follows:

  a. Claim: K-Ar dating techniques must be calibrated by uranium–lead (U-Pb) dating. Response: Some calibrations between U-Pb and K-Ar were done in the 1940s and early 1950s, but the decay rates of all the different radioisotopes involved are now known to within 1 percent, making the different dating techniques independent. With $^{40}Ar/^{39}Ar$ dating, it is possible to calibrate this dating method by using volcanic deposits created in historic volcanic eruptions—for example, the eruption of Mount Vesuvius on August 24, 79 C.E. (Renne et al. 1997). In addition, $^{40}Ar/^{39}Ar$

dating can be compared not only with U-Pb dating techniques but also with other absolute dating techniques—for example, K-Ar, Rb-Sr, and Sm-Nd dating techniques—which all provide dates consistent with each other and with associated ^{40}Ar/^{39}Ar dates. This has been demonstrated by the dating of chondrite meteorites (Dalrymple 1991) and tektites and other ejecta and deposits created by the giant meteorite impact at Chicxulub in the Yucatan Peninsula (Dalrymple et al. 1993).

b. Claim: The potassium–argon is an open system (see also CD001).
Response: The papers cited by Morris fail to probe this point. The first paper simply demonstrates that rock altered by weathering cannot be dated. This is a common-sense conclusion understood by geologists literate in the basics of their profession; it is irrelevant to the unaltered minerals that are typically dated using K-Ar, ^{40}Ar/^{39}Ar, and other techniques. The final paper claims potassium is quite mobile because potassium can be extracted from iron meteorites by using distilled water. However, K-Ar dating commonly uses potassium silicate minerals, which are very insoluble in water and resist weathering. Potassium cannot be significantly leached from the minerals used in K-Ar dating, or, conversely, the minerals from which significant potassium can be leached are not the minerals used in K-Ar dating.

c. Claim: The decay rate of potassium is subject to change.
Response: This is simply not true (see CF210).

d. Claim: Argon maybe incorporated with potassium at time of formation.
Response: See first point above.

e. Claim: K-Ar ages are extremely variable.
Response: As previously noted, K-Ar and ^{40}Ar/^{39}Ar dating both provide extremely consistent dates when the methods are used properly (Dalrymple 1991, 2000). The single paper (Engels 1971) cited by Morris clearly stated that variability resulted from presence of unwanted impurities in the specific mineral being dated. If the sample dated consisted of an absolutely pure mineral, there would not be any variability in the K-Ar dates obtained from them.

Further Reading: Attendorn, H.-G., and R.N.C. Bowen. 1997. *Radioactive and Stable Isotope Geology*; Faure, G. 1986. *Principles of Isotope Geology*; harlequin2. 2001a. Ar-Ar dating assumes there is no excess argon? http://members.cox.net/ardipithecus/evol/lies/lie024.html; harlequin2. 2001b. 200 year old lava dated 2.96 billion years old? http://members.cox.net/ardipithecus/evol/lies/lie023.html; Lindsay, D. 2000b. Fresh lava dated as 22 million years old. http://www.don-lindsay-archive.org/creation/hawaii.html; Stassen, C. 1999. Feedback response (Jan.). http://www.talkorigins.org/origins/feedback/jan99.html; Thompson, T. 2003. A radiometric dating resource list. http://www.tim-thompson.com/radiometric.html; Wiens, R.C. 2002. Radiometric dating: A Christian perspective. http://www.asa3.org/ASA/resources/Wiens.html.

by Keith Littleton

CD013.1: K-Ar dates of 1986 dacite from Mount St. Helens are very old.

The conventional K-Ar dating method was applied to the 1986 dacite flow from the new lava dome at Mount St. Helens, Washington. The whole-rock age was 0.35 ± 0.05 Ma. Ages for component minerals varied from 0.34 ± 0.06 Ma to 2.8 ± 0.6 Ma. These ages show that the K-Ar method is invalid. (Austin 1996)

1. Austin sent his samples to a laboratory that clearly states that their equipment cannot accurately measure samples less than two million years old. All of the measured ages but one fall well under the stated limit of accuracy, so the method applied to them is obviously inapplicable. Since Austin misused the measurement technique, he should expect inaccurate results, but the fault is his, not the technique's. Experimental error is a possible explanation for the older date.

2. Austin's samples were not homogeneous, as he himself admitted. Any xenocrysts in the samples would make the samples appear older (because the xenocrysts themselves would be old). A K-Ar analysis of impure fractions of the sample, as Austin's were, is meaningless.

Further Reading: Henke, K. R. n.d.c. Young-earth creationist 'dating' of a Mt. St. Helens dacite. http://home.austarnet.com.au/stear/mt_st_helens_dacite_kh.htm.

CD014: Isochron dating gives unreliable results.

Isochron dating is unreliable. The method assumes that the samples are cogenetic, that is, that they form at the same time from a reasonably homogeneous common pool. This assumption is invalid. In particular, mixing two sources with different isotopic compositions gives meaningless but apparently valid isochron plots. (Overn n.d.)

1. Mixing can usually be detected by plotting the total daughter isotopes against the ratio of daughter isotopes. These would not likely fall on a straight line if mixing occurred.

2. Isochron plots from mixing can have any slope, even negative slopes. If mixing were common, we would expect a high percentage of isochron results to show negative slopes. They do not.

3. Other factors that can produce false isochrons are

- Protracted fractionation. This requires slow cooling (over millions of years) and produces only a small error.

- Inherited ages as from partial melting. The age given by this method is the age of the source material. Furthermore, this factor requires unusual conditions and usually produces scatter in the isochron plot.

- Metamorphosism. This produces apparent ages younger than the age of the source material (Stassen 1998; Zheng 1989).

4. False isochrons can usually be avoided by choosing appropriate samples. The samples must come from an (apparently) initially homogeneous source and avoiding obvious signs of weathering and metamorphism.

Further Reading: Dalrymple, G. B. 1991. *The Age of the Earth*; Faure, G. 1986. *Principles of Isotope Geology*; Stassen, C. 1998. Isochron dating. http://www.talkorigins.org/faqs/isochron-dating.html.

CD014.1: Isochron date of young Grand Canyon lava is excessively old.

The hawaiite lava flows from the Uinkaret Plateau of the Grand Canyon are dated by potassium–argon at about 1.2 million years, an age consistent with the lava flow being younger than the canyon itself. However, a Rb-Sr isochron plot of samples from the lava flow gives an age of 1.34 billion years, which is older than even the Cardenas Basalt, some of the oldest rock in the canyon. The data points are colinear, which is supposed to indicate a valid isochron. This result shows that the isochron method is invalid. (Austin 1992)

1. One of the requirements for isochron dating is that the samples be cogenetic, that is, that they come from materials that were isotopically homogeneous (with respect to each other) when they formed. Austin's selection of samples violated this assumption. His five samples came from four different lava flows plus one phenocryst (which likely solidified in the magma chamber before the flow). Thus, Austin's conclusion, not the isochron method, is invalid.

2. Noncogenetic samples such as Austin used are sometimes used intentionally to determine the age of the common source of the samples. Austin's results confirm that the lithospheric mantle underlying the Grand Canyon (the common source for his samples) is older than the Cardenas Basalt. Geologists have known this all along.

3. Austin (1988) cited Brooks et al. (1976), showing Austin should have been aware that noncogenetic samples could produce an isochron for the age of their molten material's source. His misstatement of the significance of the isochron is just plain sloppy.

Further Reading: Stassen, C. 2003. A criticism of the ICR's Grand Canyon Dating Project. http://www.talkorigins.org/faqs/icr-science.html.

CD015: Zircons retain too much helium for an old earth.

Uranium and thorium in zircons produce helium as a by-product of their radioactive decay. This helium seeps out the zircons quickly over a wide range of temperatures. If the zircons really are about 1.5 billion years old (the age that conventional dating gives, assuming a constant decay rate), almost all of the helium should have dissipated from the zircons long ago. But there is a significant amount of helium still inside the zircons, showing their ages to be 6,000 ± 2,000 years. Accelerated decay must have produced a billion years worth of helium in that short amount of time. (D.R. Humphreys et al. 2003, 2004)

1. Subsurface pressure and temperature conditions affect how quickly the helium diffuses out of zircons. D.R. Humphreys et al. selected a rock core sample from the Fenton Hill site, which Los Alamos National Laboratory evaluated in the 1970s for geothermal energy production. The area is within a few kilometers of the Valles Caldera, which has gone through several periods of faulting and volcanism. The rocks of the Fenton Hill core have been fractured, brecciated, and intruded by hydrothermal veins. Excess helium is present in the rocks of the Valles Caldera (Goff and Gardner 1994). The helium may have contaminated the gneiss that Humphreys et al. studied. In short, the entire region has had a very complex thermal history. Based on oil industry experience, it is essentially impossible to make accurate statements about the helium-diffusion history of such a system.

2. Scientific studies, especially those with radical implications, do not mean much until the results have been replicated by others. Many scientific claims have disappeared entirely when others could not get the same results. Confidence in this particular paper is reduced by certain points:

- Most measurement errors and variabilities are not reported. Therefore, we do not know how accurate the results are.

- Humphreys et al. claimed that they studied zircons and biotites from depths of 750 and 1,490 meters in the Jemez Granodiorite. However, Sasada (1989) showed that at those depths, the samples came from a gneiss, an entirely different rock type.

- Because of math errors, the Q/Q_0 values (fraction of helium retained), used by Humphreys et al. to derive their dates, are too high.

- Humphreys et al. (2003) failed properly to total their data in Appendix C, which means that they grossly underestimated the total amount of helium released by their 750-meter-deep zircons. The amount of helium in the zircons greatly exceeds the amount that would be expected from the radioactive decay of uranium over 1.5 billion years. The high helium concentration may be due to samples that were abnormally high in uranium and/or to the presence of excess helium.

- Much is made of the fact that samples five and six retained the same amount of helium, even though the amounts are probably at the limit of what could be measured. The possibility of measurement error accounting for the results was never mentioned.

- If one discounts sample five, which is likely at the limit of measurable precision, the conclusions of Humphreys et al. (2004) rest on just three samples. Such a small data set may be the basis for further research, but not for drawing firm conclusions.

- Humphreys et al. (2003, note 9) referred to correcting "apparent typographical errors" in the raw data, casting suspicion on the validity of all the data.

The helium results could easily be due to an aberrant sample. They could be an artifact of the experimental or collecting method (e.g., defects in the zircons caused by rapid cooling) or from just plain sloppiness. We cannot know for sure until others have looked at the issue, too.

3. Producing a billion years of radioactive decay in a "Creation week" or year-long flood would have produced a billion years worth of heat from radioactive decay as well. This would pretty much vaporize the earth. Since the earth apparently has not been vaporized recently, we can be confident that the accelerated decay did not occur. (Humphreys recognizes this "heat problem" but is currently unable to provide a solution.)

4. If helium concentrations stay high around the rocks, it is possible for helium to diffuse into voids and fractures in the zircons, or at least high helium pressures could reduce the rate at which helium diffuses out. Either of these scenarios would invalidate the helium diffusion calculations of Humphreys et al. (2003, 2004). Helium concentrations within the earth become high enough for commercial mining. The sample measured by Humphreys et al. came from an area that is probably helium enriched. Helium deposits are common in New Mexico, and excess helium has been found just a few miles from where the sample was taken (Goff and Gardner 1994). To test for the presence of excess helium in their zircons, Humphreys et al. should look for ^{3}He.

5. Uranium does not decay directly to lead; rather, it proceeds through a series of multiple intermediate radioactive elements (Faure 1986, 284–287). It takes about ten half-lives of the longest lived intermediate to achieve secular equilibrium (i.e., each intermediate having the same activity). The uranium decay series contains elements with half-lives well over 10,000 years. If the decay rates changed suddenly, we would not expect the various elements to be in a secular equilibrium. Humphreys et al. should test for this in their zircons. Other uranium ores are at secular equilibrium, indicating a constant decay rate for at least the last two million years.

CD020: Consistency of radiometric dating comes from selective reporting.

The use of radiometric dating in geology involves a very selective acceptance of data. Most discrepant dates are not published. This selective reporting may account for consistencies in the data; internal consistencies, mineral-pair concordances, and agreement between differing dating methods may be illusory. (Woodmorappe 1979)

1. Geologists cannot be selective about choosing results because measurements typically cost hundreds of dollars per sample. To date multiple samples and choose a concordant set from among them would require throwing out about $100,000 worth of data if dating methods gave chance results (Henke n.d.b).

2. As creationists are fond of pointing out, radiometric dating is complicated by geological processes such as metamorphism and weathering, which can interfere with the assumptions that the dating methods use. As creationists do not point out, though, geologists know this. They examine the geological context of where their samples came from to determine whether a technique is likely to be valid, and they experiment with different techniques on different minerals subjected to different conditions to determine which combinations of techniques, minerals, and conditions are valid and which are not. Many so-called discordant dates are results from such experiments dishonestly portrayed as ordinary field measurements.

All measurement techniques, from rulers to neutrino detectors, are invalid in some contexts. That does not make them invalid in all contexts. Woodmorappe and others who cite discordant radiometric dates are claiming that the method is entirely useless because it does not apply to some contexts.

3. The factors that one must consider when doing radiometric dating were ignored by Woodmorappe. He ignored geological context and well-known limitations of dating methods. His analysis is further flawed because

- he uses obsolete data, such as data from years when the technique was still being developed
- his treatment of individual cases is extremely superficial
- his paper is written as propaganda, not as a technical analysis. If he believes what he writes, he should publish it in journals for professional geologists, not for creationists.

Further Reading: Schimmrich, S.H. 1998. Geochronology *kata* John Woodmorappe. http://www.talkorigins.org/faqs/woodmorappe-geochronology.html.

GEOLOGICAL COLUMN

CD101: Entire geological column does not exist.

The geological column is a fiction, existing on paper only. The entire geological column does not exist anywhere on the earth. (Huse 1983, 15)

1. The existence of the entire column at one spot is irrelevant. All of the parts of the geological column exist in many places, and there is more than enough overlap that the full column can be reconstructed from those parts.

Breaks in the geological column at any spot are entirely consistent with an old earth history. The column is deposited only in sedimentary environments, where conditions favor the accumulation of sediments. Climatic and geological changes over time would be expected to change areas back and forth between sedimentary and erosional environments.

2. There are several places around the world where strata from all geological eras do exist at a single spot—for example, the Bonaparte Basin of Australia (Trendall et al. 1990, 382, 396) and the Williston Basin of North Dakota (Morton 2001a).

Further Reading: Matson, D.E. 1994. How good are those young-earth arguments? http://www.talkorigins.org/faqs/hovind/howgood-gc.html#G3; Morton, G.R. 2001a. The geologic column and its implications to the Flood. http://www.talkorigins.org/faqs/geocolumn/ or http://home.entouch.net/dmd/geo.htm.

CD102: The geological column is sometimes out of order.

Strata in the geological column are sometimes out of order. The mechanisms geophysicists use to account for them are problematic. Thrust faulting would have produced great amounts of debris, which geologists do not see; folding would require great forces for which geophysicists have trouble accounting. (H.M. Morris 1985; Whitcomb and Morris 1961)

1. Folds account for out-of-order strata with sequences such as A-B-C-B-A. Faults create sequences such as B-C-A-B-C. The evidence is so overwhelming that these conclusions should be obvious. In many cases, the folds and faults can easily be seen in cross-sections of the strata. In other cases, further geological mapping verifies the presence of the fold or fault. Features such as ripple marks and mud cracks show that the strata were originally horizontal.

2. Great forces are not a problem in geophysics. First, great forces exist. Earthquakes can move many miles of crust by several feet at a time. Second, the forces act over a long period of time. Rocks which would fracture if bent suddenly will deform gradually under hundreds of millions of years of heat and constant pressure.

Faults do, in fact, produce a layer of debris along the fault line. Sometimes this layer is fairly thin. There is no reason to expect great amounts of debris along all faults.

3. The geologic column is never out of order in areas that have not been greatly disturbed.

Further Reading: Matson, D. E. 1994. How good are those young-earth arguments? http://www.talkorigins.org/faqs/hovind/howgood-gc.html#G4c.

CD102.1: Out-of-order strata occur at the Lewis Overthrust.

At the Lewis Overthrust in Alberta and Montana, Precambrian limestone rests on top of Cretaceous shales, which conventionally are dated much later. The evidence, and common sense, does not support the explanation that the discontinuity is caused by a thrust fault. (Price 1913, 86–101; Whitcomb and Morris 1961, 185–195)

1. Contrary to the claim, geologists do find convincing evidence of a thrust fault between the strata (Strahler 1987, chap. 40). This is true even of young-earth creationists with geology training. For example, Kurt Wise (1986, 136) said that "[a] close examination of the contact between the Cretaceous and Precambrian rocks leaves no doubt that the contact is a fault contact."

2. The strata on either side of the discontinuity are well ordered and have the order one would expect from a thrust fault.

3. The photo in Whitcomb and Morris's (1961) book *The Genesis Flood* showing the "Lewis Overthrust contact line" (Figure 17, p. 190) is not really a photo of the contact line, but of rocks 200 feet above it. The photographs that Whitcomb and Morris used were taken by Walter Lammerts, a botanist and geneticist, on his vacation (Numbers 1992, 216–219).

4. Whitcomb and Morris (1961, 187) quoted a description of the Lewis Overthrust out of context to give the impression that rocks along the fault are undisturbed. They quoted Ross and Rezak:

> Most visitors, especially those who stay on the roads, get the impression that the Belt strata are undisturbed and lie almost as flat today as they did when deposited in the sea which vanished so many million years ago.

The original paper (Ross and Rezak 1959, 420) continues:

> Actually, they are folded, and in certain zones they are intensely so. From points on and near the trails in the park it is possible to observe places where the beds of the Belt series, as revealed in outcrops on ridges, cliffs, and canyon walls, are folded and crumpled almost as intricately as the soft younger strata in the mountains south of the park and in the Great Plains adjoining the park to the east.

Further Reading: Weber, C. G. 1980. Common creationist attacks on geology.

CD103: The geologic column is based on the assumption of evolution.

The entire geologic column is based on the assumption of evolution. (Huse 1983, 14)

1. The geologic column was outlined by creationist geologists. For example, Adam Sedgwick, who described and named the Cambrian era, referred to the theory of evolution as "no better than a phrensied dream" (Ritland 1982). The geologic column is based on the observation of faunal succession, the fact that organisms vary across strata, and

that they do so in a consistent order from place to place. William "Strata" Smith (1769–1839) recognized faunal succession years before Darwin published his ideas on biological evolution.

2. The geologic column is validated in great detail by radiometric dating, which is based on principles of physics, not evolution. Furthermore, different dating techniques are all consistent, and all are consistent with the order established by the early pioneers of stratigraphy.

Further Reading: Matson, D. E. 1994. How good are those young-earth arguments? http://www .talkorigins.org/faqs/hovind/howgood-gc.html#G2; Ritland, R. 1982. Historical development of the current understanding of the geologic column: Part II. http://www.grisda.org/origins/09028.htm; Schneer, C. J. n.d. William "Strata" Smith on the web. http://www.unh.edu/esci/wmsmith.html; Young, D. A. 1988. *Christianity and the Age of the Earth*.

CD110: Meteorites and meteor craters are never found in deeper strata.

Meteorites and meteor craters are never found in deeper strata, as they should be if the strata were deposited over many millions of years. (W. Brown 1995, 27; Wysong 1976, 171)

1. At least 130 fossil craters have been found, ranging from the Precambrian to the Recent. (Grieve 1997).

2. Several meteorites have been found, in strata from Precambrian to Miocene (Matson 1994; Schmidt et al. 1997). There is evidence that a major asteroid disruption event about 500 million years ago caused an increase in meteor rates during the mid-Ordovician; more than forty mid-Ordovician fossil meteorites were found in one Ordovician limestone quarry (Schmitz et al. 2003).

3. In addition to meteorites and craters, we also find other evidence of meteor impacts, including shocked quartz and glass microspherules and fullerenes with extraterrestrial gases (L. Becker et al. 2000).

Further Reading: Becker, L. 2002. Repeated blows; Earth Impact Database. 2003. http://www .unb.ca/passc/ImpactDatabase; Greene, T. S. 2000, 2002. Impact craters on Earth. http://www .geocities.com/Athens/Thebes/7755/ancientproof/impactcraters.html; Matson, D. E. 1994. How good are those young-earth arguments? http://www.talkorigins.org/faqs/hovind/howgood-yea.html# proof4; Montanari, A., and C. Koeberl. 2000. *Impact Stratigraphy* (technical).

SEDIMENTATION

CD200: Uniformitarian assumption is untenable.

The evolution model is associated primarily with uniformitarianism, but evidence of catastrophism makes the uniformitarian assumption untenable. (H. M. Morris 1985, 91–100)

1. Modern uniformitarianism (actualism) differs from nineteenth-century Lyell uniformitarianism. The prevailing view in the eighteenth and early nineteenth centuries was that the earth had been created by supernatural means and had been shaped by several catastrophes, such as worldwide floods. In 1785, James Hutton published the proposal that Earth's history could be explained in terms of processes observed in the present; that is, "the present is key to the past." This was the beginning of uniformitarianism. Charles Lyell, in his *Principles of Geology*, modified Hutton's ideas and applied this philosophy to explain geological features in terms of relatively gradual everyday processes.

Geologists today no longer subscribe to Lyell uniformitarianism. Starting in the late nineteenth century, fieldwork showed that natural catastrophes still have a role in creating the geologic record. For example, in the later twentieth century, J. Harlan Bretz showed that the Scablands in eastern Washington formed from a large flood when a glacial lake broke through an ice dam (see Figure CH1); and Luis Alvarez proposed that an asteroid impact was responsible for the extinction of dinosaurs sixty-five million years ago. Actualism (modern uniformitarianism) states that the geologic record is the product of both slow, gradual processes (such as glacial erosion) and natural catastrophes (such as volcanic eruptions and landslides). However, natural catastrophes are not consistent with creationist catastrophism, such as "Flood geology." First, they are much smaller than the world-shaping events proposed as part of the creationists' catastrophism. More to the point, they still represent processes observed in the present. Meteorites, glacial melting, and flash floods still occur regularly, and we can (and do, as in the examples above) extrapolate from the observed occurrences to larger events of the same sort. The scale of events may change, but the physical laws operating today are key to the past.

Further Reading: Lyell, C. 1830. *Principles of Geology*. http://www.esp.org/books/lyell/principles/facsimile/title3.html; University of Oregon. n.d. Uniformitarianism. http://zebu.uoregon.edu/2003/glossary/uniformitarianism.html.

CD202: Sandstone and shale layers are too extensive for normal deposition.

Sandstones and shales cover large areas, larger than we observe being produced today. This is consistent with deposition by a global flood, not with uniformitarianism. (H. M. Morris 1985, 102–103)

1. Sandstones, shales, and other formations often do not have uniform ages. For example, the extensive St. Peter Sandstone of central North America was deposited at different times in different locations.

2. Shales form mostly from mud on the ocean floor, which does cover large areas. In the late Ordovician, much of North America was covered by a shallow sea. Much shale formed there over millions of years, to be exposed when the sea level lowered.

Sands occur mostly along shorelines. When a shoreline recedes gradually, sands can be left covering a large area.

3. A catastrophic flood would not be expected to produce such large amounts of shale and sandstone. The particle sizes in these sediments is uniform; the gravel, sand, and mud have been sorted apart into different areas. The high energies in the flood would mix everything together. At best, a flood could redeposit sands or muds that already existed, and it would take millions of years for such quantities to form.

Furthermore, shales are sometimes found atop sandstones. A single flood could not deposit both. Even more impossible for a single flood, we also see multiple layers of sand or shale interleaved with other materials, such as volcanic ash (Nanayama et al. 2003).

CD203: Limestone and dolomite layers are too extensive for normal deposition.

The occurrence of limestone deposits that are so great and so uniform defies explanation except by massive precipitation from chemical-rich waters, consistent with catastrophism, not uniformitarianism. Dolomite sediments are not being formed at all today; they also require an exceptional explanation. (H. M. Morris 1985, 104)

1. Uniformitarian processes explain limestone formations far better than catastrophism does:

- Limestones form continuously today over wide areas (such as the Caribbean) as calcium carbonate is precipitated from water directly and through the actions of organisms. Limestone formation easily fits within conventional geology.
- Limestones appear in strata interleaved between strata of sandstones and other rocks. A single event could not explain all the layers.
- Limestones often include fragile fossils that could not survive catastrophic transport.

Dolomites require no exceptional explanation. They form via diagenesis (a sort of chemical rearrangement in the deep subsurface) from calcite, the main ingredient of limestone. Creationism does not explain the origin of dolomite.

2. Limestone could not have formed quickly from massive precipitation, because the formation of calcite releases heat. If only 10 percent of the world's limestone were formed during the Flood, the 5.6×10^{26} joules of heat released would be enough to boil the flood waters (Isaak 1998).

CD210: The mouth of the Colorado River does not have enough sediment for the Grand Canyon.

There is nowhere near enough sediment deposited at the mouth of the Colorado River to account for ten million years worth of erosion.

1. The Colorado River delta itself is quite extensive. It covers 3,325 square miles (G. Sykes 1937) and is up to 3.5 miles deep (Jennings and Thompson 1986), containing over 10,000 cubic miles of the Colorado River's sediments from the last two to three million years. The sediments that were deposited by the river more than two to three million

years ago have been shifted northwestward by movement along the San Andreas and related faults (Winker and Kidwell 1986). Sediments have also accumulated elsewhere. Some were deposited in flood plains between the delta and the Grand Canyon.

2. Wind is a major erosional force in parts of the Colorado River basin. Some sediments from Colorado and Wyoming were blown as far as the Atlantic Ocean.

3. Much of the strata exposed in the Grand Canyon is limestone and dolomite. These rocks eventually simply would have dissolved.

CD211: Mississippi delta could have formed in 5,000 years.

The size of the Mississippi River delta divided by the sediment accumulation rate gives an age of less than 30,000 years, indicating a young earth. (Pathlights n.d.b)

1. The age of the Mississippi delta only gives a lower limit for the age of the earth.

2. The Mississippi delta is seven miles thick at the Gulf of Mexico. This is too thick to have formed suddenly by a single flood, as such a flood would have spread the sediments out, not compacted them all in one place.

3. The claimed size of the Mississippi delta considers only its current delta. The location of the delta has changed every so often due to changes in sea level and changes in the course of the Mississippi River. In the early Cenozoic, the Gulf of Mexico and the Mississippi delta extended as far north as Illinois (Weber 1980).

Further Reading: Matson, D. E. 1994. How good are those young-earth arguments? http://www.talkorigins.org/faqs/hovind/howgood-yea2.html#proof19.

CD220: There is not enough sediment in the ocean for an old earth.

At current rates of erosion, only thirty million years are needed to account for all the sediment in the ocean. If the earth were as ancient as is claimed, there should be more sediment. (H. M. Morris 1985, 155–156)

1. The thickness of sediment in the oceans varies, and it is consistent with the age of the ocean floor. The thickness is zero at the Mid-Atlantic Ridge, where new ocean crust is forming, and there is about 150 million years' worth of sediment at the continental margins. The average age of the ocean floor is younger than the earth due to subduction at some plate margins and formation of new crust at others.

2. The age of the ocean floor can be determined in various ways—measured via radiometric dating, estimated from the measured rate of sea-floor spreading as a result of plate tectonics, and estimated from the ocean depth that is predicted from the sea floor sinking as it cools. All these measurements are consistent, and all fit with sediment thickness.

Further Reading: Matson, D. E. 1994. How good are those young-earth arguments? http://www.talkorigins.org/faqs/hovind/howgood-yea2.html#proof21.

CD220.1: Much more sediment is deposited than removed by subduction.

About twenty-five billion tons of sediment are deposited in the ocean each year, but plate tectonic subduction only removes about one billion tons per year. Currently, ocean sediment thickness averages 400 meters. At the observed deposition rate, it would accumulate in only twelve million years, not the hundreds of millions of years that the oceans have been around. (Humphreys 1999)

1. Yes, more sediment is deposited in the oceans than is removed by subduction. However, subduction is not the only fate of sediment deposited into the oceans. Some sediment deposited on the continental margin can become part of the continent itself if the sea level falls or the land is uplifted. Some calcium and organic sediments become biomass or ultimately dissolve. Some sediment becomes compacted as it deepens, so its volume is not indicative of the original sediment volume. Some sediment is "scraped" off of subducting plates and becomes coastal rocks.

2. The uniformitarian assumption in the claim is not valid. Tectonics involves ocean basins forming and spreading, but it also involves them closing up again (the Wilson cycle). When the basins close, the sediment in the oceans is piled up on the edges of continents or returned to the mantle. Much of British Columbia was produced when the Pacific Ocean closed a few hundred million years ago and land in the ocean accreted to the continent.

Further Reading: Fichter, L. S. 1999. The Wilson cycle. http://csmres.jmu.edu/geollab/Fichter/Wilson/Wilson.html; Thomas, D. 1998. "Creation physicist" D. Russell Humphreys, and his questionable "evidence for a young world." http://www.cesame-nm.org/Viewpoint/contributions/Hump.html.

CD221: Oceans do not have enough dissolved minerals for an old earth.

An upper limit for the age of the oceans is obtained by dividing the amount of an element dissolved in the sea by the amount added each year by rivers. These calculations yield the following figures (H. M. Morris 1985, 153–155):

Element	Years to Accumulate
sodium	260,000,000
magnesium	45,000,000
silicon	8,000
potassium	11,000,000
copper	50,000
gold	560,000
silver	2,100,000
mercury	42,000
lead	2,000
tin	100,000
nickel	18,000
uranium	500,000

1. The numbers in the table are residence times, or the average time that a small amount of an element stays in the sea water before being removed. They are not times that it takes the element to accumulate, and individual atoms may stay much briefer or longer than those times. Elements in the ocean are in approximate equilibrium between sources adding them and mechanisms removing them.

A detailed analysis of sodium, for example, shows that 35.6×10^{10} kg/yr come into the ocean, and 38.1×10^{10} kg/yr are removed (Morton 1996). Within measurement error, the amount of sodium added matches the amount removed.

2. Morris left aluminum off the list. It would show (according to Morris's reasoning) that the earth is only 100 years old.

Further Reading: Burton J. D., and D. Wright. 1981. Sea water and its evolution; Matson, D. E. 1994. How good are those young-earth arguments? http://www.talkorigins.org/faqs/hovind/howgood-yea.html#proof13; Stassen, C. 1997. The age of the earth. http://www.talkorigins .org/faqs/faq-age-of-earth.html#ocean.

CD222: Juvenile water is added to oceans too fast for an old earth.

Juvenile water, the water added to oceans through volcanoes, hot springs, and other vents, is added at such a rate that the oceans could not be more than 320 million years old to be as full as they are now. (H. M. Morris 1985, 156)

1. Water is removed from the ocean into the earth's crust, too. Around rift systems, water is circulated underground thermally. At subduction zones, water is subducted along with the oceanic plates. In fact, the water (and carbon) that is subducted lowers the melting point of the rock, ultimately causing the volcanoes from which the juvenile water comes (Decker and Decker 1998, 84–85). In short, juvenile water is not really juvenile; it is recycled.

CD230: Natural gas and oil escape too fast to allow for long ages.

Natural gas and oil appear underground often under great pressures. If the earth were old, the high-pressure gas and oil would have seeped out and the reservoirs would have emptied. (Pathlights n.d.b)

1. The high pressures show that rocks trapping the gas and oil are impermeable enough to hold the reservoirs for many millions of years. If the assumptions (young earth and leaky rocks) behind the claim were true, the pressures never would have built up in the first place.

It is also important to remember that pressure does not mean much to seepage rate without a pressure gradient. If the pressure around a reservoir is as great as the pressure within it, the gas or oil will not be forced out.

2. A geological event that could cause oil and gas to migrate into a reservoir could have occurred relatively recently. Even if the gas reservoir is young, that does not mean the earth is young.

Further Reading: Matson, D. E. 1994. How good are those young-earth arguments? http://www .talkorigins.org/faqs/hovind/howgood-yea2.html#proof18.

CD240: Experiments show that strata can violate principles of superposition.

Experiments in a water tank showed that sedimentary layers can be laid down very quickly in patterns that violate the geological principles of superposition (that layers are deposited horizontally with younger ones on top) and continuity (that each layer has the same age at every point). In particular, lamination is sometimes the result of segregation of particles according to size, not the result of successive layering; lamination deposits can be produced on slopes. (Berthault 2000; Snelling 1997)

1. Berthault's results do not invalidate the principle of superposition. Newer layers still appear on older layers. Berthault's results duplicated a case where the layers are not laid down horizontally, but the principle of superposition does not require horizontal depositional surfaces. Berthault erred in confusing the principle of superposition with the principle of original horizontality, which was already known to have limited application. Berthault's experiments only duplicated results that were familiar to sedimentologists decades earlier. There is nothing of significance in his work.

Further Reading: Henke, K. R. n.d.a. Berthault's "stratigraphy." [Includes responses]. http:// home.austarnet.com.au/stear/henke_steno.htm, http://home.austarnet.com.au/stear/guy_response_ henke.htm, http://home.austarnet.com.au/stear/questions_berthault_k_henke.htm.

CD241: Varves can form in less than a year.

Varves (layers of silt that show seasonal differences) do not necessarily form annually. Individual varves can form in less than a year. Thus, claims that 10,000 varve layers represent 10,000 years are unwarranted. (Garner 1997)

1. The seasonal nature of varves is sometimes indicated by the systematic variation of pollen from seasonal plants (Morton 2002c).

2. There is at least one formation that contains twenty million varves. That represents more than 50,000 years even if you assume varves were formed at a rate of one per day. And the fineness of the silt precludes the possibility that they could have formed that rapidly.

3. The 45,000-year varve record of Lake Suigetsu is consistent with other dating techniques, such as carbon-14 dating and the tree ring record (Kitagawa and van der Plicht 1998; see Figure CD2).

4. Non-annual fine-grained layers are recognizably different from varves. Layers that form rapidly tend to be much more irregular, reflecting the changes in the weather conditions that cause them (Morton 1998d). Annual varves are observed forming today. They produce uniform layers seen also in the geologic record.

CD250: Stalactites can grow very rapidly.

Stalactites can grow very quickly. Some have been observed to grow more than half an inch per year. The largest stalactites and flowstones could have formed in a few thousand years. (Meyers and Doolan 1987)

1. The fast-growing stalactites form via processes very different from calcium carbonate stalactites found in limestone caves. Limestone is not soluble in water. When carbon dioxide (from decaying plants in the soil above the cave) mixes with water, it forms a very weak carbonic acid. This turns the calcium carbonate into calcium bicarbonate, which dissolves. When drips are exposed to air in the cave, a little carbon dioxide escapes from them into the atmosphere, which reverses the process and precipitates a small amount of calcium carbonate. The upper average rate for limestone stalactite growth is ten centimeters per thousand years, with lower growth rates outside of tropical areas.

Fast-growing stalactites, on the other hand, either grow from gypsum through an evaporative process, or they form from concrete or mortar. When water is added to concrete, one product is calcium hydroxide, which is about 100 times more soluble than calcite. The calcium hydroxide absorbs carbon dioxide from the atmosphere to reconstitute calcium carbonate.

2. The time for stalactite growth also has to allow for time for the cave to dissolve in the first place, which is a very slow process, sometimes on the order of tens of millions of years. Then the geological conditions have to change so that the cave is no longer under water. Only then can stalactite growth begin.

3. Direct measurement via radiometric dating gives stalactite ages over 190,000 years (Ford and Hill 1999). Other deposits in caves have been dated to several million years old. For example, $^{40}Ar/^{39}Ar$ dating of alunite (an aluminum sulfate mineral) gives an age of 11.3 million years for a cave near Carlsbad Caverns (Polyak et al. 1998).

4. Oxygen isotope measurements in stalactites give an indication of outside temperatures. They are consistent with the coming and going of ice ages back at least 160,000 years (Dorale et al. 1998; Wang et al. 2001; Zhang et al. 2004).

Further Reading: Matson, D. E. 1994. How good are those young-earth arguments? http://www.talkorigins.org/faqs/hovind/howgood-yea2.html#proof22.

EVAPORATION

CD301: Evaporites could form without evaporation.

Evaporites can precipitate from unsaturated brines; they can form without evaporation (H. M. Morris 1985, 105–106). *A mechanism is sketched by J. D. Morris (2002):*

> Many now think the salt was extruded in superheated, supersaturated salt brines from deep in the earth along faults. Once encountering the cold ocean waters, the hot brines could no longer sustain the high concentrations of salt, which rapidly precipitated out of solution, free of impurities and marine organisms.

1. Most evaporite deposits are not associated with evidence of hydrothermal activity. The huge amount of energy needed to deposit kilometers of salt in a few weeks should have left obvious evidence, such as heat-altered rocks or evidence of magma. Typical hydrothermal deposits such as iron and manganese are not often found associated with evaporites. Sea-floor basalts are a common site of hydrothermal activity, and other hydrothermal deposits are found there, but salt deposits are never found associated with them.

2. Hydrothermal systems operating today are not depositing any salt, much less the thick, laterally extensive layers we find in the sedimentary record. In fact, hydrothermal solutions contain less sodium and chlorine than normal sea water (Open University Team 1989, 100).

3. Evaporites are observed forming today in basins with no significant outflow; the water that flows in evaporates and leaves behind layers of dissolved salts. Ancient evaporites are also found in sedimentary context, and they are often associated with other evidence of being open to the air, such as footprints, dessication cracks and occasional raindrop impressions. None of these structures are consistent with an underwater hydrothermal environment.

Evaporites are also found in sabkha environments, where crystals or nodules of salt grow within fine-grained sediments as saline groundwater (usually from a nearby ocean) is drawn upward by evaporation. As the water evaporates at the surface, salt nodules grow, often forming a chicken-wire pattern. Some sabkha evaporites grow into gypsum rosettes: huge crystals resembling flowers. These features also are known from ancient evaporites. They also are inconsistent with hydrothermal deposition.

CD302: Evaporites are too thick.
Evaporite deposits are far too thick to have been formed from the evaporation of ancient seas. (H. M. Morris 1985, 106)

1. This is only a problem for creationists who cannot imagine long time spans. Thick deposits with numerous beds are consistent with cyclical sea transgression/regressions (Burke n.d.). They can also form in sinking basins that have had no outlet over geological time.

CD303: Evaporites contain no organic matter.
Evaporite deposits could not be from evaporated inland seas because they are too pure; they contain no organic matter. (H. M. Morris 1985, 106)

1. Evaporite deposits do contain impurities. The Sedom Formation evaporites in the Dead Sea Basin, which are more than 2 km thick, are about 80 percent pure halite, with 20 percent gypsum, marl, chalk, dolomite, and shale and with significant amounts of pollen (Niemi et al. 1997, 46). The Paradox Basin evaporites have many thin interbed-

ded shale layers containing brachiopods, conodonts, and plant remains (Duff et al. 1967, 204).

We should expect few impurities in evaporites because hypersaline basins are harsh environments in which few organisms can live.

GLACIATION

CD410: World War II airplanes are now beneath thousands of "annual" ice layers.
Ice cores are claimed to have as many as 135,000 annual layers. Yet airplanes of the Lost Squadron were buried under 263 feet of ice in forty-eight years, or about 5.5 feet per year. This contradicts the presumption that the wafer-thin layers in the ice cores could be annual layers. (Vardiman 1992)

1. The airplanes landed near the shore of Greenland, where snow accumulation is rapid, at about 2 m per year. Allowing for some compaction due to the weight of the snow, that accounts for the depth of snow under which they are buried. The planes are also on an active glacier and have moved about 2 km since landing. Ice core dating takes place on stable ice fields, not active glaciers. The interior of Greenland, where ice cores were taken, receives much less snow. In Antarctica, where ice cores dating back more than 100,000 years have been collected, the rate of snow accumulation is much less still.

Further Reading: Brinkman, M. 1995. Ice core dating. http://www.talkorigins.org/faqs/icecores .html; Kuechmann, F.C. 2000. Creationist comedy. http://home.austarnet.com.au/stear/ kuechmann_cretin_comedy.htm.

MOUNTAIN BUILDING

CD501: On an old earth, mountains would have eroded by now.
On an old earth, mountains would have eroded by now. At present rates of erosion, the continents would have been eroded to sea level in less than fifteen million years. (H. M. Morris 1985, 155)

1. Old mountain ranges are eroded flat. But there are also forces creating new mountains. For example, the Himalayas are still rising.

2. Present rates of erosion are particularly high due to more mountain building and higher mountains than usual in earth's history. (Erosion slows as mountains lose elevation.)

3. The reasoning behind this claim directly contradicts the reasoning behind the claim that volcanoes build too much material for an old earth (see CD502).

Further Reading: Matson, D. E. 1994. How good are those young-earth arguments? http://www
.talkorigins.org/faqs/hovind/howgood-yea.html#proof15; McPhee, J. 1998. *Annals of the Former
World.*

CD502: Volcanic mountains are built too fast for an old earth.

Volcanoes are adding material to the crust too rapidly for an earth as old as is claimed. At present rates, volcanoes could have formed the entire crust in 500 million years. (H. M. Morris 1985, 155–157)

1. Volcanic material gets eroded, too. Some also gets subducted.

2. The rate at which igneous rocks are being produced today does not necessarily reflect the rate at which they were produced in the past.

3. The reasoning behind this claim directly contradicts the reasoning behind the claim that mountains erode too quickly for an old earth (see CD501).

Further Reading: Matson, D. E. 1994. How good are those young-earth arguments? http://www
.talkorigins.org/faqs/hovind/howgood-yea.html#proof12.

CD510: Folded rocks must have been soft when folded.

Rocks do not fold without breaking. Therefore, folded rocks (found in many places) must have been soft when folded. This is easily explained by mountain building taking place soon after the sedimentary layers were deposited by a global flood. (Humphreys 1999)

1. Rocks do fold without breaking when bent very slowly under pressure. Laboratory experiments demonstrate as much (e.g., Friedman et al. 1980). Increased temperature can also increase the flow rate.
Some rocks ("weak" ones) flow more easily than others ("competent" ones). Layers of different rocks will sometimes have broken rocks in some layers and not others.

2. Deformation is not limited to sedimentary layers. There are deformed quartzite pebbles near Death Valley.

EROSION

CD610: The erosion rate of Niagara Falls' rim indicates a young earth.

The rim of Niagara Falls erodes at four to five feet per year, indicating that it could not be more than 10,000 years old. Therefore, the earth is not more than 10,000 years old. (J. D. Morris 2003; Pathlights n.d.b)

1. The age of Niagara Falls is not the age of the earth. Geologists estimate that Niagara Falls originated about 7,000 years ago, sometime after the end of the last glacial episode. This says nothing about how old the rest of the earth is, though.

Further Reading: Matson, D. E. 1994. How good are those young-earth arguments? http://www
.talkorigins.org/faqs/hovind/howgood-yea2.html#proof17.

CD620: Average soil depth is consistent with a young earth.

*Only 300 to 1,000 years are required to build an inch of topsoil. The average depth of topsoil
is about eight inches, indicating an earth less than about 8,000 years old.* (Pathlights n.d.b)

1. Soil gets eroded as well as built up, so the average depth does not mean much.
Where soil does exist under steady conditions, it does not build up continuously; there
is a maximum depth to it determined by climate, ground composition, slope, and local
ecology. The depth of a soil says very little about its age.

2. Some soils require long times to develop. R. Meyer (1997, 120) listed seven types
of soil that take more than 50,000 years to form; some took on the order of a million
years or more.

Further Reading: Matson, D. E. 1994. How good are those young-earth arguments? http://www
.talkorigins.org/faqs/hovind/howgood-yea2.html#proof16.

GEOPHYSICS AND PLATE TECTONICS

CD701: The earth's magnetic field is decaying, indicating a young earth.

The earth's magnetic field is decaying at a rate indicating that the earth must be young. (T. G.
Barnes 1973; Humphreys 1993)

1. Empirical measurement of the earth's magnetic field does not show exponential
decay. Yes, an exponential curve can be fit to historical measurements, but an exponential curve can be fit to any set of points. A straight line fits better.

2. The earth's magnetic field is known to have varied in intensity (Gee et al. 2000)
and reversed in polarity numerous times in the earth's history. This is entirely consistent
with conventional models (Glatzmaier and Roberts 1995) and geophysical evidence
(Song and Richards 1996) of the earth's interior.

3. T. G. Barnes (1973) relied on an obsolete model of the earth's interior. He viewed
it as a spherical conductor (the earth's core) undergoing simple decay of an electrical current. However, the evidence supports Elsasser's dynamo model, in which the magnetic
field is caused by a dynamo, with most of the "current" caused by convection. Barnes
cited Cowling to try to discredit Elsasser, but Cowling's theorem is consistent with the
dynamo earth (Cowling 1981).

4. Barnes measures only the dipole component of the total magnetic field, but the dipole field is not a measure of total field strength. The dipole field can vary as the total
magnetic field strength remains unchanged.

Further Reading: Brush, S. G. 1983. Ghosts from the nineteenth century; Matson, D. E. 1994. How good are those young-earth arguments? http://www.talkorigins.org/faqs/hovind/howgood-yea.html#proof11; Thompson, T. 1997. On creation science and the alleged decay of the earth's magnetic field. http://www.talkorigins.org/faqs/magfields.html.

CD740: The theory of plate tectonics is wrong.

The theory of plate tectonics is false. (W. Brown 1995; Nevins 1976)

1. Plate tectonics was uncertain as recently as the 1960s, but evidence in its favor has become overwhelming:

- Plate motions are measured directly (J. P. Davidson et al. 1997).

- The eastern edge of the continental shelves of North and South America fit closely (within 50 km) with the western continental shelves of Africa and Europe (A. C. Bishop 1981). The Mid-Atlantic Ridge has the same shape.

- Plant and animal fossil distributions, geological formations, and indications of ancient climate match up in Africa and South America as if the continents once fit together (J. P. Davidson et al. 1997).

- When new rocks are formed, they record the earth's current magnetic field, which reverses occasionally. The magnetic field pattern recorded in the sea floor rocks shows bands mirrored across a spreading center (A. C. Bishop 1981; J. P. Davidson et al. 1997; see also CD741).

- Paleomagnetic studies show different polar wandering on different continents, indicating that the continents moved relative to one another (A. C. Bishop 1981; J. P. Davidson et al. 1997).

- Oceanic sediments are young and thin, indicating that sea basins are relatively young (Graham 1981).

- Maps of earthquake locations show plate boundaries and the paths of subducting plates (J. P. Davidson et al. 1997; Graham 1981).

- Hot spots leave trails such as volcanic island chains as the plates move over them (J. P. Davidson et al. 1997).

Further Reading: Alden, A. 2003. Tectonic plate motions, Eurasia/Africa. http://geology.about .com/library/bl/maps/blplatemo_atlas.htm; Cocks, L.R.M. (ed.). 1981. *The Evolving Earth*; McPhee, J. 1998. *Annals of the Former World*. Plate tectonics will be covered in most any basic geology textbook.

CD741: Midocean magnetic anomalies are not reversals.

Plate tectonics became widely accepted when bands of reversed magnetic orientation were found mirrored on either side of the Mid-Atlantic Ridge. According to the theory, the sea floor spread gradually from the Mid-Atlantic Ridge, and periodic flips in the earth's magnetic field were preserved and recorded in the rocks as emerging magma cooled. But these bands of magnetism were misinterpreted; there are no magnetic reversals. Although magnetic intensities fluctuate, these are slight deviations around a high average. A compass needle would not change direction over these bands. (W. Brown 1995, 79)

1. The magnetic field preserved in the rocks themselves does change direction. The magnetism measured in the 1950s was measured at the ocean surface, so the earth's present magnetic field was added to the magnetism from the rocks below (A.C. Bishop 1981).

2. The main significance of the data was that the pattern was mirrored on either side of the midocean ridge. This is just the pattern one would expect from sea-floor spreading.

3. There is a great deal more evidence for plate tectonics (see CD740).

CD750: Plate tectonics occurred catastrophically and has since slowed.

Plate tectonics occurred, but catastrophically. Slabs of oceanic crust broke loose and subducted along continental margins. This lowered the viscosity of the mantle, leading to meters-per-second runaway subduction. The earth's magnetic field rapidly reversed several times. Steam caused a global rain. Flood basalts erupted. The lighter mantle material of the new ocean floors made them rise, causing the oceans to flood the continents. The flood carried and redistributed sediments. The process slowed almost to a stop when nearly all the old ocean floor had been subducted. Subsequent cooling of the ocean basins caused them to sink to where they are today. (Austin et al. 1994)

1. Much geological evidence is incompatible with catastrophic plate tectonics:

- Island chains, such as the Hawaiian islands, indicate that the ocean floor moved slowly over erupting "hot spots." Radiometric dating and relative amounts of erosion both indicate that the older islands are very much older, not close to the same age as catastrophic tectonics would require.

- Catastrophic plate tectonics says that all ocean floor should be essentially the same age. But both radiometric dating and amounts of sedimentation indicate that the age changes gradually, from brand new to tens of millions of years old.

- As sea-floor basalt cools, it becomes denser and sinks. The elevation of sea floors is consistent with cooling appropriate for its age, assuming gradual spreading.

- Guyots are flat-topped underwater mountains. The tops were eroded flat from a long time at the ocean surface, and they sank with the sea floor. Catastrophic tectonics does not allow enough time for the sea mountain to form, erode, and sink.

- Runaway subduction does not account for continent–continent collisions, such as between India and the Eurasian plate.

2. Catastrophic plate tectonics has no plausible mechanism. In particular, the greatly lowered viscosity of the mantle, the rapid magnetic reversals, and the sudden cooling of the ocean floor afterwards cannot be explained under conventional physics.

3. Conventional plate tectonics accounts for the evidence already and does a much better job of it. It explains innumerable details that catastrophic plate tectonics cannot, such as why there is gold in California, silver in Nevada, salt flats in Utah, and coal in Pennsylvania (McPhee 1998). It requires no extraordinary mechanisms to do so. Catastrophic plate tectonics would be a giant step backwards in the progress of science.

Further Reading: McPhee, J. 1998. *Annals of the Former World.*

ASTRONOMY AND COSMOLOGY

EARTH

CE001: There is not enough helium in the atmosphere for an old earth.

The radioactive decay of several elements produces helium, which migrates to the atmosphere. There is too little helium in the atmosphere to account for the amount that would have been produced in 4.5 billion years. Escape of helium into space is not sufficient to account for the lack. (H. M. Morris 1985, 150–151)

1. Helium is a very light atom, and some of the helium in the upper atmosphere can reach escape velocity simply via its temperature. Thermal escape of helium alone is not enough to account for its scarcity in the atmosphere, but helium in the atmosphere also gets ionized and follows the earth's magnetic field lines. When ion outflow is considered, the escape of helium from the atmosphere balances its production from radioactive elements (Lie-Svendsen and Rees 1996).

Further Reading: Matson, D. E. 1994. How good are those young-earth arguments? http://www.talkorigins.org/faqs/hovind/howgood-yea.html#proof14.

CE010: NASA scientists found a day missing.

NASA scientists, using computers to track planetary motions, discovered that a day of time was missing, corresponding to biblical accounts of the sun's standing still for Joshua for almost a day, plus the sun moving backwards forty minutes for Hezekiah. (Hawaii Christians Online n.d.)

1. The origin of this urban legend goes back to 1890. It is entirely baseless. Indeed, it could not be true. There is no frame of reference to measure against to determine whether a day was missing thousands of years ago.

Further Reading: Brunvand, J. H. 2000. *The Truth Never Stands in the Way of a Good Story*, pp. 137–148; Mikkelson, B., and D. P. Mikkelson. 2000. The lost day. http://www.snopes.com/religion/lostday.htm.

CE011: Earth's rotation is slowing, indicating a young earth.

Earth's rotation is slowing down, so it cannot be more than a few million years old. (Pathlights n.d.b)

1. The earth's rotation is slowing at a rate of 0.005 seconds per year per year. This extrapolates to the earth having a fourteen-hour day 4.6 billion years ago, which is entirely possible.

The rate at which the earth is slowing today is higher than average because the present rate of spin is in resonance with the back-and-forth movement of the oceans.

Fossil rugose corals preserve daily and yearly growth patterns and show that the day was about 22 hours long 370 million years ago, in rough agreement with the 22.7 hours predicted from a constant rate of slowing (Scrutton 1964; J. W. Wells 1963).

Further Reading: Matson, D. E. 1994. How good are those young-earth arguments? http://www.talkorigins.org/faqs/hovind/howgood-yea2.html#proof20. Rosenberg, G. D., and S. K. Runcorn (eds.). 1975. *Growth Rhythms and the History of the Earth's Rotation.*

CE011.1: The frequency of leap seconds indicates a young earth.

Leap seconds have had to be inserted into the year twenty-two times between 1970 and 1999, showing that the earth is slowing 0.77 seconds per year. At this rate, the earth would have slowed to a stop if it were billions of years old.

1. Adding leap seconds nearly every year does not indicate that the earth is slowing by nearly one second per year; it shows that there is a discrepancy between Coordinated Universal Time (UTC, an international standard) and astronomical time (NIST Time and Frequency Division n.d.). The earth is slowing down (see CE011), but not at such a great rate. The length of a day now is very slightly more than twenty-four hours. If the earth kept rotating at the same rate, leap seconds would need to be inserted at the same rate they have been inserted in the last thirty years.

Further Reading: NIST. Updated monthly. NIST time scale data archive. http://www.boulder.nist.gov/timefreq/pubs/bulletin/leapsecond.htm; Robinson, B. A. 2002. A failed attempt to dialog with creation scientists. http://www.religioustolerance.org/ev_dialog.htm; Thwaites, W. M., and F. T. Awbrey. 1982. As the world turns: Can creationists keep time? http://www.natcenscied.org/resources/articles/9626_issue_09__volume_3_number_3__1_3_2003.asp.

CE020: An old earth would be covered by 182 feet of meteoric dust.

The observed rate of cosmic dust influx should have produced a layer 182 feet thick over the entire surface of the earth if the earth were five billion years old. The distinctive nickel and iron content of the dust should make it easy to detect. (H. M. Morris 1985, 151–152)

1. The observed rates used in Morris's calculation are based on dust collected in the atmosphere; this measurement was contaminated by dust from the earth. More recent measurements of cosmic dust influx measured from satellites give an influx rate about 1 percent as large, corresponding to a layer 66 cm at most thick over 4.5 billion years (Kyte and Wasson 1986).

Further Reading: Thompson, T. 1996. Meteorite dust and the age of the earth. http://www.talk origins.org/faqs/moon-dust.html.

MOON

CE101: There is not enough moon dust for an old universe.

Based on measured rates of planetary dust accumulation, there is too little moon dust for an old moon. Before the moon landings, there was considerable fear that astronauts would sink in the dust. (H. M. Morris 1985, 152)

1. The high number for dust accumulation (14 million tons per year on earth) comes from the high end of a single preliminary measurement that has long been obsolete. Other higher estimates come from even more obsolete sources, although they are sometimes incorrectly cited as being more recent. The actual influx is about 22,000 to 44,000 tons per year on earth and around 840 tons per year on the moon.

The story that scientists worried about astronauts sinking in moon dust is a total fabrication. As early as 1965, scientists were confident, based on optical properties of the moon's surface, that dust was not extensive. Surveyor I, in May 1966, confirmed this.

Further Reading: Matson, D. E. 1994. How good are those young-earth arguments? http://www.talkorigins.org/faqs/hovind/howgood-yea.html#proof2; Thompson, T. 1996. Meteorite dust and the age of the earth. http://www.talkorigins.org/faqs/moon-dust.html; Thompson, T. n.d.b. Is the Earth young? http://www.tim-thompson.com/young-earth.html.

CE110: The moon is receding at a rate too fast for an old universe.

Because of tidal friction, the moon is receding, and the earth's rotation is slowing down, at rates too fast for the earth to be billions of years old. (T. G. Barnes 1982)

1. The moon is receding at about 3.8 cm per year. Since the moon is 3.85×10^{10} cm from the earth, this is already consistent, within an order of magnitude, with an earth–moon system billions of years old.

2. The magnitude of tidal friction depends on the arrangement of the continents. In the past, the continents were arranged such that tidal friction, and thus the rates of earth's slowing and the moon's recession, would have been less. The earth's rotation has slowed at a rate of two seconds every 100,000 years (Eicher 1976).

3. The rate of earth's rotation in the distant past can be measured. Corals produce skeletons with both daily layers and yearly patterns, so we can count the number of days per year when the coral grew. Measurements of fossil corals from 180 to 400 million years ago show year lengths from 381 to 410 days, with older corals showing more days per year (Eicher 1976; Scrutton 1970; J. W. Wells 1963, 1970). Similarly, days per year can also be computed from growth patterns in mollusks (Pannella 1976; Scrutton 1978) and stromatolites (Mohr 1975; Pannella et al. 1968) and from sediment deposition patterns (G. E. Williams 1997). All such measurements are consistent with a gradual rate of earth's slowing for the last 650 million years.

4. The clocks based on the slowing of earth's rotation described above provide an independent method of dating geological layers over most of the fossil record. The data is inconsistent with a young earth.

Further Reading: Matson, D. E. 1994. How good are those young-earth arguments? http://www.talkorigins.org/faqs/hovind/howgood-yea.html#proof5; Pannella, G. 1972. Paleontological evidence on the Earth's rotational history since the early Precambrian; Rosenberg, G. D., and S. K. Runcorn (eds.). 1975. *Growth Rhythms and the History of the Earth's Rotation.* Schopf, J. W. (ed.). 1983. *Earth's Earliest Biosphere*; Thompson, T. 2000. The recession of the Moon and the age of the Earth–Moon system. http://www.talkorigins.org/faqs/moonrec.html.

CE130: Lunar moonquakes, lava flows, and gas emissions indicate the moon's youth.

Lunar activity such as moonquakes, lava flows, and gas emissions indicates the moon's youth. (Pathlights n.d.a)

1. Moonquakes can be explained by tidal stresses and by lunar contraction (due to the highlands gradually sinking). Moonquakes, in fact, provide evidence that the moon has a solid core, consistent with its old age.

2. There is no evidence for recent lava flows not associated with meteor impacts.

3. Outgassing is consistent with an old moon. It can take a long time for gasses to work their way to the surface.

Further Reading: LANL. n.d. The Los Alamos built spectrometers. http://lunar.lanl.gov/pages/spectros.html.

PLANETS AND SOLAR SYSTEM

CE210: Venus's high temperature and atmosphere should have eroded its surface features.

High surface temperatures on Venus (900 degrees F, 482 degrees C), combined with its dense atmosphere, should long ago have eroded its surface features if the planet were four billion years old. Venus's features support a young age. (Pathlights n.d.a)

1. A paucity of surface features usually indicates a young surface, not an old one, because there would be less time for cratering to occur. However, a young surface does not imply a young planet.

2. In fact, there are few craters and surface features on Venus. This is taken as evidence that the surface is young. Venus shows evidence of volcanoes and rift valleys, suggesting some tectonic activity, which would help to renew its surface. Venus's thick atmosphere also reduces the impact of meteors and, as the claim notes, erodes features. The features on Venus are probably all less than a few hundred million years old.

Further Reading: Thompson, T. 1994. Is the planet Venus young? http://www.talkorigins .org/faqs/venus-young.html.

CE230: Io's great volcanic activity indicates a young age.
Jupiter's moon Io is volcanic. It is too small for its volcanism to be explained by residual heat of formation or radioactive decay, unless the moon is not millions of years old. (Pathlights n.d.a)

1. The volcanoes on Io are powered by tidal heating. Io is close to Jupiter, so it is strongly affected by Jupiter's gravity. The other moons of Jupiter exert their own gravitational forces. The resulting tides raise and lower Io's surface by about 100 m, generating frictional heat that drives the volcanoes.

Further Reading: Wood, J. S. 2003. Io: Jupiter's volcanic moon: Tidal heating. http://www.plane taryexploration.net/jupiter/io/tidal_heating.html.

CE231: Jupiter and Saturn are cooling too rapidly to be old.
Jupiter and Saturn are cooling, giving off their internal heat at a rate too great for them to be billions of years old. (W. Brown 1995, 30)

1. Jupiter is cooling slowly enough that it could still be radiating its primordial heat. Saturn's extra heat could come from gravitational potential energy as helium in its atmosphere condenses into droplets and falls toward the center.

Further Reading: Matson, D. E. 1994. How good are those young-earth arguments? http://www .talkorigins.org/faqs/hovind/howgood-yea.html#proof10.

CE240: Saturn's rings are unstable.
Saturn's rings are unstable. They gradually drift outward, and disruption from bombardment could mean that they could not last more than 10,000 years. The rings cannot be billions of years old. (W. Brown 1995, 29)

1. Saturn's rings may be less than 100 million years old (Cuzzi and Estrada 1998). However, that says nothing about the age of the planet. The rings could have formed when

Saturn captured a small moon that fell within the Roche limit (the distance below which moons will be pulled apart by tidal forces). This could have happened any time in Saturn's history.

2. Saturn's moons shepherd the particles that make up the rings, preventing them from drifting and maintaining the gaps between the rings. This shepherding may allow the rings to be much older than 100 million years. (However, the color of the rings suggests not much more than 100 million years' worth of accumulated dust.)

Further Reading: Matson, D. E. 1994. How good are those young-earth arguments? http://www.talkorigins.org/faqs/hovind/howgood-yea.html#proof9; Sobel, D. 1994. Secrets of the rings; Thompson, T. n.d.a. Answers in Genesis and Saturn's rings. http://home.austarnet.com.au/stear/aig_and_saturn's_rings.htm.

CE260: Three planets and several moons revolve backwards.

The hypothesis that the solar system formed from the collapse of a revolving nebula is contradicted by the fact that three planets and several moons revolve backwards. (W. Brown 1995, 19)

1. The "backwards" planets and moons are in no way contrary to the nebular hypothesis. Part of the hypothesis is that the nebula of gas and dust would accrete into planetessimals. Catastrophic collisions between these would be part of planet building. Such collisions and other natural processes can account for the retrograde planets and moons.

The only moons that orbit retrograde are small asteroid-sized distant satellites of giant planets such as Jupiter and Saturn, plus Triton (Neptune's large moon) and Charon (Pluto's satellite). The small retrograde satellites of Jupiter and Saturn were probably asteroids captured by the giant planets long after formation of the solar system. It is actually easier to be captured into a retrograde orbit. The Neptune system also contains one moon, Nereid, with a highly eccentric orbit. It appears that some sort of violent capture event may have taken place. The Pluto–Charon system is orbiting approximately "on its side," technically retrograde, with tidally locked rotation. As these are small bodies in the outer solar system, and binaries are likely to have been formed through collisions or gravitational capture, this does not violate the nebular hypothesis.

Uranus is rotating more or less perpendicular to the plane of the ecliptic. This may be the result of an off-center collision between two protoplanets during formation. Venus is rotating retrograde but extremely slowly, with its axis almost exactly perpendicular to the plane of its orbit. The rotation of this planet may well have started out prograde, but solar and planetary tides acting on its dense atmosphere have been shown to be a likely cause of the present state of affairs. It is probably not a coincidence that at every inferior conjunction, Venus turns the same side toward Earth, as Earth is the planet that contributes most to tidal forces on Venus.

2. Orbital motions account for 99.9 percent of the angular momentum of the solar system. A real evidential problem would be presented if some of the planets orbited the sun in the opposite direction to others, or in very different planes. However, all the planets orbit in the same direction, confirming the nebular hypothesis, and nearly in the same plane. A further confirmation comes from the composition of the giant planets, which

are similar to the sun's composition of hydrogen and helium. Giant planets could hold on to all of their light elements, but small planets like Earth and Mars could not.

by Mike Dworetsky

CE261: Comets would not have lasted in an old universe.

Comets lose material as they near the sun. If the solar system were very old, comets would long ago have evaporated. (Velikovsky 1955, 261–262)

1. The comets that entered the inner solar system a very long time ago indeed have evaporated. However, new comets enter the inner solar system from time to time. The Oort Cloud and Kuiper Belt hold many comets deep in space, beyond the orbit of Neptune, where they do not evaporate. Occasionally, gravitational perturbations from other comets bump one of them into a highly elliptical orbit, which causes it to near the sun.

Further Reading: Jewitt, D. n.d. Kuiper Belt. http://www.ifa.hawaii.edu/~jewitt/kb.html; Matson, D.E. 1994. How good are those young-earth arguments? http://www.talkorigins.org/faqs/hovind/howgood-yea.html#proof3.

CE280: Solar wind should have cleared the inner solar system of microparticles.

The sun's radiation pushes very small particles outward. Over millions of years, all particles smaller than a certain size (on the order of a micron in diameter) should have been blown out of the solar system. Yet these microparticles are abundant around the sun. (W. Brown 1995, 30)

1. Comets bring more particles into the inner solar system regularly.

CE281: The Poynting–Robertson effect would remove space dust in an old solar system.

The Poynting–Robertson effect causes orbiting particles (on the order of a centimeter in diameter) to slow and fall inward because solar radiation falls slightly more on their leading edge, like raindrops on a speeding car. If the solar system were old, the Poynting–Robertson effect would have caused all particles above a certain size to spiral into the sun, removing them from the solar system, but we still find interplanetary dust. (W. Brown 1995, 20, 30)

1. The particles are replenished by disintegrating comets and colliding asteroids. It takes hundreds of millions of years for the Poynting–Robertson effect to cause centimeter-sized particles to fall into the sun, so the replenishment need not be particularly rapid.

2. For smaller particles, there is a balance between the Poynting–Robertson effect and radiation pressure (see CE280), thereby preserving the dust in stable orbits. Gravitational effects of planets can also keep particles in stable orbits.

Further Reading: Matson, D.E. 1994. How good are those young-earth arguments? http://www.talkorigins.org/faqs/hovind/howgood-yea.html#proof7; Thompson, T. n.d.b. Is the Earth young?

http://www.tim-thompson.com/young-earth.html; Wong, M. 2001. Young-earth creationism: Pseudoscience. http://www.stardestroyer.net/Creationism/YoungEarth/Hartman-5.shtml.

SUN AND STARS

CE301: The lack of solar neutrinos indicates that the stellar model is wrong.

The number of neutrinos detected coming from the sun is only about a third of what is predicted by standard solar models. This indicates a lack of nuclear fusion in the sun, supporting gravitational collapse theory. (K. Davies 1996)

1. A new, more sensitive neutrino detector has found that the number of electron neutrinos coming from the sun is less than expected, but the total number of neutrinos coming from the sun matches predictions. Apparently, neutrinos are changing flavor en route as a result of quantum mechanical effects. (Ahmad et al. 2002).

Further Reading: Johansson, S. 1998. The Solar FAQ. http://www.talkorigins.org/faqs/faq-solar.html.

CE302: The sun has most of the mass but little angular momentum of the solar system.

The sun has 99 percent of the mass of the solar system, but less than 1 percent of the angular momentum. It is spinning too slowly to have formed naturally. (W. Brown 1995, 19)

1. Among solar-type stars, there is a strong correlation between age and rotation rate; the younger stars spin more rapidly (Baliunas et al. 1995). This implies some kind of braking mechanism that slows a star's rotation. A likely candidate is an interaction between the star's magnetic field and its solar wind (Parker 1965).

CE310: A shrinking sun indicates a young sun.

The sun is shrinking at such a rate that it would disappear completely in 100,000 years. This would make it impossibly large and hot in the distant past if the sun is millions of years old. (H.M. Morris 1985, 169)

1. This assumes that the rate of shrinkage is constant. That assumption is baseless. (In fact, it is the uniformitarian assumption that creationists themselves sometimes complain about; see CD200.) Other stars expand and contract cyclically. Our own sun might do the same on a small scale.

2. There is not even any good evidence of shrinkage. The claim is based on a single report from 1980. Other measurements, from 1980 and later, do not show any significant shrinkage. It is likely that the original report showing shrinkage contained systematic errors due to different measuring techniques over the decades.

Further Reading: Johansson, S. 1998. The solar FAQ. http://www.talkorigins.org/faqs/faq-solar.html; Matson, D. E. 1994. How good are those young-earth arguments? http://www.talkorigins.org/faqs/hovind/howgood-yea.html#proof1.

CE311: The faint young sun paradox contradicts an old earth.

According to standard models of the sun, the sun's luminosity has increased by 40 percent since the origin of the earth (the "faint young sun paradox"). This would mean that the early earth could not have supported life. Thus, an old earth is impossible; the earth must be young. (Faulkner 1998)

1. The change in luminosity is not as drastic as it sounds. Much of the change would have occurred before the origin of life; the luminosity increase since the origin of life is about 25 percent. And this translates to a 7 percent increase in temperature when the earth's heat outflow is taken into account.

2. The 7 percent change described above assumes no feedback system, but the earth's climate feedback systems are complex. In particular, the greenhouse effect and albedo could moderate the temperature further. On the early earth, it is likely that greenhouse gases such as carbon dioxide and methane were commoner than they are today.

3. Life has survived fairly large changes in climate over its history, from a near-global glaciation in the late Precambrian, to a global temperature warmer than today's in the Carboniferous.

Further Reading: Johansson, S. 1998. The solar FAQ. http://www.talkorigins.org/faqs/faq-solar.html.

CE351: Sirius was a red star 2,000 years ago and is a white dwarf now.

Astronomers from 2,000 years ago recorded that Sirius was a red star; today it is a white dwarf star. Conventional astronomy, which states that 100,000 years are required for a star to "evolve" from a red giant to a white dwarf, must be wrong. (Hovind 2003)

1. The ancient astronomers who described Sirius as red were looking at it when it was low on the horizon, so its reddening was due to the earth's atmosphere. The "red Sirius" refers to observations made at the heliacal risings and settings of the star in Greek and Roman society (Ceragioli 1996; Whittet 1999).

2. Not all ancient astronomers recorded that Sirius was red. Many ancient sources confirm that it was white or bluish white 2,000 years ago (van Gent 1984).

3. The bright star visible without a telescope, Sirius A, is not a white dwarf. Sirius A has a white dwarf companion star, Sirius B, which has nothing to do with what ancient astronomers saw. Hovind did not check the facts behind his claims.

4. Even if Sirius had changed color, it would not support creationism or a young earth in any way. It would simply mean that one observation was unexplained.

CE380: Galaxies should lose their spiral shape over millions of years.

Stars closer to the center of a spiral galaxy orbit the galaxy faster than stars farther away. Over many millions of years, the difference in orbital rates should wind the spiral tighter and tighter. We do not see any evidence for this in galaxies of different ages. (Corliss 1988)

1. Spiral arms are density waves, which, like sound in air, travel through the galaxy's disk, causing a piling-up of stars and gas at the crests of the waves. In some galaxies, the central bulge reflects the wave, giving rise to a giant standing spiral wave with a uniform rotation rate and a lifetime of about one or two billion years.

The causes of the density waves are still not known, but there are many possibilities. Tidal effects from a neighboring galaxy probably cause some of them.

The spiral pattern is energetically favorable. Spiral configurations develop spontaneously in computer simulations based on gravitational dynamics (Carlberg et al. 1999).

Further Reading: Carlberg, R., et al. 1999. Ask the experts: Astronomy: What process creates and maintains the beautiful spiral arms around spiral galaxies? http://www.sciam.com/askexpert_question.cfm?articleID=0008A68A-8C7F-1C72-9EB7809EC588F2D7.

COSMOLOGY

CE401: There are too few supernova remnants for an old universe.

If the universe is old, many supernova remnants (SNRs) should have reached the third, oldest stage. We observe no Stage 3 SNRs and few Stage 2 SNRs. Both observations are consistent with a young universe, not an old one. (K. Davies 1994)

1. Many more SNRs have been found, including many Stage 3 remnants older than 20,000 years. And the census is not over yet. If the universe is old, many SNRs should have reached the third, oldest stage, and that is what we see. The evidence contradicts a young universe, not an old one.

2. Davies's estimate of what proportion of SNRs should be visible to us is grossly oversimplified. It is impossible to say with certainty what proportion should be visible. Furthermore, he ignores data, including observations of possible old remnants, that would weaken his case.

SNRs are relatively hard to see. They would not be visible for one million years, the figure Davies used in his calculations. A million years is the theoretical lifetime of a remnant; it will be visible for a much shorter time because of background noise and obscuring dust and interstellar matter. Fewer than 1 percent of SNRs last more than 100,000 years. It may be that as few as 15–20 percent of supernova events are visible at all through the interstellar matter.

3. Supernovas are evidence for an old universe in other ways:

- Supernovas are evidence that stars have reached the end of their lifetime, which for many stars is billions of years.

- The formation of new stars indicates that many are second generation; the universe must be old enough for some stars to go through their entire lifetime and for the dust from their supernovas to collect into new stars.

- It takes time for the light from the supernovas to reach us. All supernovas and SNRs are more than 7,000 light-years from us. SN 1987A was 167,000 ± 4,000 light years away.

Further Reading: Moore, D. 2001. Supernovae, supernova remnants and young earth creationism FAQ. http://www.talkorigins.org/faqs/supernova.

by Adam Crowl

CE410: Physical constants are only assumed constant.
Physicists only assume that physical constants have been constant over billions of years. In particular, this untestable assumption underlies all radiometric dating techniques. (W. Brown 1995, 24)

1. The constancy of constants is a conclusion, not an assumption. It is tested whenever possible. For example:

- The fine structure constant affects neutron capture rates, which can be measured from products of the Oklo reactor, where a natural nuclear reaction occurred 1,800 million years ago. These measurements show that the fine structure constant has remained constant (within one part in 10^{17} per year) for almost two billion years (Fujii et al. 2000; Shlyakhter 1976).

- Despite some weak evidence that the fine structure constant may have varied slightly more than six billion years ago (Musser 1998; Webb et al. 1999), analysis of the spectra of quasars shows that it has changed less than 0.6 parts per million over the last ten billion years (Chand et al. 2004).

- Experiments with atomic clocks show that any change is less than a rate of about 10^{-15} per year (Fischer et al. 2004).

Further Reading: Ball, P. 2003. Lab tests tenets' limits. http://www.nature.com/nsu/030428/030428-20.html; SpaceDaily. 2004. Quasar studies keep fundamental physical constant—constant. http://www.spacedaily.com/news/cosmology-04i.html.

CE411: The speed of light has changed.
The speed of light was faster in the past, so objects millions of light-years away are much younger than millions of years. (T. G. Norman and Setterfield 1987)

1. The possibility that the speed of light has not been constant has received much attention from physicists, but they have found no evidence for any change. Many different measurements of the speed of light have been made in the last 180 or so years. The older measurements were not as accurate as the latest ones. Setterfield chose 120 data

points from 193 measurements available (see Dolphin n.d. for the data), and the line of best fit for these points shows the speed of light decreasing. If you use the entire data set, though, the line of best fit shows the speed increasing. However, a constant speed of light is well within the experimental error of the data.

2. If Setterfield's formulation of the changes in physical parameters was true, then there should have been 417 days per year around 1 c.e and the earth would have melted during the creation week as a result of the extremely rapid radioactive decay (Morton et al. 1983).

3. As an aside, some creationists assert that fundamental laws have not changed (H. M. Morris 1985, 18).

Further Reading: Aardsma, G. E. 1988a, Has the speed of light decayed?; Aardsma, G. E. 1988b, Has the speed of light decayed recently?

CE412: Gravitational time dilation made distant clocks run faster.

The earth is near the center of the universe, at the bottom of a deep gravitational well. Relativistic effects result in billions of years passing in the rest of the universe while only thousands pass near the earth. This explains how multibillion-year-old stars and galaxies can exist in a universe only a few thousand years old. (D. R. Humphreys 1994, 2002)

1. Gravitational time dilation, if it existed, should be observable. On the contrary, we observe (from the periods of Cepheid variable stars, from orbital rates of binary stars, from supernova extinction rates, from light frequencies, etc.) that such time dilation does not exist. There is some time dilation corresponding with Hubble's law (i.e., further objects have greater red shifts), but this is due to the well-understood expansion of the universe, and it is not nearly extreme enough to fit more than ten billion years into less than 10,000.

2. Humphreys tried to use clocks in the earth's frame of reference. But the cosmos is much older than the earth. Judging from the heavy elements in the sun and the rest of the solar system, our sun is a second-generation star at least. Billions of years must have passed for the first stars to have formed, shone, and gone nova, for the gasses from those novas to have gathered into new star systems, and for the earth to form and cool in one such system. The billions of years before the earth are not accounted for in Humphreys's model.

3. Humphreys's theory assumes that the earth is in a huge gravity well. The evidence contradicts this assumption. If the earth were in such a gravity well, light from distant galaxies should be blue-shifted. Instead, it is red-shifted.

4. See Conner and Page (1998) and Conner and Ross (1999) for several other technical objections.

5. There is a great deal of other independent evidence that the earth is very old (see CH210).

6. If there were any substance to Humphreys's proposal, at least some competent cosmologists would build on it and share in the Nobel Prize. Instead, they dismiss it as worthless.

Further Reading: Conner, S.R. and H. Ross. 1999. The unraveling of starlight and time. http://www.reasons.org/resources/apologetics/unravelling.shtml?main.

CE420: The big bang theory is wrong.

The theory of a big bang has been shaken with unresolvable inconsistencies, such as an unexpectedly uneven distribution of matter in the universe and a need for dark matter. Several astronomers think it is no longer a valid theory. (Gitt 1998)

1. The big bang is supported by a great deal of evidence:

- Einstein's general theory of relativity implies that the universe cannot be static; it must be either expanding or contracting.

- The more distant a galaxy is, the faster it is receding from us (the Hubble law). This indicates that the universe is expanding. An expanding universe implies that the universe was small and compact in the distant past.

- The big bang model predicts that cosmic microwave background (CMB) radiation should appear in all directions, with a blackbody spectrum and temperature about 3 degrees K. We observe an exact blackbody spectrum with a temperature of 2.73 degrees K.

- The CMB is even to about one part in 100,000. There should be a slight unevenness to account for the uneven distribution of matter in the universe today. Such unevenness is observed, and at a predicted amount.

- The big bang predicts the observed abundances of primordial hydrogen, deuterium, helium, and lithium. No other models have been able to do so.

- The big bang predicts that the universe changes through time. Because the speed of light is finite, looking at large distances allows us to look into the past. We see, among other changes, that quasars were more common and stars were bluer when the universe was younger.

Note that most of these points are not simply observations that fit with the theory; the big bang theory predicted them.

2. Inconsistencies are not necessarily unresolvable. The clumpiness of the universe, for example, was resolved by finding unevenness in the CMB. Dark matter has been observed in the effects it has on star and galaxy motions; we simply do not know what it is yet.

There are still unresolved observations. For example, we do not understand why the expansion of the universe seems to be speeding up. However, the big bang has enough supporting evidence behind it that it is likely that new discoveries will add to it, not overthrow it. For example, inflationary universe theory proposes that the size of the universe increased exponentially when the universe was a fraction of a second old (Guth 1997). It was proposed to explain why the big bang did not create large numbers of magnetic monopoles. It also accounts for the observed flatness of space, and it predicted quantitatively the pattern of unevenness of the CMB. Inflationary theory is a significant addition to big bang theory, but it is an extension of big bang theory, not a replacement.

Further Reading: Ferris, T. 1997. *The Whole Shebang;* Guth, A.H. 1997. *The Inflationary Universe.*

CE421: The cosmos has an axis, contrary to big bang models.

Anisotropies in the cosmic background radiation measured by the Wilkinson Microwave Anisotropy Probe show an axis. The big bang proposes no special orientations, so an axis discredits the big bang theory, but it is consistent with creationist cosmology. (D.R. Humphreys 2003)

1. Humphries referred to the work of Tegmark et al. (2003). Tegmark et al.'s map shows an axis of symmetry for the quadrapole and octopole maps, but the hexadecapole map shows no such axis of symmetry, which could indicate that the axis is an artifact of a systematic bias in the data analysis.

2. A cosmic axis is compatible with the big bang. If Tegmark et al.'s results are correct, they imply that cosmology is anisotropic (not the same in all directions) on very large length scales. There has been, to date, little evidence gathered about the universe on such scales, but anisotropic cosmologies have been seriously considered. Gödel's rotating universe (Gödel 1949) is one example. Another is a universe with one spatial dimension compacted relative to the other two.

CE440: Where did space, time, energy, and laws of physics come from?

Cosmologists cannot explain where space, time, energy, and the laws of physics came from. (W. Brown 1995, 20)

1. Some questions are harder to answer than others. But although we do not have a full understanding of the origin of the universe, we are not completely in the dark. We know, for example, that space comes from the expansion of the universe. The total energy of the universe may be zero (see CF101). Cosmologists have hypotheses for the other questions that are consistent with observations (Hawking 2001). For example, it is possible that there is more than one dimension of time, the other dimension being unbounded, so there is no overall origin of time. Another possibility is that the universe is in an eternal cycle without beginning or end. Each big bang might end in a big crunch to start a new cycle (Steinhardt and Turok 2002), or at long intervals, our universe collides with a mirror universe, creating the universe anew (Seife 2002).

One should keep in mind that our experiences in everyday life are poor preparation for the extreme and bizarre conditions one encounters in cosmology. The stuff cosmologists deal with is very hard to understand. To reject it because of that, though, would be to retreat into the argument from incredulity (see CA100).

2. Creationists cannot explain origins at all. Saying "God did it" is not an explanation, because it is not tied to any objective evidence. It does not rule out any possibility or even any impossibility. It does not address questions of "how" and "why," and it raises questions such as "which God?" and "how did God originate?" In the explaining game, cosmologists are far out in front.

Further Reading: Hawking, S. 1988. *A Brief History of Time*; Hawking, S. 2001. *The Universe in a Nutshell*; Musser, G. 2002. Been there, done that. http://www.sciam.com/article.cfm?articleID=

000D59C8-5512-1CC6-B4A8809EC588EEDF; Veneziano, G. 2004. The myth of the beginning of time.

CE441: Explosions such as the big bang do not produce order or information.

The universe was supposedly formed in the big bang, but explosions do not produce order or information. (Big-Bang-Theory 2002)

1. The total entropy of the universe at the start of the big bang was minimal, perhaps almost zero. Because it was so compact, it had considerably more order than the universe we are in now. The complexity we observe around us today can be produced from the ultimate order of the hot but cooling gas of the big bang.

2. The big bang was not an explosion. It was an expansion. Besides the fact that it got bigger over time, the big bang has almost nothing in common with an explosion.

3. Explosions do produce some order amidst their other effects:

- Large surface explosions, such as nuclear bombs, produce the familiar mushroom clouds. There are not very highly ordered, but they are not purely random either.

- Supernovae produce heavy elements, and the shock waves from them compress interstellar gases, which begins the formation of new stars.

- Powerful explosions can compress carbon into diamond crystals, the most ordered arrangement.

- Explosions of atomized gasoline produce compressed gas, which is harnessed in internal combustion engines to power automobiles and other equipment.

PHYSICS AND MATHEMATICS

SECOND LAW OF THERMODYNAMICS AND INFORMATION THEORY

CF001: The second law of thermodynamics prohibits evolution.
The second law of thermodynamics says that everything tends toward disorder, making evolutionary development impossible. (H. M. Morris 1985, 38–46)

1. The second law of thermodynamics says no such thing. It says that heat will not spontaneously flow from a colder body to a warmer one or, equivalently, that total entropy (a measure of useful energy) in a closed system will not decrease. This does not prevent increasing order because

- the earth is not a closed system; sunlight (with low entropy) shines on it and heat (with higher entropy) radiates off. This flow of energy, and the change in entropy that accompanies it, can and will power local decreases in entropy on earth.

- entropy is not the same as disorder. Sometimes the two correspond, but sometimes order increases as entropy increases (Aranda-Espinoza et al. 1999; Kestenbaum 1998). Entropy can even be used to produce order, such as in the sorting of molecules by size (Han and Craighead 2000).

- even in a closed system, pockets of lower entropy can form if they are offset by increased entropy elsewhere in the system.

In short, order from disorder happens on earth all the time.

2. The only processes necessary for evolution to occur are reproduction, heritable variation, and selection. All of these are seen to happen all the time, so, obviously, no phys-

ical laws are preventing them. In fact, connections between evolution and entropy have been studied in depth, and never to the detriment of evolution (Demetrius 2000).

Several scientists have proposed that evolution and the origin of life is driven by entropy (McShea 1998). Some see the information content of organisms subject to diversification according to the second law (D. R. Brooks and Wiley 1988), so organisms diversify to fill empty niches much as a gas expands to fill an empty container. Others propose that highly ordered complex systems emerge and evolve to dissipate energy (and increase overall entropy) more efficiently (E. D. Schneider and Kay 1994).

3. Creationists themselves admit that increasing order is possible. They introduce fictional exceptions to the law to account for it (see CF001.3).

4. Creationists themselves make claims that directly contradict their claims about the second law of thermodynamics, such as hydrological sorting of fossils during the Flood (see CH561.2).

See for Yourself:
You can see order come and go in nature in many different ways. A few examples are snowflakes and other frost crystals, cloud formations, dust devils, ripples in sand dunes, and eddies and whirlpools in streams. See how many other examples you can find.

Further Reading: Atkins, P. W. 1984. *The Second Law*; Kauffman, S. A. 1993. *The Origins of Order*; Lambert, F. L. 2003. The second law of thermodynamics. http://www.secondlaw.com.

CF001.1: Systems left to themselves invariably tend toward disorder.
Systems or processes left to themselves invariably tend to move from order to disorder. (T. Wallace 2002)

1. This is an attempt to claim that the second law of thermodynamics implies an inevitable increase in entropy even in open systems (see CF001) by quibbling with the verbiage "left to themselves." The simple fact is that, unless "left to themselves" means "not acted upon by any outside influence," disorder of systems can decrease. And since outside influence is more often the rule in biological systems, order can and does increase in them.

2. That the claim is false is not theory. Exceptions happens all the time. For example, plants around my house are left to themselves every spring, and every spring they produce order locally by turning carbon from the air into plant tissue. Drying mud, left to itself, produces orderly cracks. Ice crystals, left to themselves, produce arrangements far more orderly than they would if I interfered. How can a trend to disorder be invariable when exceptions are ubiquitous? And why do creationists argue at such length for claims that they themselves can plainly see are false?

3. Disorder and entropy are not the same. The second law of thermodynamics deals with entropy, not disorder (although disorder defined to apply to microscopic states can be relevant to thermodynamics). There are no laws about disorder as people normally use the word.

CF001.2: The second law of thermodynamics, and the trend to disorder, is universal.

The entire universe is a closed system, so the second law of thermodynamics dictates that within it, things are tending to break down. The second law applies universally. (T. Wallace 2002)

1. The second law of thermodynamics applies universally (see CF001), but, as everyone can see, that does not mean that everything everywhere is always breaking down. The second law allows local decreases in entropy offset by increases elsewhere. The second law does not say that order from disorder is impossible; in fact, as anyone can see, order from disorder happens all the time.

2. The maximum entropy of a closed system of fixed volume is constant, but because the universe is expanding, its maximum entropy is ever increasing, giving ever more room for order to form (Stenger 1995, 228).

3. Disorder and entropy are not the same. The second law of thermodynamics deals with entropy. There are no laws about things tending to "break down."

CF001.3: Instructions are necessary to produce order.

Increasing order is possible, locally and temporarily, only if there is a program to direct growth and a power converter. (H. M. Morris 1985, 43–45)

1. This claim is pure fantasy. The second law of thermodynamics says absolutely nothing about programs to direct growth, and the only "power converter" it deals with is change in entropy. Growth and order can be seen arising without a program in many places. Clouds form complex orderly patterns. Streams sort the size of the stones in their bed along their length. Cooling basalt forms a hexagonal pattern of cracks. All of these show an increase in organization and a local decrease in entropy, and none involve any program.

2. Increasing order is not a violation of the second law of thermodynamics, even temporarily (see CF001). A violation would be an increase in entropy without a greater decrease in entropy to go with it. Neither growth nor evolution violate the second law of thermodynamics because both take advantage of local differences in entropy to get work done.

3. Evolution has a program; it is called the environment. Natural selection serves to communicate information from the environment to the populations of organisms (Adami et al. 2000).

4. An increase in organized complexity is not the same as a decrease in entropy. The second law applies only to entropy; it says nothing at all about organized complexity as such.

Further Reading: Kauffman, S. A. 1993. *The Origins of Order.*

CF001.4: The second law of thermodynamics is about organized complexity, not entropy.

The second law of thermodynamics allows higher order (lower entropy) to appear locally, but it still disallows organized complexity. For example, it allows highly ordered arrangements such as "aaaaaaaaaaaaaaaaa . . ." to originate, but not complex ordered arrangements such as the words on this page. (Yahya 1999)

1. The second law of thermodynamics is about spontaneous heat flow or, more generally, about the impossibility to perform useful work indefinitely. The twists put on it by creationists, including "organized complexity," are entirely fictional.

Arrangements such as the words on this page are not prohibited by any laws of physics. If they were, they could not exist for you to be reading them now.

2. Complex ordered arrangements can and do arise naturally (see CF002).

CF001.5: Evolution needs an energy conversion mechanism to utilize energy.

Energy inflow into a system is not enough to make that energy useful. There must also be an energy conversion mechanism. Without that system, evolution cannot work. (Yahya 2003a)

1. Any atom can be an energy conversion mechanism. Atoms routinely convert between light energy, thermal energy, and chemical potential energy. The energy conversion mechanism is ubiquitous.

2. A lack of an energy conversion system would not only invalidate evolution; it would invalidate life itself. Evolution requires only reproduction, natural selection, and heritable variation, all of which are observed in life. The conversion of energy is a quality of life, so the conversion system exists for evolution to work with.

CF002: Complexity does not come from simplicity.

Complexity does not arise from simplicity.

1. Complexity arises from simplicity all the time. The Mandelbrot set is an example (D. Dewey 1996). Real-life examples include the following: A pan of water with heat applied uniformly to its bottom will develop convection currents that are more complex than the still water; complex hurricanes arise from similar principles; complex planetary ring systems arise from simple laws of gravitation; complex ant nests arise from simple behaviors; and complex organisms arise from simple seeds and embryos.

2. Complexity should be expected from evolution. In computer simulations, complex organisms were more robust than simple ones (Lenski et al. 1999), and natural selection forced complexity to increase (Adami et al. 2000). Theoretically, complexity is expected because complexity-generating processes dissipate the entropy from solar energy influxes, in accordance with the second law of thermodynamics (Wicken 1979). Ilya Prigogine

won the Nobel Prize "for his contributions to non-equilibrium thermodynamics, particularly the theory of dissipative structures" (Nobel Foundation 1977). According to Prigogine, "it is shown that non-equilibrium may become a source of order and that irreversible processes may lead to a new type of dynamic states of matter called 'dissipative structures'" (Prigogine 1977, 22).

CF002.1: Tornadoes in junkyards do not build things.

Order does not spontaneously form from disorder. A tornado passing through a junkyard would never assemble a 747. (Hoyle 1983, 18–19)

1. This claim is irrelevant to the theory of evolution itself, since evolution does not occur via assembly from individual parts, but rather via selective gradual modifications to existing structures. Order can and does result from such evolutionary processes.

2. Hoyle applied his analogy to abiogenesis, where it is more applicable. However, the general principle behind it is wrong. Order arises spontaneously from disorder all the time. The tornado itself is an example of order arising spontaneously. Something as complicated as people would not arise spontaneously from raw chemicals, but there is no reason to believe that something as simple as a self-replicating molecule could not form thus. From there, evolution can produce more and more complexity.

CF003: How could information, such as in DNA, assemble itself?

How could information, such as in DNA, assemble itself? (C. W. Brown n.d.)

1. This question is based on some major misconceptions (addressed below). Its overriding logical error, however, is that it is an argument from ignorance (see CA100). One's inability to find an answer to a question does not imply that the question has no answer.

2. Information is not meaning and does not, per se, imply any special structure or function. Any arrangement implies information; the information is how the arrangement is described. If a new arrangement occurs, whether spontaneously or from the outside, new information is assembled in the process. Even if the arrangement consists of shattering a glass into tiny pieces, that means assembling new information.

3. Nothing needs to assemble itself. Evolution and abiogenesis do not exclude outside influences; on the contrary, such outside influences are essential. In abiogenesis, it is observed that complex organic molecules easily form spontaneously due to little more than basic chemistry and energy from the sun or from the earth's interior (see CB010.2). In evolution, information from the environment is communicated to genomes indirectly via natural selection against varieties that do not do well in that environment.

Further Reading: Musgrave, I. 1998b. Re: Abiogenesis. http://www.talkorigins.org/origins/postmonth/apr98.html; Musgrave, I., et al. 2003b. Information theory and creationism. http://home.mira.net/~reynella/debate/informat.htm.

CF010: Cybernetic simulations show that Darwinian processes do not produce order.

"Darwinians and Neo-Darwinians have long maintained that randomness, plus long time spans, plus natural selection would (together) do the trick in making specific codes and molecules. However, recent progress in cybernetics has shown by simulation experiments that order sequences, specificity and coding cannot be extracted from randomness on the basis of the Darwinian postulates." (Wilder-Smith 1970, 116)

1. The claim is unequivocally false. It was made in the early days of computing, apparently on the basis of one failed simulation. Computer simulations since have shown just the opposite of what Wilder-Smith claimed. In fact, genetic algorithms, which use evolutionary principles of mutation, recombination, and natural selection, are used routinely in industry to solve complex problems (Heitkötter and Beasley 2000). Artificial life simulating evolution on a computer evolves complex features (Lenski et al. 2003).

Further Reading: Elsberry, W.R. 1997. Enterprising science needs naturalism. http://www .utexas.edu/cola/depts/philosophy/faculty/koons/ntse/papers/Elsberry.html; Holland, J.H. 1975. *Adaptation in Natural and Artificial Systems*; National Science Foundation. 2003. Artificial life experiments show how complex functions can evolve. http://www.sciencedaily.com/releases/2003/ 05/030508075843.htm.

CF011: Evolutionary algorithms smuggle in design in the fitness function.

Genetic algorithms are claimed to demonstrate that evolutionary processes can create design, but in such algorithms, the design is smuggled in in the form of the fitness function. Evolutionary algorithms do not create specified complexity. (Dembski 1999c)

1. The fitness function of genetic algorithms need not include any new information. A fitness function can be expressed as whether the algorithm performs better or worse in a particular environment. The only information is provided by the environment, which is usually modeled on the real world. The claim makes sense only if design is defined as what is in nature already.

One may argue that nature and design are inseparable (and Dembski seems to make just such an argument; Dembski 2002b, xiv), but this invalidates the design argument. Design only has meaning if contrasted with nondesign, and defining design as all of nature makes nondesign nonexistent.

2. Genetic algorithms often come up with novel solutions that sometimes even surpass direct human designs (Koza et al. 2003) and which do not rely on human expertise (Chellapilla and Fogel 2001). Humans may have told the algorithms what to do, but it is the how that defines the design.

3. Genetic algorithms are not perfect evolutionary simulations in that they have a global fitness evaluation—that is, they calculate fitness by comparison with a predefined optimum. They demonstrate the power of random variation, recombination, and selection to produce novel solutions to problems, but they are not a full simulation of evolution

(and are not intended to be). In simulations of biological evolution, fitness is evaluated only locally; survival and reproduction is based only on information about local conditions, not on ultimate goals. However, the simulations demonstrate that distant fitness peaks will be reached if there are conditions of intermediate fitness (Lenski et al. 2003). Evolutionary processes do not "search." They respond to local fitness topography only. The fact that evolution (occasionally) reaches fitness peaks is a by-product of evolving on correlated fitness landscapes using purely local fitness evaluation, not an intended outcome.

Further Reading: RBH. 2003. Untitled. http://www.iscid.org/boards/ubb-get_topic-f-6-t-000384 .html#000013.

with Richard B. Hoppe

CF011.1: The outcome of Dawkins's WEASEL program was prespecified.

Dawkins (1996) demonstrated a program that starts with a random string of letters and, via random copying errors, evolves it into the phrase "Methinks it is like a weasel" in just a few generations, demonstrating the power of natural selection unaided by intelligence. But intelligence is involved in predetermining the target sentence. (Gitt and Wieland 1998)

1. Dawkins's simulation was plainly stated in his book to demonstrate selection, not evolution. It was intended to show the difference between cumulative selection and single-step selection. Attempts to apply Dawkins's simulation to evolution as a whole are a misreading of his book.

2. Other evolution simulations do demonstrate all the salient features of evolution (Lenski et al. 2003). They do include a fitness function, but simulating fitness is part of simulating evolution (see CF011).

CF011.2: NFL theorems prove that evolutionary algorithms do not beat blind search.

The No Free Lunch (NFL) theorems (Wolpert and Macready 1997) prove that evolutionary algorithms, when averaged across fitness functions, cannot outperform blind search. This means that an evolutionary algorithm can find a specified target only if complex specified information already resides in the fitness function. Evolutionary algorithms cannot account for the complex specified information we see in life; that information has to come from design. (Dembski 2002b, 199–212)

1. The NFL theorems do not apply to biological evolution. The NFL theorems apply only when the fitness function is independent of the algorithm, but in evolution, evolving populations affect the environment and each other and therefore the fitness functions.

It should also be noted that the NFL theorems do not refer to finding a target. They can apply to problems such as finding which of several algorithms performs best; such application differs from Dembski's concept of a target.

Dembski himself later wrote that the NFL theorems are not important to his point, which is about displacement and conservation of information (Dembski 2002a; see CF011).

2. The NFL theorems consider the average of all fitness functions. Finding an above-average fitness function is not complicated and is often trivial. If you want a solution that performs well according to some metric, then a fitness function that measures that metric will usually work better than blind search. For example, if you are interested in survival and reproduction in a certain environment, then survival and reproduction in that environment is a good choice for a fitness function.

3. The ultimate test of a concept is whether it works. Evolutionary algorithms work. They find solutions to many problems that are intractable with other methods. If mathematics contradicts reliable observation, the math is misapplied, irrelevant, or wrong.

4. Complex specified information does not signify design (see CI111.2).

5. No design theorist has ever shown that complex specified information exists in life.

6. That evolution uses information from the environment (via the fitness function) is nothing new. The process is called adaptation. Darwin wrote something about the general subject (Darwin 1859). It does not imply design.

Further Reading: Perakh, M. 2003a. The No Free Lunch theorems and their application to evolutionary algorithm. http://www.talkreason.org/articles/orr.cfm; Wein, R. 2002. Not a free lunch but a box of chocolates. http://www.talkorigins.org/design/faqs/nfl; Wolpert, D. 2002. William Dembski's treatment of the No Free Lunch theorems is written in jello. http://www.talkreason.org/articles/jello.cfm.

FIRST LAW OF THERMODYNAMICS

CF101: The universe's energy cannot come from nothing.
The first law of thermodynamics says that matter/energy cannot come from nothing. Therefore, the universe itself could not have formed naturally. (W. Brown 1995, 21; see also CE440)

1. Formation of the universe from nothing need not violate conservation of energy. The gravitational potential energy of a gravitational field is a negative energy. When all the gravitational potential energy is added to all the other energy in the universe, it might sum to zero (Guth 1997, 9–12, 271–276; Tryon 1973).

Further Reading: Guth, A. H. 1997. *The Inflationary Universe.*

RADIOMETRIC DECAY

CF201: Polonium haloes indicate a young earth.
Some micas in granite have tiny haloes caused by the decay of radioactive elements. From their diameters, we know the energy of the alpha particles that caused the haloes, which tells us what

element decayed. Some of these haloes formed from isotopes of polonium, all of which have short half-lives (138 days for the longest-lived isotope). According to conventional geology, the rocks in which the polonium radio-haloes occur took millions of years to form. All of the original polonium should have decayed in that time. Thus, polonium radio-haloes indicate a sudden creation of polonium in rock. (Gentry 1986; Snelling 2000)

1. Polonium forms from the alpha decay of radon, which is one of the decay products of uranium. Since radon is a gas, it can migrate through small cracks in the minerals. The fact that Polonium haloes are found only associated with uranium (the parent mineral for producing radon) supports this conclusion, as does the fact that such haloes are commonly found along cracks. (Brawley 1992; J. R. Wakefield 1998)

2. The biotite in which Gentry (1986) obtained some of his samples (Fission Mine and Silver Crater locations) was not from granite, but from a calcite dike. The biotite formed metamorphically as minerals in the walls of the dike migrated into the calcite. Biotite from the Faraday Mine came from a granite pegmatite that intruded a paragneiss that formed from highly metamorphosed sediments. Thus, all of the locations Gentry examined show evidence of an extensive history predating the formation of the micas; they show an appearance of age older than the three minutes his polonium halo theory allows. It is possible God created this appearance of age, but that reduces Gentry's argument to the omphalos argument (see CH220), for which evidence is irrelevant. (J. R. Wakefield 1998)

3. Stromatolites are found in rocks intruded by (and therefore older than) the dikes from which Gentry's samples came, showing that living things existed before the rocks that Gentry claimed were primordial (J. R. Wakefield 1998).

Further Reading: Brawley, J. 1992. Evolution's tiny violences: The Po-halo mystery. http://www .talkorigins.org/faqs/po-halos.html; Collins, L. G. 1997. Polonium halos and myrmekite in pegmatite and granite. http://www.csun.edu/~vcgeo005/revised8.htm; Wakefield, J. R. 1998. The geology of Gentry's "tiny mystery." http://www.csun.edu/~vcgeo005/gentry/tiny.htm.

CF210: Radiometric dating falsely assumes that rates are constant.

Radiometric dating assumes that radioisotope decay rates are constant, but this assumption is not supported. All processes in nature vary according to different factors, and we should not expect radioactivity to be different. (H. M. Morris 1985, 139)

1. The constancy of radioactive decay is not an assumption, but is supported by evidence:

- The radioactive decay rates of nuclides used in radiometric dating have not been observed to vary since their rates were directly measurable, at least within limits of accuracy. This is despite experiments that attempt to change decay rates (Emery 1972). Extreme pressure can cause electron-capture decay rates to increase slightly (less than 0.2 percent), but the change is small enough that it has no detectable effect on dates.

- Supernovae are known to produce a large quantity of radioactive isotopes (Nomoto et al. 1997a, 1997b; Thielemann et al. 1998). These isotopes produce gamma rays with frequencies and fading rates that are predictable according to present decay rates. These predictions hold for supernova SN1987A, which is 169,000 light-years away (Knödlseder 2000). Therefore, radioactive decay rates were not significantly different 169,000 years ago. Present decay rates are likewise consistent with observations of the gamma rays and fading rates of supernova SN1991T, which is sixty million light-years away (Prantzos 1999), and with fading rate observations of supernovae billions of light-years away (Perlmutter et al. 1998).

- The Oklo reactor was the site of a natural nuclear reaction 1,800 million years ago. The fine structure constant affects neutron capture rates, which can be measured from the reactor's products. These measurements show no detectable change in the fine structure constant and neutron capture for almost two billion years (Fujii et al. 2000; Shlyakhter 1976).

2. Radioactive decay at a rate fast enough to permit a young earth would have produced enough heat to melt the earth (Meert 2002).

3. Different radioisotopes decay in different ways. It is unlikely that a variable rate would affect all the different mechanisms in the same way and to the same extent. Yet different radiometric dating techniques give consistent dates. Furthermore, radiometric dating techniques are consistent with other dating techniques, such as dendrochronology, ice core dating, and historical records (e.g., Renne et al. 1997).

4. The half-lives of radioisotopes can be predicted from first principles through quantum mechanics. Any variation would have to come from changes to fundamental constants. According to the calculations that accurately predict half-lives, any change in fundamental constants would affect decay rates of different elements disproportionally, even when the elements decay by the same mechanism (Greenlees 2000; Krane 1987).

Further Reading: Johnson, B. 1993. How to change nuclear decay rates. http://math.ucr.edu/ home/baez/physics/ParticleAndNuclear/decay_rates.html; Matson, D. E. 1994. How good are those young-earth arguments? http://www.talkorigins.org/faqs/hovind/howgood-c14.html#R2.

CF220: Short-lived isotopes ^{230}Th and ^{236}U exist on the moon.
Thorium-230 and Uranium-236 exist on the moon. But these isotopes are so short-lived that we would not expect any to still be around if the moon were billions of years old. (Wysong 1976, 177–178)

1. Both nucleotides are formed by ongoing processes. ^{230}Th is an intermediate decay product of ^{238}U, which has a half-life of 4.468 billion years. ^{236}U is produced in trace amounts by the capture of slow neutrons.

Further Reading: Matson, D. E. 1994. How good are those young-earth arguments? http://www .talkorigins.org/faqs/hovind/howgood-yea.html#proof6.

MISCELLANEOUS ANTIEVOLUTION

HISTORY

CG001: Darwin recanted on his deathbed.
Darwin renounced evolution on his deathbed.

1. The story of Darwin's recanting is not true. Shortly after Darwin's death, Lady Hope told a gathering that she had visited Darwin on his deathbed and that he had expressed regret over evolution and had accepted Christ. However, Darwin's daughter Henrietta, who was with him during his last days, said Lady Hope never visited during any of Darwin's illnesses, that Darwin probably never saw her at any time, and that he never recanted any of his scientific views (R. W. Clark 1984, 199; Yates 1994).

2. The story would be irrelevant even if true. The theory of evolution rests upon reams of evidence from many different sources, not upon the authority of any person or persons.

Further Reading: Clark, R. W. 1984. *The Survival of Charles Darwin*; Greig, R. 1996. Did Darwin recant? http://www.answersingenesis.org/docs/1315.asp; Yates, S. 1994. The Lady Hope story. http://www.talkorigins.org/faqs/hope.html.

CG010: The oldest living thing is younger than 4,900 years.
The oldest living thing (a bristlecone pine) is younger than 4,900 years, supporting a recent date for a worldwide cataclysm. (H. M. Morris 1985, 193)

1. The age of the oldest living thing does not indicate dates of events happening before it. It merely shows that no global cataclysm happened less than 4,900 years ago.

2. Tree rings give an unbroken record back more than 11,000 years (Becker and Kromer 1993; Becker et al. 1991; Stuiver et al. 1986). A worldwide cataclysm during that time would have broken the tree ring record.

3. The King Clone creosote bush in the Mojave Desert is 11,700 years old.

Further Reading: Matson, D. E. 1994. How good are those young-earth arguments? http://www.talkorigins.org/faqs/hovind/howgood-yea2.html#proof27.

LINGUISTICS

CG101: Chinese glyph for ark is literally "8 mouths."
The Chinese glyph for ship is made up of pictographs for "vessel," "eight," and "mouth," indicating the eight passengers on Noah's ark. (Kang and Nelson 1979; van Arnhem 2002; see Figure CG1)

1. The Chinese character for boat (chuan 2) consists of the boat radical on the left and a phonetic element on the right. The phonetic element has two parts. The upper part is a primitive ideograph for "divide," though it looks the same as the character for "eight." The lower part is the pictograph for "mouth." However, these two elements have only phonetic significance (M. Wright 1996, n.d.).

2. The "vessel" on the left side of the glyph is a pictograph of a dugout canoe, nothing like an ark.

3. According to the Bible, Noah's ark carried very many more than eight mouths.

4. No flood myths from China include an ark with eight passengers.

Further Reading: Wright, M. n.d. Do Chinese characters tell us something about Genesis? http://www.coastalfog.net/languages/chinchar/chinchar.html.

CG110: The first known languages are highly complex.
The first known human languages were already very complex. Languages do not show the evolutionary progression we would expect if humans evolved gradually. (Skjaerlund n.d.)

1. The first known languages were written languages (else they would not be known). Since most cultures in the world have had no written language, and most people have been illiterate even where written language existed, written language is a poor metric to use to measure language in general. Language had been developing for an unknown period of time before written language evolved.

2. The earliest known writing is simpler than written languages today. There are very simple, nonlinguistic precursors (no grammar) to cuneiform writing (Coulmas 1989).

boat   vessel   eight   mouth

Figure CG1

FOLKLORE

CG201: There are flood myths from all over the world.

Many cultures around the world have flood myths, indicating the universality of the Flood. (W. Brown 1995, 35; LaHaye and Morris 1976, 231–241)

◇◇◆◇◇

1. Flood myths are widespread, but they are not all the same myth. They differ in many important aspects, including

- reasons for the flood. (Most do not give a reason.)
- who survived. (Almost none have only a family of eight surviving.)
- what they took with them. (Very few saved samples of all life.)
- how they survived. (In about half the myths, people escaped to high ground; some flood myths have no survivors.)
- what they did afterwards. (Few feature any kind of sacrifice after the flood.)

The biblical flood myth has close parallels only to other myths from the same region, with which it probably shares a common source, and to versions spread to other cultures by missionaries (Isaak 2002a).

2. Flood myths are likely common because floods are common; the commonness of the myth in no way implies a global flood. Myths about snakes are even more common than myths about floods, but that does not mean there was once one snake surrounding the entire earth.

Further Reading: Dundes, A. (ed.). 1988. *The Flood Myth.*

BIBLICAL CREATIONISM

BIBLICAL CREATIONISM GENERALLY

CH001: Creationism has explanatory power.
Creationism is explanatory. It can accommodate all the results of evolution and more. In particular, it can also explain the results of a designer. (Dembski 2001b)

1. Accommodation is very different from explanation. An explanation tells why something is one way and not another. A theory that accommodates anything explains nothing, because it does not rule out any possibilities. Accommodating all possibilities also makes a theory exactly useless. Since creationism accommodates all possibilities, it is not explanatory.

CH010: Creationism, being Bible-based, is good.
Creationism, because it is based on the Bible, is moral. The denial of creationism is a denial of the Bible and is therefore immoral.

1. Many evils in the past have been justified by claiming biblical support. Claiming a biblical basis has no bearing whatsoever on whether something is good or not.

2. Creationism is not based on the Bible. Most people who accept the Bible do not accept creationism. Biblical creationism (we will not deal here with creationism based on the Qur'an or Vedas) is based on one particular interpretation of the Bible. It is a form of religious bigotry; it declares that a particular religious interpretation applies not just to people of that religion, but to everybody everywhere, and that the religion of any-

one who believes otherwise is wrong. This bigotry is overt from many creationists (Tparents n.d.). In fact, creationism claims to apply an individual's religious opinion to the whole universe. That is not merely bigotry; it is also hubris. Since bigotry and hubris are immoral, creationism is immoral at its very foundation.

3. Morals are properly judged on the basis of deeds, not claims. There are several indications that creationist deeds are below average morally:

- out-of-context quotes (see CA113)
- bogus credentials (Vickers 1998)
- fraudulent claims, such as the Paluxy footprints (see CC101), Moab and Malachite man (see CC110 and CC111), and the Lady Hope story (see CG001)
- repeated use of refuted claims, such as moon dust (see CE101) and the second law of thermodynamics (see CF001)
- vilification of opponents; for example, the persistent comparison of them to mass murderers (Elsberry and Perakh 2004).

The examples above are not indicative of all creationists. Most creationists, like most people of any category, are good people on the whole. But creationists, unlike evolutionists or most other people, have a strong ideological commitment. Strong commitments such as theirs can, and judging by the examples above probably do, lead people into questionable morality if they think it will support what they consider a higher cause. Objective study is still necessary to determine definitively whether creationists are any less moral than average, but theory and what evidence there is suggests that such is the case.

4. The Bible is not a consistent guide to morality; it describes several actions that would generally be considered immoral if not downright repugnant:

- In Numbers 31:17–18, Moses commands his troops to kill every woman and child in Midian, except young virgins, who the troops may keep for themselves.
- In Exodus 32:27, Moses commands his people to kill their brothers, sons, and neighbors who worshiped improperly.
- According to 2 Kings 2:23–24, Elisha called down a curse on some youths, resulting in forty-two of them being killed by a bear, simply because they mocked his baldness.
- In 1 Chronicles 13:7–11, Uzza is killed for trying to keep the Ark of the Covenant from harm.

See Robinson (2000) for more examples.

Further Reading: Carrier, R. 1998. Does the Christian theism advocated by J. P. Moreland provide a better reason to be moral than secular humanism? http://www.infidels.org/library/modern/richard_carrier/moreland.html; Drange, T. M. 1998. Why be moral? http://www.infidels.org/library/modern/theodore_drange/whymoral.html; Grünbaum, A. 1995. The poverty of theistic morality; Vickers, B. 1998. Some questionable creationist credentials. http://www.talkorigins.org/faqs/credentials.html.

CH010.1: Learning creationism stimulates mental health, joy, and morals.

Learning creationism stimulates mental health, because it is consistent with people's innate thoughts; joy, from scientific discovery; and morals, because it promotes awareness of a creator to whom one must give account. (H. M. Morris 1985, 14)

1. These claims have no support. There are healthy, joyful, and moral people who accept evolution as well as who are creationists. There are unhealthy, unhappy, and immoral people among creationists.

The claims have no logical basis, either:

- Intellectual honesty demands that evidence be followed whether we like the conclusion or not. Creationism demands that certain conclusions not be overturned by any amount of evidence. Some people might think that intellectual dishonesty and living in denial are incompatible with mental health.

- Joy is more likely to come from noncreationist scientific discovery for the simple reason that creationism discourages scientific discovery; anything incompatible with the creationists' foregone conclusions is not pursued.

- Many people consider it more moral to do good for the sake of doing good rather than for fear of punishment. In Kohlberg's stages of moral development, fear of punishment is the very first stage (Barger 2000).

CH050: Genesis is foundational to the Bible.

True science and true religion are founded on Genesis. All biblical doctrines have their foundations laid there, and the book of Genesis itself is founded on the events of its first chapter. (H. M. Morris 1983)

1. This claim is an instance of religious bigotry. Lots of religions, including Buddhism, Hinduism, Druidism, and many more, have no connection with Genesis at all. For a person to say that these are not true religions is

- a gross insult to the people who practice the religions. Many of these people are highly devout, with a spiritual relationship at least as great as any creationist.

- a gross insult to God. The person is saying that God's revelation must coincide with his own opinion to be valid, that God cannot reveal himself differently to different people. Anyone making this claim places themself above God.

- a disservice to oneself. Bigotry is hateful and will prevent good relationships with good people.

2. If Genesis is so all-important, why do creationists reject serious study of it? Modern (and even not-so-modern) scholarship has revealed much about the authors of Genesis (called J, E, P, and R) and other books of the Old Testament, including their motivations and places in history. For example, the Flood account is an interleaving of two different flood stories by J and P (R. E. Friedman 1987). Creationists studiously avoid any such knowledge. (Creationists are not alone in this; most Christians generally are woefully ignorant of biblical scholarship.)

3. Ideas in other parts of the Bible stand on their own. Creationists themselves frequently quote them out of context. The Old Testament itself refers to documents that no longer exist: the Book of the Wars of the Lord (Num. 21:14); the Book of Jasher (Josh. 10:13, 2 Sam. 1:18); and others (1 Kings 11:41; 14:29, 19; 16:5; 1 Chron. 29:29; 2 Chron. 20:34, 13:22). Knowledge of earlier scriptures is helpful but not critical. Jesus sometimes rejected the letter of some Old Testament laws, so the letter of the Old Testament cannot be too important, and Jesus exemplified the spirit. The reason creationists find Genesis so important is because they depend on it, not because other parts of the Bible depend on it.

4. If one believes that God created the earth and heavens, then surely the earth and heavens are God's primary work. Study of the earth and heavens should be foundational. Placing an object such as the Bible before them is idolatry.

5. No accepted science has ever been based on the Bible. That is not for lack of trying. Up to the nineteenth century, serious scientists tried to accomodate literal readings of the Bible to what they saw in nature. Young-earth creationism failed early on, so scientists tried gap creationism, day-age creationism, and other attempted reconciliations. But purely Bible-based science has always failed. True science is based on reality as expressed in the world (D. A. Young 1988).

Further Reading: Friedman, R. E. 1987. *Who Wrote the Bible?*

CH055: Noncreationist Christians are compromisers.
People who call themselves Christians but who do not accept creationism, such as theistic evolutionists, are seriously misguided. The same applies even to those who reject a particular interpretation of creationism, such as young-earth creationism. The Bible does not allow such compromises. (H. M. Morris 1984; Sarfati 2004)

1. Those who disparage the Christianity of Christian theistic evolutionists (many creationists deny that theistic evolutionists can even be Christian) show only their own arrogance and hubris. The claimants are saying that they, not God, get to dictate the proper way for another person to relate to God and the Bible. A religious relationship with God is a personal matter; it cannot effectively be decided by total strangers.

2. Many creationists are themselves compromisers. They compromise on major factual matters, rejecting the flat earth and solid firmament that a plain literal reading indicates. They compromise on details—for example, allowing insects and other "creeping things with the breath of life" not to be on Noah's ark as the Bible says. They compromise on biblical teachings, such as the numerous dietary and other laws in Exodus through Deuteronomy. Most of all, they compromise on Christian theology, rejecting the spirit of Christ's teaching to be humble in one's religion (e.g., Matt. 6:1–6).

BIBLICAL ACCURACY

CH100: The Bible says it; I believe it; that settles it.
God's word, the Bible, must be our ultimate authority. The Bible says it, I believe it, and that settles it. (Sarfati 2004, 17)

1. This claim is dogmatism. It suggests no reason for its conclusion. The views of others, that the Bible is not God's word or is not an ultimate authority, have just as much validity.

2. The Bible says different things to different people. Beliefs that creationists take as gospel today, such as the fixity of "kinds" and the impossibility of life from nonlife, would have seemed absurd to creationists of centuries past, and those past creationists would have cited the Bible to support their views (such as Moses's staff changing to a snake and the plagues of Egypt appearing from nowhere), just as today's creationists quote the Bible to support their own views (Brewster 1927).

3. In practice, this claim really means, "My view of the Bible is the ultimate authority." (Since there are so very many different interpretations of the Bible, not to mention other religions, the claim would be meaningless otherwise.) In practice, then, this claim displays a great deal of arrogance, hubris, and closed-mindedness. It says that the final word on how the universe operates depends on one's personal decision of what to believe.

4. This belief, when applied as a standard for others, is religious bigotry in its purest form. It shows contempt to others who believe that God's influence may be seen elsewhere than the Bible and the select few who are defined to interpret it correctly. This claim has started wars.

5. The Bible says several things that you probably do not believe:
- Slavery is acceptable (Skeptic's Annotated Bible n.d.b).
- You should kill your child if he strikes you (Exod. 21:15).
- If you work on Sunday, you should be put to death (Exod. 35:2–3).
- If you curse, you should be stoned to death (Lev. 24:14–15).
- Happiness is smashing children upon the rocks (Ps. 137:9).
- Women should be subjugated by their husbands (1 Pet. 3:1–7).

CH101: The Bible is inerrant.
The Bible, being God's revealed word, is without error or fault in everything it teaches, including what it says about creation, historical events, and its own origin. Scientific study of the earth cannot be used to overturn scriptural accounts of creation and the Flood. (International Council on Biblical Inerrancy 1978)

1. Inerrancy cannot be trusted. Errors can only be corrected if they are first recognized and admitted. Inerrancy makes that impossible. Therefore, errors in an inerrant interpretation of the Bible can never be fixed.

2. Inerrancy is a contempt that breeds hate. Inerrantists take it as divinely certain that other people's religious views are inferior to their own. One reaps what one sows, so when inerrantists show their contempt, contempt for their own religious views is returned. History is bloodied by the consequences. Jews, Muslims, heathens, and other Christians have been subjugated, tortured, and slaughtered in the name of the "true" god. Jacob Bronowski (1973, 374), speaking of Auschwitz, wrote,

Into this pond were flushed the ashes of some four million people. And that was not done by gas. It was done by dogma. It was done by arrogance. When people believe that they have absolute knowledge, with no test in reality, this is how they behave. This is what men do when they aspire to the knowledge of gods.

The contempt also shows up as intolerance—against women's roles, in attitudes about sex, and through a variety of other different views. Even those who do not commit atrocities, when they display such intolerance, are guilty of fomenting the atmosphere that makes the atrocities possible.

3. Inerrancy rejects much study of the Bible (not infrequently to the point of persecuting the studier). One who accepts inerrancy generally

- ignores textual criticism. Most inerrantists accept the King James version as authoritative, but analysis of the earliest biblical manuscripts shows that the King James version includes numerous errors. For example, the story of Jesus chiding those who would stone an adulteress (John 8:1–11) does not appear until about 300 years after the Gospel of John was written.

- ignores source criticism. Many stories in the Bible are repeated, but with different emphasis, different details, and different language. These differences show that the Bible was written by different people at different times for different purposes, and their accounts were redacted by people with still different motives (R. E. Friedman 1987).

- ignores the reality of syncretism, the process by which rituals and concepts from one religion are adapted by another. Many biblical stories show Sumerian and Canaanite influence, for example.

- ignores the values of the writers of the Bible, who likely did not distinguish literalism or consider it important. The Bible was not written to record accurate histories, but to convey and persuade spiritual ideas. Those ideas may not even be the same to all people.

It is ironic that people who purport to hold the Bible in such high esteem reject serious, objective study of it.

4. Jesus himself said that religious laws are not absolute. In Matthew 5:38, he rejects the "eye for an eye" law (Exod. 21:23–25, Lev. 24:19–20, Deut. 19:21). Jesus rejected all dietary law (Mark 7:19; cf. Lev. 11). He rejected the commandment about working on the Sabbath (Mark 2:27). If Jesus considered that even the laws of Moses were not inerrant, why should we consider any part of the Bible inerrant?

5. Ultimately, there is no authority for inerrancy except oneself:

- God cannot be the authority because God has not said anything on the subject directly. The whole point of inerrancy is to attribute God's authority to an indirect vehicle.

- The Bible cannot be an authority to its own authoritativeness; that would be circular reasoning.

- The church cannot be an authority for inerrancy because there is no one church. There are over 10,000 different Christian denominations, all with different ideas about the Bible. In fact, there are at least three significantly different Bibles (the Catholic, Protestant, and Ethiopian Orthodox versions).

- For the same reason, historical tradition cannot be the authority for inerrancy. Views about the Bible have changed over history.

6. Claiming inerrancy in the Bible is pointless unless one also claims inerrancy in one's interpretation of it. Some people believe that the earth is flat and is covered by a solid dome because the Bible says so and the Bible is inerrant (Schadewald, 1987). Most people, including most biblical inerrantists, would say they are wrong. Claiming inerrancy for a particular view of creation or the flood is no different in principle. Claiming that the Flood account is a true literal account is an error if it was written as an allegory; claiming that it is a true allegory is an error if it was a literal account. To claim that a particular interpretation of any part of the Bible is inerrant is to claim that you yourself are inerrant.

7. There are several aspects of the Bible that show it is not inerrant. These include factual errors:

- Leviticus 11:4 states that rabbits chew their cud.
- Leviticus 11:20–23 speaks of four-legged insects, including grasshoppers.
- 1 Chronicles 16:30 and Psalm 93:1 state that the earth is immobile; yet it not only revolves and orbits the sun but is also influenced by the gravitational pull of other bodies.

and contradictions:

- In Genesis 1, Adam is created after other animals; In Genesis 2, he appears before animals.
- Matthew 1:16 and Luke 3:23 differ over Jesus's lineage.
- Mark 14:72 differs from Matthew 26:74–75, Luke 22:60–61, and John 18:27 about how many times the cock crowed.
- 2 Samuel 24:1 and 1 Chronicles 21:1 differ over who incited David to take a census.
- 1 Samuel 17:23, 50 and 2 Samuel 21:19 disagree about who killed Goliath.
- 1 Samuel 31:4 and 2 Samuel 1:8–10 differ over Saul's death.
- The four Gospels differ about many details of Christ's death and resurrection (Barker 1990). For example, Matthew 27:37, Mark 15:26, Luke 23:38, and John 19:19 have different inscriptions on the cross.
- Matthew 27:5–8 differs with Acts 1:18–19 about Judas's death.
- Genesis 9:3 and Leviticus 11:4 differ about what is proper to eat.
- Romans 3:20–28 and James 2:24 differ over faith versus deeds.
- Exodus 20:5, Numbers 14:18, and Deuteronomy 5:9 disagree with Ezekiel 18:4, 19–20 and John 9:3 about sins being inherited.

Inerrantists are familiar with these and find rationalizations for these and other errors and contradictions, but they are unconvincing. The rationalizations merely make the point that what the Bible seems to say is not what it means, which defeats the whole concept of scriptural inerrancy.

Further Reading: Bringas, E. 1996. *Going by the Book*; Hildeman, E. J. 2004. *Creationism: The Bible Says No!* Straight Dope. 2002. Who wrote the Bible? (Part 5). http://www.straightdope.com/mailbag/mbible5.html.

CH101.1: If part of the Bible is wrong, none of it can be trusted.

If the Bible cannot be trusted on scientific and historical matters, then it cannot be trusted on matters of salvation and spirituality. (H. M. Morris 2000b)

1. The Bible was not intended to teach matters of science and history. Therefore, those areas should not be held to standards of literal accuracy.

2. The claim is a non sequitur. That something is wrong in one area does not prevent it from being perfectly accurate in another.

3. Theologians through the ages have considered parts of the Bible suspect but accepted the rest as canon. In fact, it was exactly such a process by which canon was determined. Even Martin Luther considered some Old Testament passages suspect (Armstrong 1996; Engwer n.d.; Shea 1997).

4. A logical consequence of this claim is that the Bible cannot, in fact, be trusted, because parts of it (not only Genesis) are known to be wrong if interpreted literally (see CH101).

5. Creationists themselves sometimes make claims that contradict the Bible. For example, Whitcomb and Morris (1961, 69) claimed, contrary to Genesis 7:21–23, that some land animals not aboard Noah's ark survived.

Further Reading: Charles, R. H. (ed.). 1913. The Apocrypha and Pseudepigrapha of the Old Testament in English. http://www.ccel.org/c/charles/pseudepigrapha; Seghers, J. 1998. Sola Scriptura. http://totustuus.com/solascri.htm; Sungenis, R. A. 1997. *Not by Scripture Alone*.

CH102: The Bible is literal.

The Bible should be read literally. (H. M. Morris 1985, 204)

1. A literal reading of the Bible misses the meaning behind the details. It is like reading Aesop's Fables without trying to see the moral of the stories. Finding the meaning in a figurative reading requires more thought, but is thinking about the Bible a bad thing?

2. There are many inconsistencies and inaccuracies in the Bible that cannot be resolved without excessive pseudological contortions unless one does not take them literally. Augustine said,

> It is a disgraceful and dangerous thing for an infidel to hear a Christian, presumably giving the meaning of Holy Scripture, talking nonsense on these topics; and we should take all means to prevent such an embarrassing situation, in which people show up vast ignorance in a Christian and laugh it to scorn. (Augustine 1982, 42–43)

Augustine's warning has merit. The invalid "proofs" necessary to support antievolution, a global flood, and a young earth have pushed people away from Christianity (Hildeman 2004, Morton n.d.).

3. There are several passages of the Bible itself that indicate that it should not be taken literally:

- 2 Corinthians 3:6 says of the new covenant, "the letter kills, but the Spirit gives life."

- 1 Corinthians 9:9–12 says that one of the laws of Moses is figurative, not literal.
- Galatians 4:24 says that the story of Abraham is an allegory.
- Jesus frequently taught in parables, with the obvious intention that the lesson from the story, not the details of the story, was what was important.

4. There is extensive tradition in Christianity, including Catholicism and Protestantism, of accepting nonliteral interpretations (Rogerson 1992). Biblical literalism is not a requirement; it is a fashion.

5. Nobody reads the Bible entirely literally anyway. For example, when God says, "into your hands they [all wild animals] are delivered" (Gen. 9:2), the phrase is obviously meant metaphorically.

6. Even reading the Bible literally requires interpretation. For example, what does "fountains of the deep" (Prov. 8:28) mean?

CH102.1: Genesis must be literal; it is straightforward narrative.
Genesis must be literal; it is straightforward narrative. (H. M. Morris 1998c)

1. Straightforward narrative does not imply literalness. Aesop's fables are also straightforward narrative, but they are not literal.

2. It is far from clear that Genesis is straightforward narrative. Genesis 1 has a formulaic and poetic structure.

CH102.2: Genesis must be literal; later writers refer to it as fact.
Genesis must be literal because writers of later books of the Bible refer to it as fact. (H. M. Morris 1985, 244–247)

1. Referring to something as fact does not mean it is fact. Writers often use metaphors. I have often seen writers refer to the story of blind men describing an elephant as if it were fact, for example, even though it is a fictional story.

2. Even if the later writers thought what they were referring to was true, it may not have been. People mistake folklore for fact all the time.

CH103: Bible claims inspiration.
The Bible is special because it is inspired; its writers claim that their thoughts came from God. (Watchtower 1989, 10)

1. Other works claiming divine inspiration include the Qur'an, *Mahabharata*, and *The Book of Mormon*. Many traditions of North American Indians claim to be divinely inspired (Vecsey 1991). Some contemporary New Age books claim divine inspiration, such as *The Urantia Book* and *Talking with Angels: A Document from Hungary*.

2. Claiming divine inspiration does not mean having it. I could easily claim divine inspiration even for this book.

3. Even if the Bible is divinely inspired, one's interpretation of it can still be entirely wrong. The Bible has been invoked to justify burning innocent young women (supposedly witches), maintaining slavery, and other atrocities. Divine inspiration, even if true, is no reason to accept ideas that someone gets from the Bible.

CH110: Prophecies prove the accuracy of the Bible.

The Bible contains many prophecies that have accurately been fulfilled, proving it is a divine source. (Watchtower 1985, 216–223)

1. There are several mundane ways in which a prediction of the future can be fulfilled:

1. Retrodiction. The "prophecy" can be written or modified after the events fulfilling it have already occurred.

2. Vagueness. The prophecy can be worded in such a way that people can interpret any outcome as a fulfillment. Nostradomus's prophecies are all of this type. Vagueness works particularly well when people are religiously motivated to believe the prophecies.

3. Inevitability. The prophecy can predict something that is almost sure to happen, such as the collapse of a city. Since nothing lasts forever, the city is sure to fall someday. If it has not, it can be said that according to prophecy, it will.

4. Denial. One can claim that the fulfilling events occurred even if they have not. Or, more commonly, one can forget that the prophecy was ever made.

There are no prophecies in the Bible that cannot easily fit into one or more of those categories.

2. In biblical times, prophecies were not simply predictions. They were warnings of what could or would happen if things did not change. They were meant to influence people's behavior. If the people heeded the prophecy, the events would not come to pass; Jonah 3 gives an example. A fulfilled prophecy was a failed prophecy, because it meant people did not heed the warning.

3. The Bible contains failed prophecies, in the sense that things God said would happen did not (Skeptic's Annotated Bible n.d.a). For example, Isaiah 17:1–3 says that Damascus will cease to be a city and be deserted forever, yet it is inhabited still.

4. Other religions claim many fulfilled prophecies, too (Prophecy Fulfilled n.d.).

5. Divinity is not shown by miracles. The Bible itself says true prophecies may come elsewhere than from God (Deut. 13:1–3), as may other miracles (Exod. 7:22, Matt. 4:8). Some people say that to focus on proofs is to miss the whole point of faith (John 20:29).

Further Reading: Festinger, L., et al. 1956. *When Prophecy Fails.*

CH120: The Bible must be accurate because archaeology supports it.

Archaeology supports the accuracy of the Bible. The Bible's historical account has many times been substantiated by new archaeological information. (Watchtower 1985, 207–214)

1. Archaeology supports at most the general background of the Bible and some relatively recent details. It does not support every biblical claim. In particular, archaeology does not support anything about creation, the Flood, or the conquest of the Holy Land.

If a few instances of historical accuracy are so significant, then an equal claim for accuracy can be made for the *Iliad* and *Gone with the Wind*.

2. Archaeology contradicts significant parts of the Bible:

- Luke 2:4 describes Nazareth as being Joseph's home, but the archaeological evidence indicates that the town did not exist at the time (K. Humphreys 2003).

- The Bible contains anachronisms. Details attributed to one era actually apply to a much later era. For example, camels, mentioned in Genesis 24:10, were not widely used until after 1000 B.C.E. (Finkelstein and Silberman 2001).

- The Exodus, which should have been a major event, does not appear in Egyptian records. There are no traces in the Sinai that one would expect from forty years of wandering of more than half a million people. And other archaeological evidence contradicts it, showing instead that the Hebrews were a native people (Finkelstein and Silberman 2001; Lazare 2002).

- There is no evidence that the kingdoms of David and Solomon were nearly as powerful as the Bible indicates; they may not have existed at all (Finkelstein and Silberman 2001; Lazare 2002).

Many claims that archaeology supports the Bible, especially earlier ones, were based on the scientists trying to force the evidence to fit their own preconceptions.

Further Reading: Finkelstein, I., and N. A. Silberman. 2001. *The Bible Unearthed*; Lazare, D. 2002. False testament: Archaeology refutes the Bible's claim to history; Miller, L. 2001. King David was a nebbish. http://dir.salon.com/books/feature/2001/02/07/solomon/index.html; Moorey, P.R.S. 1991. *A Century of Biblical Archaeology*.

CH130: The Bible's accuracy on other scientific points shows overall accuracy.

The Bible's accuracy on various scientific and historical points shows its overall accuracy. For example, it says that the earth is spherical and unsupported, that the water cycle is correct, and that certain practices are good hygiene. (Jeffrey 1996; Watchtower 1985, 200–206)

1. The accuracy of the Bible is not remarkable. All of its accurate points can be explained by

- simple observation of nature. People would not need special knowledge to see that streams are replenished by rain and do not run dry or that certain practices were unhealthy and should be avoided.

- selective interpretation of scriptures. The Bible says not that the earth is spherical, but that it is round and covered by a tentlike canopy (Isa. 40:22). A passage cited as knowledge of the importance of washing hands refers only to ritual sprinkling, not washing (Num. 19:11–22; Watchtower 1985, 205).

2. Accuracy on individual points does not indicate overall accuracy. Just about every thesis that is wrong overall still has some accurate points in it.

3. Claims about accuracy assume that the purpose of the Bible is to document scientific data. There is not the slightest indication that the Bible was ever intended as a scientific textbook. It is intended to teach people about God; even those who claim scientific accuracy for it use it with that intent. For at least some of the Bible's teachings, scientific accuracy is unnecessary and perhaps even counterproductive.

4. The Bible is not entirely accurate. If its value is made to depend on scientific accuracy, it becomes valueless when people find errors in it, as some people invariably will (see CH101).

5. If occasional scientific accuracy shows overall accuracy of the Bible, then the same conclusion must be granted to the Qur'an (see CJ530), Zend Avesta, and several other works from other religions, all of which can make the same claims to scientific accuracy.

Further Reading: Meyers, S. 2000. The Signature of God by Grant R. Jeffrey (review). http://members.aol.com/ibss2/gospel.html; Till, F. 1990. What about scientific foreknowledge in the Bible? http://www.infidels.org/library/magazines/tsr/1990/4/4scien90.html.

CH190: The Bible is harmonious throughout.

The Bible's internal harmony around a central theme testifies to its divine authorship. It is sixty-six books written over sixteen centuries by some forty different writers of diverse backgrounds, but every part follows the same theme. (Watchtower 1985, 215)

1. The Bible's harmony can also be attributed to the fact that its contents were selected and edited, by people, to make it harmonious (R. E. Friedman 1987).

2. The Bible is not harmonious on some very important points:

- Many people have noticed the difference between the Old Testament God, who is vengeful and bloodthirsty (e.g., Gen. 6–8; Exod. 7–11) and commands and aids the slaughter of one's enemies (e.g., Exod. 32:27–28; Deut. 3:6; Num. 31:1–18), versus the New Testament God, who preaches peace and commands people to love their enemies.

- Much of the Bible emphasizes the unity of God. Genesis 1, for example, stresses that all of creation came from the same God, not different gods as other contemporary religions taught. However, the New Testament, particularly Revelation, introduces a good/evil dualism akin to Zoroastrianism, which has become particularly common in Christian tradition.

Further Reading: Friedman, R. E. 1987. *Who Wrote the Bible?*

YOUNG-EARTH CREATIONISM

Age of the Universe

CH200: The universe is 6,000–10,000 years old.

The universe is relatively young, only 6,000 to 10,000 years old.

1. The age of the earth is 4.5 billion years (see CH210).

2. The universe is shown to be old by several independent types of measurements:

- We can measure the distances to some types of stars from their apparent brightness. (We know their absolute brightness from nearby stars of the same type whose distances can be measured geometrically.) We find distances more than fifty million light-years away, which means the universe must be at least fifty million years old for the light to reach us. Measurements based on the brightness of supernovae and galaxies, although less accurate, give distances up to billions of light-years.

- The Large Magellanic Cloud is 153,000 light-years away, as measured by an eclipsing binary star (A. A. Cole 2000). This method gives a relatively direct measurement from simple observations. A star's absolute brightness is determined from its temperature and diameter, which can be determined from its spectrum and length of eclipse. Distance is then determined from the apparent brightness.

- The orbits of thirteen of the Koronis family of asteroids were traced back and found to match 5.8 million years ago, suggesting that they formed then from a collision of larger asteroids (Nesvorný et al. 2002).

- There are white dwarf stars found to be twelve to thirteen billion years old, based on their cooling rate.

Further Reading: Ferris, T. 1997. *The Whole Shebang.*

CH210: The earth is 6,000–10,000 years old.
The earth is relatively young, about 10,000 years old or less. (H. M. Morris 1985, 158)

1. Radiometric dating shows the earth to be 4.5 billion years old (see CD010 regarding the reliability of radiometric dating).

2. If the earth is old, then radioactive isotopes with short half-lives should have all decayed already. That is what we find. Isotopes with half-lives longer than eighty million years are found on earth; isotopes with shorter half-lives are not, the only exceptions being those that are generated by current natural processes (Dalrymple 1991, 376–378).

3. Loess deposits (deposits of wind-blown silt) in China are 300 m thick. They give a continuous climate record for 7.2 million years. The record is consistent with magnetostratigraphy and habitat type inferred from fossils (Russeau and Wu 1997; Sun et al. 2002).

4. Varves are annual sediment layers that occur in large lakes. They are straightforward to measure, cover millions of years, and correlate well with other dating mechanisms:

- In seasonal areas, sedimentation rates vary across the year, so sediments often show annual layers (varves) distinguished by texture and/or composition. We can be confident that the layers are seasonal because we see the same sorts of layers occurring today. Even if they were not seasonal, the fineness of the sediments is often such that each layer would require several days, at least, to form. Some formations have millions of layers, such as the varve record from Lake Baikal with five million annual layers (D. F. Williams et al. 1997); and the 20,000,000 layers in the Green River formation. They must have taken hundreds of thousands of years to form at the very least.

- Dates obtained by counting annual layers of varves match dates obtained from radiometric dating. One varve formation, covering 45,000 years, was used to calibrate carbon-14 dating using terrestrially produced leaves, twigs, and insect parts that also appeared in the sediments. The varves were easy to count because they included an annual diatom bloom (Kitagawa and van der Plicht 1998; see Figure CD2).

- Varves record climate changes, too, since climate affects the amount of sediments. Climate is affected by orbital cycles known to occur at about 400,000-, 600,000-, and million-year periods (the so-called Milankovitch cycles). Climate cycles of these durations occur in the varve records. For example, Lake Baikal contains annual layers from twelve million years ago to the present. These sediments contain periodic changes matching the orbital cycles. (Kashiwaya et al. 2001).

Further Reading: Dalrymple, G. B. 1991. *The Age of the Earth;* Strahler, A. N. 1987. *Science and Earth History;* Young, D. A. 1988. *Christianity and the Age of the Earth.*

CH220: The universe was created with apparent age.
The universe was created mature, with apparent age. Light from the sun and stars fell on the earth from its beginning. (H. M. Morris 1985, 209–210) *This is sometimes called the omphalos argument after the title of an early book expounding it.* (Gosse 1857)

1. Apparent age is indistinguishable from real age. Why not forget the distinction and just call it age?

2. The universe has an appearance of history as well as an appearance of age. Stars of all ages are seen, including the supernovae of dying stars. Geological evidence shows that the early earth went through millions of years with little oxygen in the atmosphere (see CB035.1). Many other examples could be added. The appearance of age asks us to accept that light from supernovas came from stars that never actually existed, and that the evidence for low oxygen was also faked. This makes God into a deceiver, since he created an appearance different from reality. Romans 1:20 says that God is to be "understood from what has been made." The apparent age claim says we cannot trust what has been made.

Death and the Fall

CH301: There was no death or decay before the Fall.
There was no death or decay before the Fall of Adam and Eve, when sin came into the world. (H. M. Morris 1985, 211)

1. The biblical references to death refer to spiritual death. Romans 6:23, for example, says God, through Christ, already gave us eternal life. Obviously, physical death still occurs, so Christ did not save mankind from it. If the death from which Christ was supposed to save us was physical death, then Christ was a failure.

2. A world without death would be far from Edenic. Either there would be no possibility of new birth, in which case none of us could be alive today, or the world would become so overpopulated with living things that most people would be buried alive under others. Death is a necessary part of life.

Further Reading: Larson, G. 1998. *There's a Hair in My Dirt!*

CH310: The pre-Flood vapor canopy would have made the world Edenic.

In the pre-Flood earth, the vapor canopy, a translucent layer of water vapor above the atmosphere, caused a greenhouse effect that kept the climate moderate all over the planet, minimized winds and storms, and prevented rainfall. (H. M. Morris 1985, 210–211)

1. A vapor canopy with more than twelve inches of precipitable water would raise the temperature of the earth above boiling (Morton 1979). A vapor canopy of only four inches of water would raise the temperature of the earth to 144 degrees F. It is worth noting that several prominent creationists agree with this conclusion, yet their close colleagues continue to teach that there was a vapor canopy (Morton 2000a).

2. A vapor canopy capable of producing the global flood would have increased earth's atmospheric pressure from 15 PSI to 970 PSI.

3. Some creationists try to solve the vapor canopy problems by moving the canopy out of the earth's atmosphere and into orbit. A canopy of orbiting ice would have been unstable (it could only exist in a ring much like Saturn's). It would have cooled the climate (probably just slightly) until it somehow collapsed to cause the flood. Then the release of its gravitational potential energy would have converted all the ice into superheated steam, not into a flood.

Further Reading: Farrar, P., and B. Hyde. n.d. The vapor canopy hypothesis holds no water. http://www.talkorigins.org/faqs/canopy.html; Matson, D. E. 1994. Water and vapor and Noah's flood. http://www.talkorigins.org/faqs/hovind/howgood-add.html#A2; Morton, G. R. 2000a. The demise and fall of the water vapor. http://home.entouch.net/dmd/canopy.htm.

CH311: The pre-Flood vapor canopy would have extended human lifetimes.

In the pre–Flood earth, the vapor canopy, a translucent layer of water vapor above the atmosphere, filtered out radiation from space, decreasing mutations and drastically increasing life spans. (H. M. Morris 1985, 210–211)

1. The vapor canopy as proposed would make the environment on earth highly inimical at best (see CH310).

2. Filtering out radiation from space has no appreciable effect on life spans. Life spans are determined mostly by programmed cell death, not by mutations. Mutations themselves are mostly caused by factors other than radiation. People who live at high altitudes, and thus who have significantly less atmospheric shielding from cosmic radiation, do not live noticeably shorter lives as a result.

Further Reading: Morton, G.R. 2000a. The demise and fall of the water vapor. http://home .entouch.net/dmd/canopy.htm.

CH320: Life is deteriorating.

Before sin entered the world, there were no net deteriorating effects of the second law of thermodynamics. The decay that we see today, including organisms such as fungi, predators, and parasites, is a result of Adam's sin. (H.M. Morris and Clark 1987)

1. Parasites have advanced derived features, which are necessary for them to survive as parasites (see CH321). The same is true of fungi, predators, and other supposedly "deteriorated" organisms. Even the Institute for Creation Research (ICR) recognizes that features such as thorns, fangs, and poisons evolved after the Fall. The evolution of features such as thorns and poisons where there were none before is not a deterioration of the organisms that get them.

2. The record of life on earth shows advancement, not deterioration. The earliest records show only bacteria. This was followed in the late Precambrian by primitive multicellular organisms, then animals with hard parts, then land animals, and so forth. The largest plants and animals that ever lived are living today. We know of no other time when biodiversity was greater than in historic times. A few mass extinctions have occurred, but these provided fresh starts for new forms of life to evolve.

3. The amount of evolution that the ICR attributes to "deterioration" would certainly be regarded as macroevolution by any rational standard. The evolution of a death-based ecology after the Fall would be almost as big a change as the original creation.

CH321: Parasites are degenerations of free-living species.

Parasites are degenerate forms of free-living or mutualistic organisms. They became parasites when something went wrong as a result of the Fall. For example, the parasite came to invade the wrong host or the wrong organ within the host, or it changed to harm the host where it did not before. (Mace et al. 2003)

1. Parasites are far from degenerate. They have lost features that are familiar to us as nonparasites, but they also have acquired many other highly sophisticated features and abilities, allowing them to find their hosts, to survive their hosts' immune systems (often multiple hosts for one parasite), and to survive some otherwise hostile environments within their hosts. Creationists themselves tout the complexity of the immune system (see CB200.4); does not circumventing an immune system deserve at least as much credit? Fast-evolving viruses like the common cold show that such adaptations are evolving all the time.

Here are just a few features that parasites have. Similar adaptations are common (Hajek and St. Leger 1994; Zimmer 2000b):

- *Sacculina*, a parasitic barnacle, infests crabs. It prevents the crab from molting and reproducing and induces the crab to care for the parasite's brood as if it were the crab's. Even male crabs are feminized to groom as if they had a female's brood pouch (Zimmer 2000b, 79–82).

- The larva of the *Hymenoepimecis* wasp parasitizes an orb–weaving spider. When the larva is ready to pupate, it modifies the spider's behavior to make it spin a cocoon for the wasp (Eberhard 2000).

- The fungus *Entomophthora muscae* infects and kills house flies, but before it kills them, it manipulates the fly's behavior to make it crawl to a high place and adopt a sexually receptive pose, behaviors that increase the likelihood of the fungus spreading to other flies (Møller 1993).

2. Evolution often goes the other way; parasites that initially are very harmful become more benign to their host over time. The virulence of a pathogen is generally predictable on the basis of evolutionary principles. For example, parasites are less virulent at low host population densities where the parasites risk destroying available hosts and themselves with them (Nesse and Williams 1994, 57–61; Zimmer 2000b, 151–155).

3. Why do organisms have defenses against pathogens in the first place? They would not have been needed in a pre-Fall world without pathogens, and their complexity and effectiveness show that features such as immune systems are not degenerate forms themselves.

Further Reading: Nesse, R. M., and G. C. Williams. 1994. *Why We Get Sick*; Zimmer, C. 2000b. *Parasite Rex.*

Created Kinds

CH350: Organisms come in discrete kinds.

The created kinds are distinct; evolution between them is impossible. "Creation of distinct kinds precludes transmutation between kinds." (H. M. Morris 1985, 13, 216–218; quote on 216)

1. Creationists have been unable to specify what the created kinds are. If kinds were distinct, it should be easy to distinguish between them. Instead, we find a nested hierarchy of similarities, with kinds within kinds within kinds. For example, the twelve-spotted ladybug could be placed in the twelve-spotted ladybug kind, the ladybug kind, the beetle kind, the insect kind, or any of dozens of other kinds of kind, depending on how inclusive the kind is. No matter where one sets the cutoff for how inclusive a kind is, there will be many groups just bordering on that cutoff. This pattern exactly matches the pattern expected of evolution. It does not match what creationism predicts.

2. Fixity of kinds is based on the philosophy of Plato, not the Bible (J. Dewey 1910). Nowhere does the Bible say that kinds themselves cannot change and diversify. Reproduction "according to their kind" is entirely consistent with evolution, as long as it is recognized that kinds are not fixed.

3. Although major changes from one kind to another do not normally happen, except gradually over hundreds of thousands of generations, a sudden origin of a new kind has been observed. A strain of cancerous human cells (called HeLa cells) have evolved to become a wild unicellular life form (Van Valen and Maiorana 1991).

4. According to Morris, fungi were not part of the original creation. They were not among the categories listed in Genesis 1, and as decayers they would not have their form

until after the Fall. Thus, Morris's own theology requires new kinds to originate after the creation.

CH370: The stars and galaxies are unchanging.

The creation model predicts galaxies constant and stars unchanging, in the main. They may decay, but they were created entire and did not build up over time. (H. M. Morris 1985, 13, 24–25)

1. The claim is baseless. The formation of stars takes on the order of millions of years, so we cannot expect to see major changes as we watch, but that does not mean the stars are unchanging.

Given our knowledge of physical laws and our observations of stars and interstellar matter, we expect star formation to be occurring continuously, as molecular clouds form and condense. And we see molecular clouds, protostars, and young stars in all stages of formation, in close agreement with what we expect (Pudritz 2002; Ward-Thompson 2002).

The existence of stars with differing amounts of heavy elements is also in good agreement with star formation over time, since the heavy elements come only from supernovae of earlier stars.

2. Galaxies have changed over time, too. Quasars were more common in the earlier universe; there are no recent ones. We also see galaxies in various stages of colliding.

Flood

CH400: Global flood

A global flood occurred as described, literally, in Genesis 6–9. Waters covered the entire earth for about a year. All animals not preserved on Noah's ark perished. (Whitcomb and Morris 1961)

1. See CH401–CH590 below. See also CG201 (flood myths worldwide) and much of section CD (geology).

CH401: Flood from vapor canopy

A "vapor canopy" (water vapor above the troposphere) was the source for much of the flood waters. (H. M. Morris 1985, 124–125, 210–211)

1. There is no way to hold up a vapor canopy of any great extent. (It could orbit, like the rings of Saturn, but that would not cover the earth like a canopy.)

2. If the canopy began as vapor, any water from it would be superheated. This scenario essentially starts with most of the flood waters boiled off. If the water began as ice in orbit, the gravitational potential energy would likewise raise the temperature past boiling. When the canopy fell to form the Flood, everything on earth would be poached.

If the canopy began as solid ice in orbit above the atmosphere, its gravitational potential energy would be converted to heat as it fell, superheating the atmosphere well beyond the ability of any life to survive (W. Brown 1995, 175).

3. A vapor canopy would have made the earth unliveable before the Flood (see CH310).

Further Reading: Farrar, P., and B. Hyde. n.d. The vapor canopy hypothesis holds no water. http://www.talkorigins.org/faqs/canopy.html.

CH420: Hydroplate theory

The Flood's waters came from a layer of water about ten miles underground, which was released by a catastrophic rupture of the earth's crust, shot above the atmosphere, and fell as rain. (W. Brown 1995, 87–98)

1. The rock that makes up the earth's crust does not float. The water would have been forced to the surface long before Noah's time, or before Adam's time for that matter.

2. Even two miles deep, the earth is boiling hot, and thus the reservoir of water would be superheated. Further heat would be added by the energy of the water falling from above the atmosphere. As with the vapor canopy model, Noah would have been poached.

3. The escaping waters would have eroded the sides of the fissures, producing poorly sorted basaltic erosional deposits. These would be concentrated mainly near the fissures, but some would be shot thousands of miles along with the water. Such deposits would be quite noticeable but have never been seen.

CH430: Runaway subduction

Baumgardner's computer model shows that runaway subduction explains how the global flood occurred. The cold, heavy crust of the ocean floor sinks into the lighter, hotter mantle, releasing gravitational potential energy as heat. Runaway subduction posits that this process greatly accelerated: "As the plates deform the surrounding rock, the mechanical energy of deformation is converted into heat, creating a superheated 'envelope' of silicate around the sinking ocean floor. Silicate is very sensitive to heat, so it becomes weaker, allowing the plates to sink faster and heating the envelope still further, and so on, faster and faster. As the plates pull apart, the gap between them grows into a broadening seam in the planet. This sends a gigantic bubble of mantle shooting up through these ridges; [w]hich displaces the oceans; [w]hich creates a huge flood." (Burr 1997, 57) God "caused an enormous blob of hot mantle material to come rushing up at incredible velocity through the underwater midocean ridges. The material ballooned, displacing a tidal wave of sea water over the continents. . . . Then, after 150 days (Genesis 7:24), the bubble retreated with equal speed into the Earth." (Burr 1997, 56)

1. Baumgarder's theory still does not work without miracles, as Baumgardner himself admitted (Baumgardner 1990a, 1990b). The thermal diffusivity of the earth would have to increase ten thousandfold to get the subduction rates proposed, and something

would have to cause the advance and retreat of the magma bubble (Matsumura 1997). Miracles would also have been necessary to cool the new ocean floor and to raise sedimentary mountains in months rather than in the millions of years it would ordinarily take.

2. The miraculously lowered viscosity would likely also lower frictional heating, removing the heat source that the model needs to accelerate the subduction (Matsumura 1997).

3. A series of events such as the magma bubble Baumgardner described would create "an enormous volcanic province in a single region. So, where is it?" (Geissman, quoted in Matsumura 1997, 30). The incredible amount of subduction proposed would also have produced much more vulcanism around plate boundaries than we see (Matsumura 1997).

4. Baumgardner estimated a release of 10^{28} joules from the subduction process. This is more than enough to boil off all the oceans. In addition, Baumgardner postulated that the mantle was much hotter before the Flood (giving it less viscosity); that heat would have to go somewhere, too.

5. Baumgardner's own modeling shows that during the Flood, currents would be faster over continents than over ocean basins (Baumgardner and Barnette 1994), so sediments should, on the whole, be removed from continents and deposited in ocean basins. Yet sediments on the ocean basin average 0.6 km thick, while on continents (including continental shelves), they average 2.6 km thick (Poldervaart 1955).

6. Cenozoic sediments are post-Flood according to this model. Yet fossils from Cenozoic sediments alone show a sixty-five-million-year record of evolution, including a great deal of the diversification of mammals and angiosperms (Carroll 1997, chaps. 5–6, 13).

7. Terra, the computer program that Baumgardner created, is a useful computer program for modeling convection, but the program adds no credibility. Unreal assumptions of runaway subduction will produce unreal conclusions.

Further Reading: Isaak, M. 1998. Problems with a global flood. http://www.talkorigins.org/faqs/faq-noahs-ark.html; Matsumura, M. 1997. Miracles in, creationism out. http://www.ncseweb .org/resources/rncse_content/vol17/4787_miracles_in_creationism_out_12_30_1899.asp; Strahler, A.N. 1987. *Science and Earth History.*

with Reynold Hall

CH500: Noah's ark has been found.

There have been many sightings of Noah's ark, including the following:

- Berosus, ca. 275 B.C.E., reported remains of it in the mountains of the Gordyaeans in Armenia (LaHaye and Morris 1976, 15).
- According to a 1952 story by Harold Williams, Haji Yearam helped guide three scientists to the ark in 1856. Upon finding the ark sticking out of a glacier near the summit, the scientists flew into a rage and tried futilely to destroy it. Then they took an oath to keep the discovery a secret and murder anyone who revealed it. About 1918, Williams saw a newspaper article giving a scientist's deathbed confession, which corroborated Yearam's story (LaHaye and Morris 1976, 43–48).

- In 1908 and again in 1910, a local Armenian, Georgie Hagopian, then just a boy, visited the ark with his uncle. The ark was on the edge of a cliff; its wood was like stone (LaHaye and Morris 1976, 69–72).

- In 1916, a story by Vladimir Roskovitsky told how he and other Russian aviators sighted the ark, nearly intact, grounded on the shore of a lake on Ararat. An expedition reached the ark about a month later. Photographs and plans were sent to the czar, but the Bolsheviks overthrew the czar a few days later, and the evidence was lost. Later testimony revealed that that account was 95 percent fiction, but other Russian soldiers have told of hearing of an expedition that discovered Noah's ark in 1917 (LaHaye and Morris 1976, 76–87).

- While lost on Ararat in 1936, Hardwicke Knight found timbers of dark, soft wood (LaHaye and Morris 1976, 98–101).

- Two American pilots saw the ark several times and once brought a photographer along. The photograph appeared in the Tunisian edition of *Stars and Stripes* in 1943. Many people remembered the article, but no copies remain (LaHaye and Morris 1976, 102–107).

- George Green photographed the ark from a helicopter in 1953, but his pictures aroused no serious interest, and they are now lost (LaHaye and Morris 1976, 135–137).

1. What the reports of ark sightings have in common is that none has been corroborated. Most have few if any witnesses. Photographs and newspaper articles disappear, sometimes inexplicably, or they are too vague to be meaningful. Physical evidence either is not retrieved, is faked, or comes from recent wood carried up the mountain. They have the appearance of fables, not fact.

2. The reports are inconsistent. The ark has been found in different places on the mountain (and on different mountains, if you include earlier accounts). Its condition varies from almost intact to broken in half to only isolated timbers. The character of the wood varies from too hard to cut to falling apart at a touch. Early accounts make it sound like local residents visited the ark routinely, while other accounts stress the hardships encountered.

3. Noah's ark is the sort of subject that people would tell stories about. Some people might be motivated by misplaced piety to make up stories. Some have been motivated by money. Others might elaborate a story simply to get attention. Since the ark story is so famous, some people might conclude they have found the ark on the basis of ambiguous evidence. For example, they might misinterpret a blurry photograph or a shape seen through fog, or they might conclude that any wood they find is from the ark, although wood has been carried up Ararat in historical times for building crosses and huts (Bailey 1989, 105).

Further Reading: Bailey, L. R. 1989. *Noah: The Person and the Story in History and Tradition.*

CH501: We can expect to find Noah's ark on Mount Ararat.
We can expect to find Noah's ark on Mount Ararat. (LaHaye and Morris 1976)

1. Mount Ararat (known locally as Agri Dagi) is very likely the wrong place to look. Genesis says only that the ark landed on the mountains of Ararat, where Ararat is not a single mountain but a region (2 Kings 19:37; Isa. 37:38; Jer. 51:27). That region, known in Assyrian records as Urartu, is, roughly, bounded on the west by the Euphrates River, on the south by the western Taurus Mountains (northern Iraq), somewhat east of Lake Urmia, and north to include the plain of the Araxes River. "The mountains of Ararat" implies not a single peak, but a mountainous region within this area, such as the Qardu region (northern Kurdistan) west of Lake Urmia. Early reports of the ark place it on several different mountains, including some in the Qardu region. Mount Ararat is not mentioned as a landing site until the eleventh or twelfth century (Bailey 1989, 61–82).

2. No wooden structure, including the ark, should be expected to survive intact after 4,500 years. The weather on Ararat is harsh and changeable; it would have destroyed the ark if it were exposed. Some people claim that the ark could have survived in a glacier, protected by the covering ice, but this also is unlikely. First, the ice would have crushed the ark. Second, glaciers flow, carrying along whatever is inside them. Different parts often flow at different rates, which would deform the ark. And the ark started on the ground, so it would have been eroded as the glacier dragged it over the rocks.

Further Reading: Bailey, L. R. 1989. *Noah: The Person and the Story in History and Tradition.*

CH503: Noah's ark has been found near Dogubayazit, Turkey.

Noah's ark has been found near Dogubayazit, Turkey. A symmetrical streamlined stone structure near there has the right dimensions, and interior structure and symmetrically arranged traces of metal are consistent with the ark. Also, anchor stones have been found near there. (Wyatt 1989; see also CH503.1)

1. The metal traces that were interpreted as iron brackets were actually goethite, a hydrated iron oxide. This mineral was thoroughly mixed with clay, calcite, quartz, and anthophyllite particles, and it showed a large amount of chemical variability across the sample. Neither of these properties would occur in smelted iron.

The purported walls of the ark are limonite concentrations. Their boatlike shape is consistent with an eroded doubly plunging syncline. The stresses of such folding commonly cause fractures that cut across the layers. Water moving through these fractures would have produced the limonite concentrations that were interpreted as dividing walls.

In short, the structure is consistent with the following geological history:

1. Rocks formed when sediments eroded from nearby volcanic rocks and were compacted.

2. These layers were folded into a doubly plunging syncline.

3. A marine sea eroded a channel into the rocks and deposited fossiliferous limestone in it.

4. The land was uplifted, and erosion removed most of the limestone and exposed the fold.

5. A landslide carried blocks of rock and mud around the synclinal structure.

This interpretation is consistent with the structure itself and with the surrounding geology (L. G. Collins and Fasold 1996).

2. No fossilized wood or traces of wood, reed, or elemental carbon were found associated with the structure (L. G. Collins and Fasold 1996).

Further Reading: Bailey, L. R. 1989. *Noah: The Person and the Story in History and Tradition;* Collins, L. G., and D. F. Fasold. 1996. Bogus "Noah's Ark" from Turkey exposed as a common geologic structure. http://www.csun.edu/~vcgeo005/bogus.html.

CH503.1: Anchor stones of Noah's ark have been found.

Giant anchor stones found in the Durupinar area of the Middle East are too large and too far from water to have been transported by normal means. They are evidence for Noah's ark. (Wyatt 1989, 21–22, 24)

1. The "anchor" stones likely had nothing to do with Christianity or the ark. Such stones were known to have been crafted by pagans for their worship before Christianity came to Armenia. The "rope holes" were niches for lamps. When Christianity came to the region, the stones were Christianized by inscribing Christian symbols on them (Merling n.d.).

2. The rock from which the anchor stones are made is volcanic rock found around Mount Ararat, where the anchor stones were found, but not found in Mesopotamia (Iraq), from which Noah is alleged to have departed (L. G. Collins and Fasold 1996). If the stones were crafted by Noah, they would have come from the region where Noah came from, not where he landed.

Further Reading: Merling, D. n.d. Has Noah's Ark been found? http://www.tentmaker.org/WAR/HasNoahsArkBeenFound1.html.

CH511: Insects survived on floating vegetation mats.

Insects and other invertebrates were not taken aboard the ark during Noah's flood. They survived on vegetation mats. (Whitcomb and Morris 1961, 69; Woodmorappe 1996, 3)

1. Many insects could not survive for a year on vegetation mats. Most insects are specialized at least somewhat for their food or environment. Some of the requirements of various insects include

- living vegetation or flowers to feed on
- dry wood
- soil
- dung
- animal corpses
- shallow streams

In particular, a global flood would have caused the extinction of most aphids, drywood termites, dung beetles, burying beetles, black flies, mayflies, ground beetles, and many more, unless special care were taken to ensure their survival.

2. A global flood would cause the extinction of millions of species of insects and other invertebrates simply as a result of the reduced quantity of habitat. Insect species are going extinct today simply from the cutting down of sections of forests. A global flood would be many orders of magnitude more devastating. Given the fact that insects are alive today, if there was a flood, Noah must have gathered them and saved them with the rest of the animals.

3. The Bible says that Noah took "every creeping thing on the ground" and that these were distinct from animals (Gen. 6:20, 7:8, 14). It further says that all life that was not aboard the ark was killed, including creeping and swarming things (Gen. 7:21–23). There is not the slightest bit of biblical support for anything living on vegetation mats and a great deal of biblical contradiction of the idea. Obviously, the only reason to put insects on mats is so the ark apologists do not have to worry about them.

CH512: All kinds could fit.

Noah's ark could have carried pairs of all kinds of animals for a year. (Woodmorappe 1996, 1–44)

1. Woodmorappe (1996) has done a detailed analysis of the possibility of fitting all animals aboard the ark. He found that the animals, together with the food and water they require, would fit in about 90 percent of the available space. However, he made several invalid assumptions that, when corrected, fill the ark past overflowing (Isaak 1998).

- The "kinds" used in Woodmorappe's calculations were genera. Taking individual species, which is a much more reasonable definition of kind in the context of the ark, increases the load three- or fourfold.

- Woodmorappe did not account for the extra clean animals, considering their number negligible. However, he believed that the only clean animals would be thirteen domestic ruminants traditionally considered clean. But if the Bible is taken literally, all ruminants would be considered clean. Under Woodmorappe's assumption, the extra clean animals would increase the load by 1.5 percent, or 3 percent if you include seven pairs of the animals. Taking all ruminants increases the load by 14 or 28 percent.

- Woodmorappe included only juveniles of animals larger than about 10 kg (see CH512.1). This assumption, however, is unbiblical and, for some animals, impractical. Taking adult animals would increase the total mass more than thirteenfold. Taking even some of these animals as adults or taking older juveniles could easily fill the ark beyond capacity.

- According to the creation model, dinosaurs and other animals now extinct would have been alive at the time of the flood and therefore would be aboard the ark. The only extinct animals that Woodmorappe included in his calculations were the ones that were known at the time. Since then, many other dinosaur genera have been discovered, and no doubt there are many more as yet undiscovered.

- Woodmorappe excluded land invertebrates from his calculations, despite the fact that they must have been aboard the ark (see CH511). These animals are small enough that they alone would not have increased the load significantly, but they are numerous enough and have many special requirements, so the infrastructure needed to house and care for them would have been significant.

- Woodmorappe made no allowance for food spoilage or water wasted from spilling, although the conditions he described aboard the ark guarantee that both of these problems would have been severe.

Further Reading: Isaak, M. 1998. Problems with a global flood. http://www.talkorigins.org/faqs/faq-noahs-ark.html.

CH512.1: Juveniles of large animals were taken aboard.
There was room for all the animals aboard Noah's ark because juveniles of the largest animals (animals larger than about 10 kg = 22 lb.) were taken. (Woodmorappe 1996, 13)

1. Even assuming the largest animals were juveniles, the ark would still have been overcrowded (see CH512).

2. Noah was instructed to take "the male and its mate," implying sexual maturity of the animals (Gen. 7:2). Juveniles do not have mates.

3. Juveniles must still be old enough to be weaned and, in some cases, socialized to learn behaviors from their parents. This would make many of the animals old enough to be mostly grown already.

CH512.2: The average land animal is the size of a sheep.
The average size of the animals aboard the ark was the size of a sheep or less, so Noah's ark was not overly crowded. (Whitcomb and Morris 1961, 69; Woodmorappe 1996, 13)

1. It is the total size of all animals aboard the ark that matters, not the average size. The total size is greater than would have been practical for the ark to carry (Isaak 1998).

2. The average size of vertebrate aboard the ark was 347 kg (765 lb.), according to the values used by Woodmorappe (1996). Woodmorappe said that Whitcomb and Morris's sheep comparison is overly generous because the median size is that of a large rat, but the median statistic is not useful for anything in this context. Woodmorappe placed 15,746 animals aboard the ark, for a total weight of 5,464,000 kg. This is equivalent to 347 kg per animal.

Further Reading: Isaak, M. 1998. Problems with a global flood. http://www.talkorigins.org/faqs/faq-noahs-ark.html.

CH514: The crew could feed and care for the animals.
The eight-person crew aboard Noah's ark was sufficient to feed and care for all the animals. (Woodmorappe 1996, 71–81)

1. Three hundred and twenty full-time employees are needed to care for fewer than 3,000 animals at the Washington National Zoo (Grimaldi and Barker 2003). Granted,

many of these would be working on administration and visitor concerns that would not have existed on the ark. Still, assuming that only a quarter of them cared for animals, that is still eighty people to care for 3,000 animals. On the ark, there were eight people to tend more than 15,000 animals (assuming Noah's crew were not needed to do maintenance and bail water). They would have had to work more than fifty times harder than professional zookeepers. Double shifts would not have been enough to make up the difference.

Accepting Woodmorappe's number of 15,754 animals aboard the ark, and assuming the crew attended to them sixteen hours per day (a very generous assumption), each animal would receive an average of about thirty seconds of attention per day for all its needs.

2. Labor-saving mechanisms proposed by Woodmorappe are unrealistic. For example:

- Watering many animals at once via troughs would not work on a ship. Most of the water would slosh out as the ark rolled with the waves.

- Automatic feeders would allow pests to infest the food. Animals with automatic feeders would probably eat more and waste more food, too, increasing the amount of food that must be stored. Woodmorappe did not account for the extra space required.

- At least one third, and probably two thirds, of the manure could not be disposed of by simply pushing it overboard, since it would be below the water line. The manure would have to be carried up a deck or two.

3. Woodmorappe did not consider some time-consuming tasks:

- The ark itself had to be maintained. It would be a miracle if bailing alone were less than a full-time job.

- All of the hoofed animals would need to have their hooves trimmed several times during the year (R. P. Batten 1976, 39–42).

CH514.1: Many animals do not require fresh or live food.
Although a few animals have specialized food needs, such animals are rare, the needs are often exaggerated, and the specialized diets are not labor or space intensive. In short, the specialized dietary needs of animals do not prevent the voyage of Noah's ark from being feasible. (Woodmorappe 1996, 111–117)

1. Woodmorappe did not consider all animals with special needs. Just because some snakes can be coaxed to eat nonliving food, for example, does not mean all can.

Most problematical, Woodmorappe did not consider terrestrial invertebrates, especially insects, which must have been on the ark (see CH511). Many insects only eat a single species of plant. Keeping all the plants alive for a year would have taken considerable resources.

2. Some animals' needs may be exaggerated, but Woodmorappe grossly exaggerated how easy it would be to deal with them. Many animals, such as the platypus, are difficult to keep alive during transport even in the best of conditions (Fleay 1958). Noah could give hardly any attention to individual animals and would have had to keep them in nigh intolerable conditions. Modern livestock shipping often results in high casualty

rates, even though only domestic animals are shipped, and they are at sea only a few weeks.

3. Woodmorappe noted that some animals can be fed artificial diets. He failed to note that the artificial diets were developed by the work of hundreds of researchers working over tens to hundreds of years. Noah would not have had that knowledge to draw upon.

4. Some of Woodmorappe's solutions to feeding problems have problems of their own. He proposed feeding insectivores by breeding insects on grain in special compartments, and letting the insects escape into the cages of the insectivores through perforated pipes. Some of the escaping insects, however, would escape into the general grain stores, reducing a great deal of the food to waste before the voyage was over.

CH521: Animals' exacting needs could have evolved after the Flood.

The specialized dietary needs of many animals might have come about only after the Flood via microevolution. Microevolution could also account for climate preferences, lack of dormancy, wild temperament, and other traits, meaning that Noah never would have had to face many of the challenges that would be posed by animals in their present form. (Woodmorappe 1996, 61, 116–117, 125, 134)

1. It is ironic that someone opposed to evolution would invoke evolution as a magic wand to solve so many problems. The rates of evolution proposed by Woodmorappe are far greater than the evolution rates that biologists propose to account for common descent of all plants and animals from a common ancestor.

Woodmorappe (1996, 5–7) further proposed that all species evolved after the Flood from representative genera or families aboard the ark. Since the evolution Woodmorappe proposed involves speciation and has no barriers to change, it is unquestionably macroevolution, not microevolution.

2. Rapid evolution requires populations that include lots of variation already; the evolution then proceeds via selection of existing variation. If there is little or no variation in the population already, nonharmful mutations must first occur to provide some variation, and evolution is much slower. According to the Flood story, all populations would have begun from just two individuals, making variation virtually nil. (Few populations would have had the capacity even to survive normal environmental fluctuations; Simberloff 1988). The populations would not have had the genetic variation to allow microevolution of specialized traits to be common.

CH541: Aquatic organisms could have survived the Flood.

Present-day fish and other aquatic organisms could have survived the Flood. Many freshwater fish can survive in salt water, and many saltwater fish can tolerate fresh water. The floodwaters may have been layered by salinity, allowing others to find their preferred habitat. (Woodmorappe 1996, 140–152)

1. Layering of the floodwaters contradicts the Flood model, which proposes that the Flood was turbulent enough to stir up sediments on an incredible scale. The model pro-

poses that the floodwaters became the present oceans, so all the water flowing into the oceans would have ensured that they were well mixed. The freshwater fish would have had no place to find fresh water.

2. The fact that many fish can tolerate wide ranges in salinity does not mean that all can. Furthermore, the problem applies to more than fish. Freshwater invertebrates are commonly used as indicators of the health of streams. Even a tiny amount of pollution can cause many species to disappear from the stream.

3. Aquatic organisms would have more than salinity to worry about, such as the following:

- Heat. All mechanisms proposed to cause the Flood would have released enough heat to boil the oceans. The deposition of limestone would release enough heat to boil them again. Meteors and volcanoes that occurred during the Flood, as implied by their presence in layers attributed to the Flood by flood geologists, would probably have boiled them again (Isaak 1998; Woodmorappe [1996, 140] dismissed the problem of volcanoes but ignored all the other sources of heat).

- Acid. The volcanoes that erupted during the Flood would also have produced sulfuric acid, enough to lower the pH of the ocean to 2.2, which would be fatal to almost all marine life (Morton 1998b).

- Substrate. Many freshwater and marine invertebrates rely on a substrate. They anchor themselves on the substrate and rely on currents to carry their food to them. During the Flood, substrates would have been uninhabitable at least part of the time, especially on land. Woodmorappe (1996, 141) suggested floating pumice as a substrate, but it would float with the currents, so currents would not bring nutrients to animals on them.

- Pressure. The Flood would have caused great fluctuation in sea pressures. Many deep-sea creatures invariably die from the decompression when brought to the surface. Other surface animals would die from too much pressure if forced deep underwater.

4. Woodmorappe predicted a sudden extinction of fish caused by the Flood: "[P]resent-day marine life is but an impoverished remnant of that which had originally been created and had existed before the Flood" (1996, 142). However, the actual pattern of extinction we see shows convincing disproof of the Flood. Living genera become decreasingly represented in fossils as one goes deeper in the geological column, until there are no recent genera in the Triassic, and only about 12 percent of recent genera have any fossil record. Extinct genera continue back to the Cambrian (Morton 1998a). This pattern exactly matches what one would expect from evolution. It contradicts a global flood, which should include modern fish more-or-less uniformly throughout the flood-deposited sediments.

CH542: Plants could have survived the Flood.
All existing kinds of plants could have survived Noah's Flood. (Woodmorappe 1996, 153–162)

1. Not all plants could survive the Flood for some of the following reasons:

- Many plants (seeds and all) would be killed if soaked for several months in water, especially salt water.

- Some plants do not produce seeds; they would have been killed when the Flood either uprooted or covered them.

- Not all seeds could survive a year before germinating (Benzing 1990; Densmore and Zasada 1983; Garwood 1989).

2. The Flood was an ecological catastrophe. Creationists credit it with eroding and redepositing sediments miles thick, raising mountains, carving immense canyons, and even repositioning continents. This alone would doom many plants to extinction, even if they or their seeds survived the Flood, for some of the following reasons:

- Most of the world's seeds would have been buried under many feet—even miles—of sediment. This would keep them from sprouting.

- Many plants require particular soil conditions to grow. The Flood would have eroded away all the topsoil which provides the optimum conditions for most plants.

- Some seeds will germinate only after being exposed to fire. After the Flood, there would have been nothing to burn.

- Most flowering plants are pollinated by insects, but the only insects around after the Flood would have been those Noah carried aboard the ark. The surviving seeds would have had to find the proper conditions of soil type and burial depth in a small area around where the ark landed.

- Plants live not as individuals, but as communities. If you cut down the redwoods, you kill not only the redwoods but also dozens of other plants that depend on the community structure. After the Flood, there would have been no ecological communities, only bare land. Any plant that depends on a mature community (for shade, shelter, humidity, or support, for example) could not survive until such a community matures, which usually takes years to decades.

Woodmorappe (throughout his book, not just regarding plants) made two fundamental errors:

1. He noted that "many" could survive the flood conditions, disregarding the significant number that could not but that are alive anyway.

2. He assumed that plants and animals could live in isolation, ignoring that life lives in, and depends upon, ecologies. Simply preserving plants and animals would keep them alive for a very short time. Noah would have had to rebuild many entire ecologies to maintain the life we see today.

3. Evolution predicts the geographical distribution of plant kinds that we observe, with many species occurring on one continent and not others. Flood geology predicts that this pattern would not occur. Flood theory fails.

CH550: The geologic column was deposited by the Flood.
The Flood deposited the geologic column.

1. Varves within the geologic column show seasonal layers over many, many years. In many cases, such as the Green River formation, these layers are too fine to have settled

out in less than several weeks per layer. Varves in New England show evidence of climate change 17,500 to 13,500 years ago, which matches climate patterns in other parts of the world (Rittenour et al. 2000). These layers prove that the geological record was not produced in just one event.

2. There are many different kinds of surface features preserved in the middle of the geological column. These features include soils, mud cracks, evaporite deposits, footprints, raindrop impressions, meteor craters, worm burrows, wind-blown sediments, stream channels, and many others. For example:

- The Loess Plateau in China has a layer of loess more than 300 m thick. Loess is wind-blown sediment that would not occur during a global flood. The Loess Plateau occurs around the downwind edges of the Ordos Desert, its source of sediments, and the grain size of the loess decreases the further one gets from the desert (Vandenberghe et al. 1997).

- The Loess Plateau includes paleosols within it. These are buried fossil soils, some of which would require tens of thousands of years to form (Kukla and An 1989; Liu et al. 1985).

Further Reading: Harding, K. 1999. What would we expect to find if the world had flooded? http://www.creationism.ws/what_if_flood.htm; Isaak, M. 1998. Problems with a global flood. http://www.talkorigins.org/faqs/faq-noahs-ark.html; Miller, H. 1857. *The Testimony of the Rocks.* See also MacRae, A. n.d. Hugh Miller—19th-century creationist geologist. http://home.tiac.net/~cri/1998/miller.html; Strahler, A.N. 1987. *Science and Earth History.*

CH561.1: Fossils are sorted by ecological zonation.

Patterns of fossil deposition in Noah's Flood can be explained by ecological zonation. The lower strata, in general, would contain animals that lived in the lower elevations. Thus, marine invertebrates would be buried first, then fish, then amphibians and reptiles (who live at the boundaries of land and water), and finally mammals and birds. Also, animals would be found buried with other animals from the same communities. (H.M. Morris 1985, 118–120)

1. The fossil record does not show such a pattern of organisms sorted ecologically:

- Many animals that appear in the lower strata appear in all strata, even recent ones. Corals and clams, for example, appear at all levels.

- Whales do not appear until much later than fish, despite living in the same ecological zones.

- Birds do not appear until after flying reptiles.

- Dinosaurs consistently appear in strata before modern land animals.

- Grasses live in virtually all land areas, but they appear in the geological record only near the top, long after other land animals and plants.

2. Even if ecological zonation could explain how deeply various faunal zones are buried, it does not explain how they came to be buried atop one another. How did a terrestrial ecology come to be transported on top of a marine ecology, such that fine details such as footprints, burrows, and paleosols were undisturbed and such that the layer extends over hundreds of square miles? How did many such layers get stacked on top of

each other? Ecological zonation implies that the ecological zones got buried in place. What we see is ecological zones forming and living for awhile on top of the fossils of older ecological zones, repeatedly.

3. Fossil strata often appear in orders that contradict ecological zonation (and other flood deposition explanations). For example, North American midcontinent outcrops record at least fifty-five cycles of marine inundation and withdrawal (Boardman and Heckel 1989; Heckel 1986). That is, marine ecologies are interleaved with terrestrial ecologies.

CH561.2: Fossils are sorted hydrologically.
The order of fossils deposited by Noah's Flood, especially those of marine organisms, can be explained by hydrologic sorting. Fossils of the same size will be sorted together. Heavier and more streamlined forms will be found at lower levels. (Whitcomb and Morris 1961, 273–274)

1. Fossils are not sorted according to hydrodynamic principles. Ammonites, which are buoyant organisms similar to the chambered nautilus, are found only in deep strata. Turtles, which are rather dense, are found in middle and upper strata. Brachiopods are very similar to clams in size and shape, but brachiopods are found mostly in lower strata than are clams. Most fossil-bearing strata contain fossils of various sizes and shapes. Some species are found in wide ranges, while others are found only in thin layers within those ranges. Hydrologic sorting can explain none of this.

2. The sediments in which fossils are found are not hydrodynamically sorted. Coarse sediments are often found above fine sediments. Nor are the sediments sorted with the fossils. Large fossils are commonly found in fine sediments.

3. A catastrophic flood would not be expected to produce much hydrologic sorting. A flood that lays down massive quantities of sediment would jumble up most of them.

CH561.3: Fossils are sorted by the ability to escape.
The order of fossils in the fossil record is explained by the animals' ability to escape the rising floodwaters. Slow animals, such as clams, are found low in the fossil record, while quicker animals, such as mammals and birds, appear higher. (H.M. Morris 1985, 119)

1. Fossils are not sorted according to their ability to escape rising floodwaters. If they were, we would expect to see slow-moving species like sloths and tortoises and every low-elevation plant at the bottom of the fossil record, while fast-moving species, such as velociraptors, pterosaurs, and giant dragonflies, would be at the top. But this is nothing like what we actually observe; in many cases we find just the opposite. For example, in undisturbed strata there has not been a single sloth fossil found below even the highest velociraptor remains, and flowering plants do not appear in the fossil record until after winged insects and reptiles.

2. Even common present-day floods trap all manner of people and animals. The violence of a flood that could cover the entire earth in forty days would be bound to trap many individuals from even fast-moving species, especially those that were old and infirm, crippled, or trapped in low-lying areas. Therefore, we would expect to find the occasional member of fast-moving species near the bottom of the fossil record. However, the vast majority of fossilized species are only found within certain relatively narrow ranges within the fossil record. For example, human fossils are only found at the very top of the fossil record (Pleistocene period and later), and tyrannosaurs are only found at the end of the Cretaceous period.

3. The fossil record preserves entire ecosystems, not just individual species. Fossils of one species are found in association with fossils of other species common to their ecosystem. If fossil distribution is dependent on the ability to escape rising floodwaters, then all the species within an ecosystem must be equally capable of escape for them to be preserved together. But since these associated species include both highly motile animals and completely nonmotile plants, this is obviously not the case.

Further Reading: Cuffey, C. A. 2001. The fossil record. http://www.gcssepm.org/special/cuffey_00.htm.

by Derek Mathias

CH561.4: Fossils are sorted by a combination of these factors.
The order of fossils deposited by Noah's Flood can be explained by a combination of hydrologic sorting, differential escape, and ecological zonation. (H. M. Morris 1985, 118–120)

1. Even this combination of forces fails to explain many aspects of the sorting of fossils. In particular, the problems with ecological zonation (see CH561.1) are not significantly mitigated by the other two sorting methods.

2. Conventional geology explains the geological record, including its fossils, mineral content, geomagnetism, and radioisotopes. It does so in great detail and with great consistency in many, many places. Flood geology does not give a detailed explanation anywhere; what little explanation it offers is extremely vague hand waving, inconsistent with observations.

3. The geological record contains more than biological fossils. Sediment patterns also record planetary rhythms from which we can determine the length of the day and long-term changes in climate. From these, we find that the geological record shows that the moon is slowing consistent with tidal friction (see CE110; see also Sonett et al. 1996) and that climate changes often follow the Milankovitch cycles (Krumenaker 1995). These "fossils" depend only on astronomical forces; they could not be explained by the Flood.

4. There are innumerable other observations that contradict a global flood (Isaak 1998).

Further Reading: Isaak, M. 1998. Problems with a global flood. http://www.talkorigins.org/faqs/faq-noahs-ark.html#georecord; Morton, G. R. 2001a. The geologic column and its implications for the flood. http://www.talkorigins.org/faqs/geocolumn.

CH570: High mountains were raised during the Flood.

The earth was relatively flat before the Flood. Most of the world's high mountains were formed during the Flood. This explains how all the waters in the oceans could cover all the mountains at the time. It also explains how mountains formed (from the violence accompanying the Flood) and the existence of marine fossils on mountains. (Whitcomb and Morris 1961, 127–128; see also CC364)

1. This claim originated before the theory of plate tectonics existed as an explanation for mountain building. Plate tectonics, however, solved the problem in terms of relatively gradual processes we see working (and still building mountains) today. All the major mountain ranges have been studied in detail, the plate movements that caused them have been mapped, and their histories have been worked out for millions of years in the past. The problem of mountain formation has been solved, and a flood had no part in the solution.

2. The catastrophic formation of mountains and subsequent return of the sea into its basin would have released tremendous amounts of heat and mechanical energy, enough to boil the oceans and metamorphose the minerals in the mountains. No trace of such a catastrophe exists.

3. Formation of mountains during the Flood does not explain why different mountains are different ages. The Appalachians are much older than the Rockies, for example, as one can immediately see just from how the two ranges are differently eroded.

Further Reading: McPhee, J. 1998. *Annals of the Former World.*

CH580: The Flood shaped the earth's surface in other ways.

All geological features are consistent with a global flood, including plateaus, overthrusts (see also CD102.1), canyons, submarine trenches, and geosynclines. (W. Brown 1995; Whitcomb and Morris 1961)

1. A major flood leaves distinctive features, including
- a wide, relatively shallow bed, not a deep, sinuous river channel.
- anastamosing channels (i.e., a braided river system), not a single, well-developed channel.
- coarse-grained poorly sorted sediments, including boulders and gravel, on the floor of the canyon.
- giant ripple marks.
- streamlined relict islands.

Such features are observed on a local scale in the Channeled Scablands (McMullen 1998; Parfit 1995) and on Mars (NASA Quest n.d.; see Figure CH1). They are far from global, though.

2. Almost all features of the earth can be explained by conventional geology, including processes such as plate tectonics and glaciation. A global flood does not help to explain any of the exceptions.

Further Reading: Harding, K. 1999. What would we expect to find if the world had flooded? http://www.creationism.ws/what_if_flood.htm.

CH581: The Grand Canyon was carved by retreating flood waters.
The Grand Canyon was created suddenly by the retreating waters of Noah's Flood. (Austin 1995)

1. We know what to expect of a sudden massive flood, such as a wide and shallow bed, braided channels, and streamlined islands (see CH580). The Scablands in Washington state were produced by such a flood and show such features (Allen et al. 1986; Baker 1978; Bretz 1969; Waitt 1985). Such features are also seen on Mars at Ares Vallis (Baker 1978; NASA Quest n.d.). They do not appear in the Grand Canyon (see Figure CH1).

2. The same flood that was supposed to carve the Grand Canyon was also supposed to lay down the miles of sediment (and a few lava flows) from which the canyon is carved. A single flood cannot do both. Creationists claim that the year of the Flood included several geological events, but that still stretches credulity.

3. The Grand Canyon contains some major meanders. Upstream of the Grand Canyon, the San Juan River (around Gooseneck State Park, Southeast Utah) has some of the most extreme meandering imaginable. The canyon is 1,000 feet high, with the river flowing five miles while progressing one mile measured in a straight line (American Southwest n.d.). There is no way a single massive flood could carve this.

4. Recent flood sediments would be unconsolidated. If the Grand Canyon were carved in unconsolidated sediments, the sides of the canyon would show obvious slumping.

5. The inner canyon is carved into the strongly metamorphosed sediments of the Vishnu Group, which are separated by an angular unconformity from the overlying sedimentary rocks, and also in the Zoroaster Granite, which intrudes the Vishnu Group. These rocks, by all accounts, would have been quite hard before the Flood began.

6. Along the Grand Canyon are tributaries, which are as deep as the Grand Canyon itself. These tributaries are roughly perpendicular to the main canyon. A sudden massive flood would not produce such a pattern.

7. Sediment from the Colorado River has been shifted northward over the years by movement along the San Andreas and related faults (Winker and Kidwell 1986). Such movement of the delta sediment would not occur if the canyon were carved as a single event.

8. The lakes that Austin proposed as the source for the carving floodwaters are not large compared with the Grand Canyon itself. The flood would have to remove more material than the floodwaters themselves.

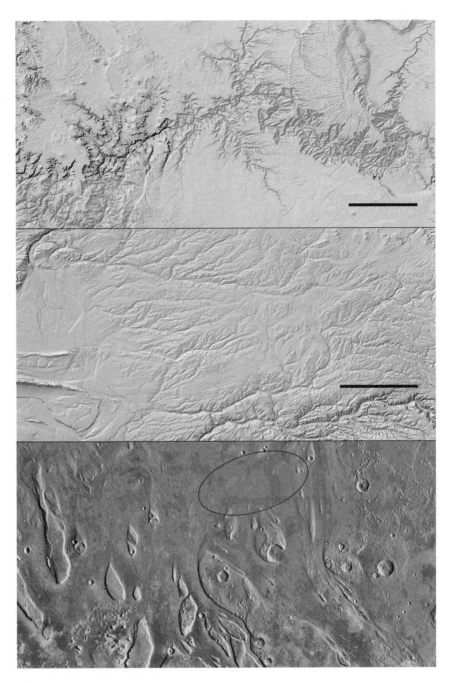

Figure CH1

Nonflood versus flood-shaped terrain. *Top*: Grand Canyon, Arizona. *Middle*: Scablands (southwest of Spokane), Washington. *Bottom*: Ares Vallis, Mars. Scale bars are 20 miles (32.2 km); ellipse is 124 miles (200 km) long. Note that the Scablands and Mars have braided stream beds and relict streamlined islands (evidence of a flood), whereas the Grand Canyon has sharp relief and a single meandering channel (evidence against a flood). Relief maps of Grand Canyon and Scablands from USGS, http://nmviewogc .cr.usgs.gov/viewer.htm; Mars photo from NASA Quest.

9. If a brief interlude of rushing water produced the Grand Canyon, there should be many more such canyons. Why are there not other grand canyons surrounding all the margins of all continents?

10. There is a perfectly satisfactory gradual explanation for the formation of the Grand Canyon that avoids all these problems. Sediments deposited about two billion years ago were metamorphosed and intruded by granite to become today's basement layers. Other sediments were deposited in the late Proterozoic and were subsequently folded, faulted, and eroded. More sediments were deposited in the Paleozoic and Mesozoic, with a period of erosion in between. The Colorado Plateau started rising gradually about seventy million years ago. As it rose, existing rivers deepened, carving through the previous sediments (D. V. Harris and Kiver 1985, 273–282).

Further Reading: Allen, J. A., et al. 1986. *Cataclysms on the Columbia*; Beus, S. S., and M. Morales (eds.). 2002. *Grand Canyon Geology*; Chronic, H. 1983. *Roadside Geology of Arizona*; Elders, W. A. 1998. Bibliolatry in the Grand Canyon; Woolf, J. 1999. Young-earth creationism and the geology of the Grand Canyon. http://my.erinet.com~jwoolf/gc_intro.html.

CH581.1: Rapid erosion on Mount St. Helens shows that the Grand Canyon could form suddenly.

Rapid erosion of sediments along the north fork of Toutle River, flowing out of Spirit Lake on Mount St. Helens, carved a canyon like a miniature Grand Canyon, showing that the Grand Canyon could form suddenly. (Austin 1986)

1. The sediments on Mount St. Helens were unconsolidated volcanic ash, which is easily eroded. The Grand Canyon was carved into harder materials, including well-consolidated sandstone and limestone, hard metamorphosed sediments (the Vishnu schist), plus a touch of relatively recent basalt.

2. The walls of the Mount St. Helens canyon slope 45 degrees. The walls of the Grand Canyon are vertical in places.

3. The canyon was not entirely formed suddenly. The canyon along Toutle River has a river continuously contributing to its formation. Another canyon also cited as evidence of catastrophic erosion is Engineer's Canyon, which was formed via water pumped out of Spirit Lake over several days by the U.S. Army Corps of Engineers.

4. The streams flowing down Mount St. Helens flow at a steeper grade than the Colorado River does, allowing greater erosion.

5. The Grand Canyon (and canyons further up and down the Colorado River) is more than 100,000 times larger than the canyon on Mount St. Helens. The two are not really comparable.

CH590: The Flood caused an ice age.

The great release of energy during the Flood caused much water from the new oceans to enter the atmosphere. This moisture fell at the poles as snow and caused the Ice Age. (H. M. Morris 1985, 126–127)

1. Adding heat to a system tends to make it hotter. The falling moisture would have been a hot rain, not snow.

Creationists invoke evaporation as a cooling method. They forget that all the heat lost to evaporation returns when the water condenses again and that more latent heat is then released in the freezing.

2. A proper ice age cannot fit into a young-earth timescale. For a continent-scale glacier to form, advance enough to change the landscape, and retreat takes centuries or more, not a decade.

Cores from ice sheets reveal annual layers that date back 160,000 years in places. Volcanic eruptions recorded in the top few thousand years match historic records. The top 4,000 or so layers have to be annual layers. It is unlikely that the other 156,000 layers were laid down in just a few years (Brinkman 1995).

3. The earth under the ice sheets is isostatically adjusted to the mass of ice. Even if 10,000 or more feet of ice were dropped on Greenland and Antarctica in only a few years about 4,000 years ago, it would take over 12,000 years to reach the observed (today) degree of adjustment. Scandinavia and Canada are still rebounding from the disappearance of glaciers covering them at the end of the last ice age (Strahler 1987, chap. 27). It would have taken thousands of additional years for the weight of the ice to push them down in the first place.

4. There are multiple lines of evidence for many glacial advances and retreats in the last two million years (Shackleton 2000):

- Species of foraminifera vary with ocean temperature. The variation is recorded in deep ocean sediments, showing many long-term changes (Strahler 1987, 252).

- Oxygen isotope ratios ($^{18}O/^{16}O$) indicate when more water is held in glaciers (because ^{16}O evaporates more easily and so is disproportionately common in snow). This ratio is recorded in carbonate shells in seafloor sediments; it shows the same sort of variation (Strahler 1987, 253).

- Formations caused by glaciers on Mauna Kea show that the volcano experienced at least four glaciations. Lava flows between the formations show that the glaciations were separate (Strahler 1987, 255).

- The water level of Lake Bonneville (of which Great Salt Lake is a remnant) rose in times of glaciation, leaving different fossil shorelines. Simultaneously, glaciers from the Rockies advanced, leaving superposed morainal material. Research on these deposits reveals glaciations at about 125,000, 200,000, 300,000, 400,000, and 440,000 years ago, plus several more cycles between 500,000 and 800,000 years ago (Strahler 1987, 255).

Furthermore, there is evidence for ice ages in the late Ordovician and in the late Carboniferous to early Permian (Strahler 1987, 265). These ice ages would have had to occur in the middle of the Flood. They cannot be easily discounted because they are indicated by just the same kind of evidence that causes creationists (and mainstream geologists) to recognize a recent ice age.

5. Changes in climate are correlated with Milankovitch cycles, long-term cycles in the earth's orbit (Lindsay 1997a).

Further Reading: Alley, R. B., and M. L. Bender. 1998. Greenland ice cores.

Miscellaneous Young-Earth Creationism

CH710: Man and dinosaurs coexisted.
Humans and dinosaurs once lived together. (Woodmorappe 1996)

1. There are no human fossils or artifacts found with dinosaurs, and there are no dinosaur fossils found with human fossils (except birds, which are descended from dinosaurs; out-of-place human traces, such as the Paluxy footprints [see CC101], do not withstand examination). Furthermore, there is an approximately sixty-four-million-year gap in the fossil record when there are neither dinosaur nor human fossils. If humans and dinosaurs coexisted, traces of the two should be found in the same time places. At the very least, there should not be such a dramatic separation between them.

2. If dinosaurs and humans were found together, it would not be evidence against evolution (see CB930).

CH710.1: Ica stones show that humans and dinosaurs coexisted.
Ica stones, collected by Dr. Javier Cabrera Darquea near the village of Ica in Peru, show ancient drawings of humans hunting or otherwise interacting with living dinosaurs. (Berlitz 1984, 179–181)

1. The stones are almost certainly modern, created by local villagers to sell to gullible tourists. Two peasants from Callango, Basilio Uchuya and his wife, Irma Gutierrez de Aparcana, have admitted to carving the stones they sold to Cabrera, basing their designs on illustrations from comic books, school books, and magazines (Polidoro 2002).

2. The stones cannot be dated without knowing their source, and their source has never been revealed.

3. The Ica stones reputedly give evidence for a highly advanced, very ancient civilization, but no other trace of such a civilization exists. (The Nazca drawings are nearby, but they do not depict any dinosaurs or evidence for advanced technology.)

Further Reading: Mathews, D. 2000. Domesticated dinosaurs? http://www.geocities.com/athens/agora/3958/weekly/weekly56.htm.

CH710.2: Dinosaur figurines from Acambaro show a human–dinosaur association.
Thousands of clay and stone figurines discovered in Acambaro, Mexico, include figurines of dinosaurs. They are apparently from the preclassical Chupicuaro Culture (800 B.C.E. to 200 C.E.). Radiocarbon and thermoluminescent dating gives them even older ages. These figurines

show that the ancient people were familiar with dinosaurs. (Berlitz 1984, 181; Swift et al. n.d.)

1. The figurines show every evidence of being recent folk art, fraudulently buried in an archeological excavation. Di Peso (1953) made the following observations:

- The surfaces of the figurines were new, not marred by a patina characteristic of genuinely old artifacts from the same area. Edges of depressions were sharp and new. No dirt was packed into crevices.

- Genuine archeological relics of fragile items are almost always found in fragments. Finding more than 30,000 such items in pristine condition is unheard of. The excavators of the artifacts were "neither careful nor experienced" in their field technique, yet no marks of their shovels, mattocks, or picks were noted in any of the 32,000 specimens. Some figurines were broken, but the breaks were unworn and apparently deliberate to suggest age. No parts were missing.

- "The author spent two days watching the excavators burrow and dig; during the course of their search they managed to break a number of authentic prehistoric objects. On the second day the two struck a cache and the author examined the material *in situ*. The cache had been very recently buried by digging a down sloping tunnel into the black fill dirt of the prehistoric room. This fill ran to a depth of approximately 1.30 m. Within the stratum there were authentic Tarascan sherds, obsidian blades, tripod metates, manos, etc., but these objects held no concern for the excavators. In burying the cache of figurines, the natives had unwittingly cut some 15 cms below the black fill into the sterile red earth floor of the prehistoric room. In back-filling the tunnel they mixed this red sterile earth with black earth; the tracing of their original excavation was, as a result, a simple task" (Di Peso 1953, 388).

- Fresh manure was found in the tunnel fill.

- Fingerprints were found in freshly packed earth that filled an excavated bowl.

2. The story of their discovery gives a motive for fraud. Waldemar Julsrud, who hired workers to excavate a Chupicuaro site in 1945, paid workers a peso apiece for intact figurines. It very well may have been more economical for the workers to make figurines than to discover and excavate them. Given the quantity that he received, the contribution to the peasants' economy would have been substantial.

3. The figurines are not from the Chupicuaro. They came from within a single-component Tarascan ruin. The Tarascan are post-classical and historical, emerging between 900 and 1522 C.E.

4. If authentic, the figurines imply even more archeological anomalies:

- If the figurines really were based on actual dinosaurs, why have no dinosaur fossils been found in the Acambaro region?

- Why did no other Mexican cultures record any dinosaurs?

- What caused the dinosaurs to disappear in the last 1,100 years?

Further Reading: Di Peso, C.C. 1953. The clay figurines of Acambaro, Guanajunto, Mexico.

CH711: Behemoth, from the book of Job, was a dinosaur.

Behemoth, from Job 40:15–24, was a dinosaur. Job 40:17 says, "His tail sways like a cedar."
Such tails only existed on dinosaurs. (T. Willis 1997)

1. There is no evidence to support such a claim. Fantastic creatures appear in folklore from all times and places. There is no reason to believe that the ancient Hebrews would be different.

2. The "tail like a cedar," which creationists think indicates a large dinosaur, is not even a real tail. "Tail" was used as a euphemism in the King James version. A more likely translation for the phrase is "his penis stiffens like a cedar" (Mitchell 1987). The behemoth was probably a bull, and the cedar comparison referred to its virility.

CH711.1: Leviathan, from the book of Job, was a dinosaur.

Leviathan, described in Job 41 and mentioned in Psalms 104:26, describes a dinosaur, perhaps
Parasaurolophus *or* Corythosaurus. (Gish 1977, 30, 51–54)

1. Leviathan appears also in Ugaritic texts, where it is described as a twisting serpent (echoing language from Isa. 27:1) with seven heads. It personifies the waters of the primeval chaos. The rousing of Leviathan in Job 3:8 implies an undoing of the process of creation (J. Day 1992).

It has also been suggested that Leviathan was a crocodile or whale, but its multiple heads (referred to also in Ps. 74:14) make it clear that it is a fantastic creature, such as appear in folklore from all times and places.

2. Leviathan is clearly described as a sea creature in the Bible. *Parasaurolophus* and *Corythosaurus* were terrestrial.

3. The message of Job 41 is that part of nature is indomitable, that "no purpose of [God's] can be thwarted" (Job 42:2). That message would lose its meaning if Leviathan was an ordinary animal that humans would be able to kill. The larger message of Job is that God's ways cannot always be understood. That message is best served by leaving Leviathan mythical.

CH712: Dragons were dinosaurs.

Legends about dragons are really actual accounts of man meeting up with dinosaurs. (P. Taylor 1998)

1. Folklore does not require a physical basis. Leprechauns, the Loch Ness Monster, djinni, the tooth fairy, and other creatures have long survived in folklore without any bodies to examine.

2. Men never met up with living dinosaurs (see CH710).

3. Dinosaurs need not be living to inspire myth and legend. In China, fossil bones (of all kinds of creatures, not just dinosaurs) have long been called dragon bones. Fossils of

Protoceratops inspired legends of griffins (Mayor 2000). In Lakota myth, dinosaur fossils in the Badlands of South Dakota are attributed to river monsters (Erdoes and Ortiz 1984, 220–222). The Pawnee attribute fossils to a former race of giants (Grinnell 1961, 355–356).

Further Reading: Jones, D. E. 2000. *An Instinct for Dragons.*

CH712.1: Some dinosaurs breathed fire.

Job 41 describes Leviathan (see CH711.1), a dinosaur-like creature, as fire breathing. This suggests that some dinosaurs could breathe fire. Humans lived at the time as these dinosaurs and preserved the memory as fire-breathing dragons. (Gish 1977, 51–54)

1. If dinosaurs could breathe fire, they would have had adaptations around their mouths to protect their mouth and throat from flame. Nothing resembling such an adaptation has ever been seen.

2. Fire breathing in myth and legend is not limited to dragons. There are also fire-breathing snakes from the Chippewa (H. Norman 1990, 127–131), fire-breathing bulls from the Greeks (Ovid 1958, bk.7), and fire-breathing horses from the Bible (Rev. 9:17–18). Fire breathing is a folkloric motif not to be taken literally.

3. Not uncommonly, dragons have other fantastic properties not found in dinosaurs, such as multiple heads (such as in the Grimms' story "The Two Brothers," and the *basmu* from Akkadian myth; Dalley 1989, 323). Legendary creatures are poor evidence for biblical literalism.

DAY–AGE CREATIONISM

CH801: Genesis 1 got the order of events right.

The creation account in Genesis 1 lists ten major events in this order: (1) a beginning; (2) a primitive earth in darkness and enshrouded in heavy gases and water; (3) light; (4) an expanse or atmosphere; (5) large areas of dry land; (6) land plants; (7) sun, moon, and stars discernible in the expanse, and seasons beginning; (8) sea monsters and flying creatures; (9) wild and tame beasts and mammals; (10) man. The odds of getting that order correct by chance are one in 3,628,800. (Watchtower 1985, 37)

1. The real order is: (1) a beginning; (2) light; (3) sun and stars; (4) primitive earth, moon, and atmosphere; (5) dry land; (6) sea creatures; (7) some land plants; (8) land creatures and more plants and sea creatures; (9) flying creatures (insects) and more plants and land and sea creatures; (10) mammals and more land and sea animals, insects, and plants; (11) the first birds, (12) fruiting plants (which is what Genesis talks about) and more land, sea, and flying creatures; (13) man and more of the various animals and plants. That is nothing like the order endorsed by Jehova's Witnesses.

2. The odds of choosing that particular order are not one in 3,628,800. Much of the order is constrained. For example, the beginning must have been first, and land had to

exist before land animals and plants. When these are taken into account, the chance of getting that order are one in 5,760 at worst.

3. The claim contradicts what Genesis says. Genesis does not say when the sun and moon became visible (which would not have been until after eyes were created in any event); it tells when they were created. Genesis also refers to fruiting plants, which came after the first sea and land animals.

with John Harshman

GEOCENTRISM

CH901: Bible says the sun goes around the earth.
The earth is fixed at (or near) the center of the universe. The sun and other planets travel around it. That is what the Bible plainly says (Ps. 93:1; Ps. 19:1–6; Josh. 10:12–14) and what the evidence indicates. (T. Willis 2000)

1. A rotating earth produces observable, and observed, effects:
- The most noticeable is the Coriolis effect, the apparent deflection of the path of an object that moves in a rotating coordinate system. This affects ocean currents, wind patterns (including the path and direction of the spin of hurricanes), and iceberg drift. It must be taken into account when aiming long-range missiles.
- The rotation of the earth is also demonstrated by a Foucault pendulum, the swing of which rotates in relation to the earth's surface as the earth rotates beneath it. (The rate of rotation equals the rate of earth's rotation times the sine of the latitude.)

2. The orbit of the earth around the sun is also observable:
- The nearest stars show a parallax. Their apparent position shifts relative to more distant stars as the earth moves from one side of its orbit to the other. (The effect is the same as the apparent movement of a nearby telephone pole relative to distant mountains as you move a few feet to the side.)
- Stellar aberration shows up as the need to point the telescope slightly ahead of the star's true position, due to the earth's motion perpendicular to the star. It was first measured by James Bradley in 1728.
- Stars near the plane of the earth's orbit show a radial velocity, a slight red shift as the earth moves away from them in its orbit, and six months later, a slight blue shift (Herrick 1935).
- Related to radial velocity, the "light time" effect affects the timing of pulsars and short-term variable stars. General relativistic calculations are needed to correct for it.
- Since the earth's orbit is elliptical, it is closer to the sun in January than in June. The difference in the apparent size of the sun can be observed.

If the earth were stationary, these effects could only be explained if every star in the universe were moving in unison relative to the earth with a periodic variation that matched the earth's year.

3. Heliocentrism falls out naturally from the law of universal gravitation.

4. Heliocentrism is useful. As implied above, it is used for predicting hurricane and iceberg paths and for aiming missiles. The space program would be impossible without it. (The Cassini probe, for example, used the earth's motion around the sun to slingshot the probe to Jupiter.) As with all of creationism, strict geocentrism is useless.

5. To the vast majority of Christians, the Bible is not plainly saying that the earth is stationary. They have accepted that reality is more important than their interpretation of what is "plainly" said.

See for Yourself
You can make your own Foucault pendulum with a weight on a long, thin cable in a room with a high ceiling. It must be long enough so that air resistance does not stop it before the rotation is evident, and it should be sheltered from winds and drafts.

CH910: Relativity shows geocentrism is true.
Relativity states that there is no favored frame of reference, so a geocentric frame is as good as a heliocentric one (T. Willis 2000). *"We know that the difference between a heliocentric theory and a geocentric theory is one of relative motion only, and that such a difference has no physical significance."* (Hoyle 1975, cited in T. Willis 2000, 2)

1. The fact that different frames of reference all work does not mean that one frame makes as much sense as any other in any application. For navigating city streets, a geocentric frame makes sense; we would not want constantly to recalculate our position relative to the sun. For considering the solar system as a whole, however, a heliocentric frame makes sense. Figuring the calculations of the rest of the universe spinning and wobbling around the earth would be possible in theory, but prohibitive in practice.

Another frame of reference and mathematical transformations put the universe on the inside of a hollow earth. That model is mathematically equivalent to standard cosmology (S. Morris 1983). If physical significance is the only criterion, it is just as good as a heliocentric frame, too.

2. The claim cuts both ways. It also says that heliocentrism is just as good as geocentrism, as is the frame of reference relative to any other planet around any other star. It says that geocentrism is correct only if you choose to interpret it that way. Some people may want to interpret it that way, but they cannot then claim that others are wrong. The earth still moves.

INTELLIGENT DESIGN

ID AS SCIENCE

CI001: Intelligent design theory is scientific.

Intelligent design theory is science. (Dembski 1998a)

1. The terms used in design theory are not defined. "Design," in design theory, has nothing to do with "design" as it is normally understood. Design is normally defined in terms of an agent purposely arranging something, but such a concept appears nowhere in the process of distinguishing design in the sense of "intelligent design." Dembski defined design in terms of what it is not (known regularity and chance), making intelligent design an argument from incredulity (see CA100); he never said what design is.

A solution to a problem must address the parameters of the problem, or it is just irrelevant hand waving. Any theory about design must somehow address the agent and purpose, or it is not really about design. No intelligent design theorist has ever included agent or purpose in any attempt at a scientific theory of design, and some explicitly say they cannot be included (Dembski 2002b, 313). Thus, even if intelligent design theory were able to prove design, it would mean practically nothing; it would certainly say nothing whatsoever about design in the usual sense.

Irreducible complexity also fails as science because it, too, is an argument from incredulity that has nothing to do with design (see CB200).

2. Intelligent design is subjective. Even in Dembski's mathematically intricate formulation, the specification of his specified complexity can be determined after the fact, making "specification" a subjective concept. Dembski now talks of "apparent specified complexity" versus "actual specified complexity," of which only the latter indicates design. However, it is impossible to distinguish between the two in principle (Elsberry n.d.).

3. Intelligent design implies results that are contrary to common sense. Spider webs apparently meet the standards of specified complexity, which implies that spiders are intelligent. One could instead claim that the complexity was designed into the spider and its abilities. But if that claim is made, one might just as well claim that the spider's designer was not intelligent but was intelligently designed, or maybe it was the spider's designer's designer that was intelligent. Thus, either spiders are intelligent, or intelligent design theory reduces to a weak Deism where all design might have entered into the universe only once at the beginning, or terms like "specified complexity" have no useful definition.

Further Reading: Elsberry, W. 2000. The anti-evolutionists: William A. Dembski. http://www .antievolution.org/people/dembski_wa/sc.html; Pennock, R. T. 2003. Creationism and intelligent design.

CI001.1: Intelligent design theory is not religious.
Intelligent design (ID) is scientific, not religious.

1. The ID movement is motivated by and inseparable from a narrow religious viewpoint. In the words of its founders and leaders:

> Our strategy has been to change the subject a bit so that we can get the issue of intelligent design, which really means the reality of God, before the academic world and into the schools (P. Johnson 2003).

> Father's words, my studies, and my prayers convinced me that I should devote my life to destroying Darwinism, just as many of my fellow Unificationists had already devoted their lives to destroying Marxism. When Father chose me (along with about a dozen other seminary graduates) to enter a Ph.D. program in 1978, I welcomed the opportunity to prepare myself for battle (J. Wells, n.d.).

> Intelligent design is the Logos of John's Gospel restated in the idiom of information theory (Dembski 1999b, 84).

> Johnson said he and most others in the intelligent design movement believe the designer is the God of the Bible (Maynard 2001).

See Poindexter (2003) for more such quotes.

2. Intelligent design is explicitly religious as a motive for legislative change of educational standards. Legislation introduced in Michigan attempts to add "intelligent design of a Creator" to the science standards of middle and high school (Michigan House Bill 4946).

3. Several books on intelligent design are published by InterVarsity Press, which says of itself,

> Who is InterVarsity Press? We are a publisher of Christian books and Bible studies. As an extension of InterVarsity Christian Fellowship/USA, InterVarsity Press serves those in the university, the church and the world by publishing resources that equip and encourage people to follow Jesus as Savior and Lord in all of life (IVP n.d.).

The video "Unlocking the Mystery of Life" purportedly "tells the story of contemporary scientists who are advancing a powerful but controversial idea—the theory of intelligent design." But it was produced by and promoted almost exclusively by fundamentalist Christian organizations (Evans 2003b).

4. The ID movement attempts to hide its religious basis in order to give the appearance of secular objectivity (Branch 2002). Their attempt is dishonest propaganda. "The trend among many Christian groups these days is to camouflage their creationism as 'Intelligent Design' or 'Progressive Creationism'" (J. D. Morris 1999b). And despite their claims, the movement has no basis in science (see CI001).

5. Intelligent design is blatantly antireligious if the religion is one they disagree with. For example, Phillip Johnson equates theistic evolution (which would include most of Christianity) with atheism because of its acceptance of evolution.

Further Reading: Poindexter, B. 2003. The horse's mouth. http://home.kc.rr.com/bnpndxtr/download/HorsesMouth-BP007.pdf.

CI001.2: Intelligent design is not creationism.
Intelligent design (ID) is quite different from creationism, because

1. "Intelligent design creationism" is a pejorative term, not a term used by members of the ID movement.
2. Creationists and fair-minded critics recognize a difference between intelligent design and creationism.
3. Intelligent design is scientific.
4. Intelligent design's religious implications are distinct from its science program (West 2003).

1. The reasons given for intelligent design not being creationism fail:
1. The term "intelligent design creationism" is used because it is descriptive. The fact that the ID movement does not use it themselves means nothing, because the movement is based on propaganda and image manipulation (Branch 2002; CRSC 1998; Forrest 2002).

Claiming this reason is also blatant hypocrisy. ID members are relentless in referring to evolution as Darwinism and evolutionary scientists as Darwinists, despite the fact that evolutionary scientists do not use those labels in such a way.

2. There are differences between ID creationism, young-earth creationism, old-earth creationism, gap creationism, Vedic creationism, and other forms of creationism. Still, they are all creationism.
3. ID is anything but scientific (see CI001).
4. Since ID has no science program at all, their last point is meaningless.

2. Intelligent design is defined and treated as a form of creationism by its supporters. For example:

- The internet domain www.creation-science.com (as of Sept. 17, 2004) is registered by Access Research Network, a major ID organization, and directs you to their Web site.
- One prominent ID book captures the idea of creation in its definition:

> Intelligent design means that various forms of life began abruptly through an intelligent agency, with their distinctive features intact—fish with fins and scales, birds with feathers, beaks, and wings, etc. (P. Davis and Kenyon 1989, 99–100).

The ID movement rejects naturalistic explanations for origins and seeks to replace them with one or a few sudden creations by a supernatural agent who almost everyone in the movement identifies as the Christian God. That is creationism, plainly.

Further Reading: Thomas, D. 2003. The C-Files: The smoking gun—"intelligent design" IS creationism! http://www.nmsr.org/smkg-gun.htm.

CI001.3: Intelligent design is mainstream.

Intelligent design has been accepted as a mainstream scientific theory. It is argued before school boards; top scientists have published articles about it; its promotional videos have even been shown on public television. (Winn 2003)

1. Intelligent design is not mainstream science. (In fact, it is not science at all; see CI001.) It is not generating any research. Zero scientific research articles have been written about it. Most of the few articles that mention it at all are critical of it (Gilchrist 1997; Lane 2003).

2. The ID movement has been designed as a propaganda machine for achieving the appearance of respectability (Forrest 2002; Forrest and Gross 2004). The movement relies on deception to become accepted as mainstream. For example:

- A 2003 poll reported support for teaching intelligent design, but the poll was falsely reported and worthless to begin with (NMSR 2003; see also Mooney 2003).

- Discovery Institute fellows presented a bibliography of reputable scientists whose "publications represent dissenting viewpoints" to the Ohio Board of Education. But the scientists, when contacted, said that their work did not support intelligent design or challenge evolution. Many said that their work is evidence against intelligent design (NCSE 2002a).

- In order to get their promotional video onto television, the ID movement deliberately hid the fact that it was about intelligent design (Evans 2003a; see also Evans 2003b).

The resources of the Discovery Institute and other proponents of intelligent design are devoted to speaking engagements, popular publishing, and political lobbying. There is a lot of hot air surrounding intelligent design, but no substance.

3. Intelligent design may, in some sense, be mainstream in the public. Most people believe in some kind of divine creation. However, this in itself cannot be considered acceptance of intelligent design because it includes theistic evolution, which most ID proponents find distasteful. Roughly half of Americans believe in creationism, which might qualify as "mainstream intelligent design," but that number has probably fallen since the ID concept was popularized 200 years ago. In one poll, 84 percent of the Ohio public did not know what "intelligent design" is (G. Bishop 2002).

More to the point, intelligent design's popularity with the public is a logical fallacy. Astrology, for example, is at least as mainstream and just as wrong.

Further Reading: Forrest, B. 2002. The wedge at work; Forrest, B., and P. R. Gross. 2004. *Creationism's Trojan Horse.*

CI002: Intelligent design has explanatory power.

Intelligent design has explanatory power, especially given Dembski's "explanatory filter." It accounts for a wide range of biological facts. This makes it scientific. (Dembski 2001b)

1. Merely accounting for facts does not make a theory scientific. Saying "it's magic" can account for any fact anywhere but is as far from science as you can get. A theory has explanatory power if facts can be deduced from it. No facts have ever been deduced from ID theory. The theory is equivalent to saying "it's magic."

2. Dembski's explanatory filter (see CI111) requires the examination of an infinite number of other hypotheses—even unknown ones—to reject the design hypothesis. Thus, it is impossible to apply. Intelligent design remains untestable and impossible to use in practice. Dembski himself has never rigorously applied his filter (Elsberry 2002).

3. "Intelligent" and "design" remain effectively undefined. A theory cannot have explanatory power if it is uncertain what the theory says in the first place.

Further Reading: Pennock, R. T. 1999. *Tower of Babel.*

CI009: Evidence for design disproves evolutionary mechanisms.

"The Darwinian mechanism claims the power to transform a single organism . . . into the full diversity of life that we see both around us and in the fossil record. If intelligent design is correct, then the Darwinian mechanism of natural selection and random variation lacks that power." (Dembski 2001a)

1. The claim is a non sequitur. Intelligent design says nothing at all about the Darwinian mechanism of evolution. Whether or not evolution lacks power must be decided by looking at evolution. If intelligent design is correct, it provides an alternative explanation, but it does not automatically falsify other alternatives.

2. The Darwinian mechanism of natural selection is part of the normal design process. If designers were not able to abandon old designs in favor of better modified versions of the same design, then intelligent design would have very little power itself.

CI010: There is a law of conservation of information.

Information cannot be created by either natural processes or chance, so there is a law of conservation of information. (Dembski 1999a)

1. Dembski defines his information as Shannon uncertainty, which is equivalent to entropy. We know that entropy can and does increase. Dembski's law of conservation of information is simply wrong.

2. No recognized theory of information (i.e., the statistical theory of Shannon et al, and the algorithmic theory of Kolmogorov, Chaitin, and Solomonoff) has a law of conservation of information. William Dembski and Werner Gitt have each invented their own nonstandard information theories, but neither of these theories is used in science or engineering, and their claims are not supported by the vast body of research into information theory.

3. Even if there were a law of conservation of information, it would not necessarily invalidate evolution. Information is transferred from the environment to organisms by natural selection and other processes.

4. Normally, physical laws get to be considered laws after they are tested and verified by independent sources under very many various conditions. For Dembski to claim a new physical law without any testing whatsoever is hubris of the highest magnitude.

Further Reading: Elsberry, W., and J. Shallit. 2003. Information theory, evolutionary computation, and Dembski's "complex specified information." http://www.talkreason.org/articles/eandsdembski.pdf; Stenger, V. J. 2000. The emperor's new designer clothes. http://www.csicop.org/sb/2000-12/reality-check.html.

DETECTING DESIGN

CI100: Design is detectable.
Life looks intelligently designed because of its complexity and arrangement. As a watch implies a watchmaker, so life requires a designer. (P. Davis and Kenyon 1989; W. Paley 1802)

1. According to the definition of design, we must determine something about the design process in order to infer design. We do this by observing the design in process or by comparing with the results of known designs. The only example of known intelligent design we have is human design. Life does not look man-made.

2. Nobody argues that life is complicated. However, complexity is not the same as design. There are simple things that are designed and complex things that originate naturally. Complexity does not imply design; in fact, simplicity is a design goal in most designs.

3. In most cases, the inference of design is made because people cannot envision an alternative. This is simply the argument from incredulity (see CA100). Historically, supernatural design has been attributed to lots of things that we now know form naturally, such as lightning, rainbows, and seasons.

4. Life as a whole looks very undesigned by human standards, for several reasons:

- In known design, innovations that occur in one product quickly get incorporated into other, often very different, products. In eukaryotic life, innovations generally stay confined in one lineage. When the same sort of innovation occurs in different lineages (such as webs of spiders, caterpillars, and web spinners), the details of their implementation differ in the different lineages. When one traces lineages, one sees a great difference between life and design. (Eldredge has done this, comparing trilobites and cornets; G. Walker 2003.)

- In design, form typically follows function. Some creationists expect this (H. M. Morris 1985). Yet life shows many examples of different forms with the same function (e.g., different structures making up the wings of birds, bats, insects, and pterodactyls; different organs for making webs in spiders, caterpillars, and web spinners; and at least eleven different types of insect ears), the same basic form with different functions (e.g., the same pattern of bones in a human hand, whale flipper, dog paw, and bat wing) and some structures and even entire organisms without apparent function (e.g., some vestigial organs, creatures living isolated in inaccessible caves and deep underground).

- As noted above, life is complex. Design aims for simplicity.
- For almost all designed objects, the manufacture of the object is separate from any function of the object itself. All living objects reproduce themselves.
- Life lacks plan. There are no specifications of living structures and processes. Genes do not fully describe the phenotype of an organism. Sometimes in the absence of genes, structure results anyway. Organisms, unlike designed systems, are self-constructing in an environmental context.
- Life is wasteful. Most organisms do not reproduce, and most fertilized zygotes die before growing much. A designed process would be expected to minimize this waste.
- Life includes many examples of systems that are jury-rigged out of parts that were used for another purpose. These are what we would expect from evolution, not from an intelligent designer. For example:
  - Vertebrate eyes have a blind spot because the retinal nerves are in front of the photoreceptors.
  - On orchids that provide a platform for pollinating insects to land on, the stem of the flower has a half twist to move the platform to the lower side of the flower.
- Life is highly variable. In almost every species, there is a spread of values for anything you care to measure. The "information" that specifies life is of very low tolerance in engineering terms. There are few standards.

5. Life is nasty. If life is designed, then death, disease (see CH321), and decay also must be designed since they are integral parts of life. This is a standard problem of apologetics. Of course, many designed things are also nasty (think of certain weapons), but if the designer is supposed to have moral standards, then it is added support against the design hypothesis.

6. The process of evolution can be considered a design process, and the complexity and arrangement we see in life are much closer to what we would expect from evolution than from known examples of intelligent design. Indeed, engineers now use essentially the same processes as evolution to find solutions to problems that would be intractably complex otherwise.

7. Does evolution itself look designed? When you consider that some sort of adaptive mechanism would be necessary on the changing earth if life were to survive, then if life were designed, evolution or something like it would have to be designed into it.

8. Claiming to be able to recognize design in life implies that nonlife is different, that is, not designed. To claim that life is recognizably designed is to claim that an intelligent designer did not create the rest of the universe.

9. As it stands, the design claim makes no predictions, so it is unscientific and useless. It has generated no research at all.

Further Reading: Aulie, R. P. 1998. A reader's guide to *Of Pandas and People*. http://www.nabt .org/sub/evolution/panda1.asp; Isaak, M. 2003. What design looks like; Miller, K. R. n.d. Of pandas and people: A brief critique. http://www.kcfs.org/pandas.html; Pennock, R. T. 1999. *Tower of Babel*; Perakh, M. 2003b. *Unintelligent Design.*

CI100.1: Look—is design not obvious?
Design is self-evident. You just need to open your eyes and see it.

1. This claim lacks any substance. It is nothing more than a subjective assertion. That design is far from self-evident is demonstrated by the difficulty people have in trying to describe the objective evidence for it.

2. There are good reasons why people should see design that is not there:

- Humans anthropomorphize. We tend to attribute our humanlike qualities to all sorts of things. Since design is what humans do, we attribute it far and wide.

- Evolution and some human design both involve complex systems dealing with the same physical constraints (Csete and Doyle 2002).

- Evolution has much in common with a design process. It generates trial-and-error modifications of existing forms and discards the inferior versions.

Further Reading: Isaak, M. 2003. What design looks like.

CI101: Complexity indicates design.
Complexity indicates intelligent design.

1. This is a quintessential argument from incredulity (see CA100). Complexity usually means something is hard to understand. But the fact that one cannot understand how something came to be does not indicate that one may conclude it was designed. On the contrary, lack of understanding indicates that we must not conclude design or anything else.

Irreducible complexity (see CB200) and complex specified information (see CI111) are special cases of the "complexity indicates design" claim; they are also arguments from incredulity.

2. In the sort of design that we know about, simplicity is a design goal. Complexity arises to some extent through carelessness or necessity, but engineers work to make things as simple as possible. This is very different from what we see in life.

3. Complexity arises from natural causes: for example, in weather patterns and cave formations.

4. Complexity is poorly defined.

CI102: Irreducible complexity indicates design.
Systems are irreducibly complex if removing any one part destroys the system's function. Irreducible complexity in organisms indicates they were designed. (Behe 1996a)

1. Irreducible complexity is claimed to indicate (but does not; see CB200) that certain systems could not have evolved gradually. However, jumping from there to the con-

clusion that those systems were designed is an argument from incredulity (see CA100). There is nothing about irreducibly complex systems that is positive evidence for design.

2. Irreducible complexity suggests a lack of design. For critical applications, such as keeping an organism alive, you do not want systems that will fail if any one part fails. You want systems that are robust (Steele 2000).

CI111: Complex specified information indicates design.

Design can be recognized by the following filter:

1. If an event E has high probability, accept regularity as an explanation; otherwise move to the next step.
2. If the chance hypothesis assigns E a high probability or E is not specified, then accept chance; otherwise move down the list.
3. Having eliminated regularity and chance, accept design.

This filter is equivalent to detecting complex specified information (Dembski 1998a).

1. The filter is useless in practice because the probabilities it asks for can never be known. Step 1, in particular, does not ask us to accept or reject just one regularity hypothesis, it asks us about all regulatory hypotheses, even ones that nobody has thought of before. Similarly, rejecting chance requires a complete list of all chance processes that might apply to the event.

2. The filter is based on the premise that regularity, chance, and design are mutually exclusive and exhaustive categories. But they are not mutually exclusive. R. A. Fisher, for example, included mutations in all three categories. Individually, they were due to chance, but collectively they were governed by laws, and all of this was planned by God (Ruse 2001, 121).

3. Although the filter claims to detect design, it really says nothing about design. The filter defines design as the elimination of regularity and chance, not, as most people would define design, as purposeful, intelligent arrangement. The two definitions are not equivalent. Dembski himself noted that some intelligent design will be eliminated in the first two steps. And what the filter actually detects is copying, not intelligent agency (see CI111.2).

4. Since the filter does not say anything about design, there is no intelligent design hypothesis that can be used scientifically or for any practical purposes.

5. Key terms in the filter, especially "chance hypothesis" and "specified," are poorly defined.

6. Dembski does not consider that design is a process. The process that produces design is itself not regularity (or the resulting design would have high probability) or chance (or the design would likely not result), so the filter says the process must itself be design. Thus, the design process must have another design process to produce it, which needs a design process of its own, ad infinitum, or somewhere along the way there must be no process at all and design must come out of nowhere. In actuality, design is typically done as an iterative process involving lots of trial and error. Regularity and chance are both parts of the process, as is selection. Evolution uses the same processes.

Further Reading: Elsberry, W. R., and J. Shallit. 2003. Information theory, evolutionary computation, and Dembski's "complex specified information." http://www.talkreason.org/articles/eandsdembski.pdf; Fitelson, B., et al. 1999. How not to detect design. http://philosophy.wisc.edu/sober/dembski.pdf; Pennock, R. T. 1999. *Tower of Babel*; Wilkins, J. S., and W. R. Elsberry. 2001. The advantages of theft over toil. http://www.talkdesign.org/faqs/theftovertoil/theftovertoil.html.

CI111.1: Specified information criterion produces no false positives.

Specified complexity is a reliable criterion for detecting design (see CI111). The complexity-specification criterion successfully avoids false positives—in other words, whenever it attributes design, it does so correctly. (Dembski 2002b, 24–25)

1. Complexity-specification allows false positives because it does not consider the combination of regularity and chance acting together, and it does not consider unknown causes.

2. Specific examples of false positives are irreducibly complex structures for which plausible evolutionary origins have been found (see CB200).

Another false positive is canals on Mars. Percival Lowell saw that many Martian canals meet at each of several points. The odds of this happening by chance, he calculated, are less than 1 in 1.6×10^{260}, proving that Mars must be inhabited (Lowell, 1907). We now know that the canals were optical illusions caused by the human mind connecting indistinct features.

3. Dembski himself admitted the possibility of error in the same book in which he claimed reliability:

> Now it can happen that we may not know enough to determine all the relevant chance hypotheses. Alternatively, we might think we know the relevant chance hypotheses, but later discover that we missed a crucial one. In the one case a design inference could not even get going; in the other, it would be mistaken. But these are the risks of empirical inquiry, which of its nature is fallible. Worse by far is to impose as an *a priori* requirement that all gaps in our knowledge must ultimately be filled by non-intelligent causes. (Dembski 2002b, 123)

What Dembski failed to appreciate is that his complexity specification criterion imposes an a priori requirement that all gaps must be filled by supernatural causes.

Dembski also said, "On the other hand, if things end up in the net that are not designed, the criterion will be useless" (Dembski 1999a, 142).

Further Reading: Ratzsch, D. 2001. *Nature, Design and Science.* (The appendix addresses false positives and other problems with Dembski 2002b.)

CI111.2: Specified complexity characterizes what intelligent agents do.

An intelligent agent is one that chooses between different possibilities. Specified complexity (also called complex specified information) detects design because it detects what characterizes intelligent agency; it detects the actualization of one among many competing possibilities. (Dembski 2002b, 28–30)

1. Specified complexity does not indicate intelligence agency; it merely indicates copying. When a pattern matches a specification, that can happen only by coincidence, by the causes of both patterns following the same constraints, or by some kind of copying of information. The specified complexity criterion explicitly rules out the first two possibilities (Dembski 2002b, 6–13), leaving only copying.

Consider the following scenario: A person accidentally spills some ink and creates a complex inkblot on a page of a report. The spill goes unnoticed until several copies of the report have been made. The inkblot images in the copies of the report exhibit specified complexity, as they are complex, and they match a specification (the original spill). But they achieved specified complexity by copying, not by deliberate choosing.

2. Nonintelligent processes also select between different possibilities. The machines that select lotto numbers are an example.

CI113: Genetic algorithms require a designer to specify desired outcome.
Genetic algorithms and computer evolution simulations do not show that intelligent design is unnecessary. On the contrary, those programs must be designed themselves, and they require a designer to specify the outcome.

1. See the response to the claim that evolutionary algorithms smuggle in design in the fitness function (CF011). Modeling evolution requires modeling fitness differences. Fitness functions often are not designed but are taken from the real world. The exact outcome is not specified in genetic algorithms, only the general requirements. And evolutionary simulations have no intended outcome for the simulated organisms at all.

2. The fact that the programs are designed themselves is not relevant so long as they accurately model real phenomena. They will not model everything, of course; that is why they are called models. But the fact that modeling evolution is done by designers has no more implications for intelligent design than the fact that painting seascapes is done by designers, too.

CI120: Purpose indicates design.
A purpose for an object indicates that the object is designed. (W. Paley 1802, 2)

1. When somebody designs something, he or she usually has a purpose for it, but the purpose is that of the designer, not the object designed. For example, people have a purpose for windows and airbags in automobiles, but the automobile itself has no such purpose. When the purpose argument is applied to life, though, the designer is intentionally left entirely unknowable, and thus the purpose of the designer is not part of the picture. We know only the object's purpose for part of the object, which is not relevant unless you want to claim that the object designed itself.

2. To the extent that traits of living things have a purpose, that purpose, ultimately, is the reproductive success of the organism's genes. Such purpose is entirely consistent with evolution.

3. It is not uncommon for undesigned objects to have a purpose. The North Star, for example, has a purpose in navigation, but it got that purpose entirely through the chance of its being in a certain spot.

Even with designed things, it is common for purposes to come and go. The same object can have different purposes at different times or even multiple purposes at the same time. It will gain and lose its purposes as conditions change.

4. Some life forms have no apparent purpose. There have been species in isolated caves discovered quite by chance (Decu et al. 1994). Very likely, there have been species similarly isolated that were never discovered.

Some parts of life forms also appear to have no purpose: junk DNA, for example (see CF011).

Life also exists at cross-purposes. A bobcat's purpose for a rabbit is likely to be quite different from the rabbit's purpose.

CI130: Functional integration indicates design.
Design is indicated by functional integration, which is multiple parts working together to produce a particular function or end. (Lumsden 1995)

1. Functional integration can be produced naturally. Parts come together all the time and, with the addition of energy, act on each other. It is inevitable that sometimes such workings will create an end that someone or something considers functional. An example of such functional integration occurring naturally is climate (which has even been proposed as evidence of intelligent design; Morton 2001b). Climate is produced by the interactions of terrain, trade winds, bodies of water, and latitude. These can occur arbitrarily, but they are bound to produce some kind of functional climate.

2. Functional integration is what we would expect from evolution. Evolution requires units that reproduce. "Units" implies at least some amount of integration, and reproduction is a function. (And natural selection will favor innumerable other functions that help with survival and reproduction.)

3. Biological systems often include imperfections and jury-rigging that argue against intelligent planning of their integration. For example:

- DNA sometimes includes a "wrong" sequence that, if not edited after the DNA is transcribed, would be fatal. For example, a gene involved in mouse neurons contains a codon specifying glutamine. In normal mice, an enzyme edits the RNA that is transcribed from the DNA, so the codon will specify arginine. Mice die if this editing enzyme is disabled. But mice that are genetically engineered with DNA for arginine in the first place are healthy even without the editing enzyme (Keegan et al. 2000).

- In men, the urethra passes through the prostate gland. When the prostate gets infected and swells, as it is prone to do, it causes problems not only with reproduction but also with the excretory system (Colby et al. 1993).

- Although baleen whales are toothless as adults, their fetuses grow teeth that are resorbed as they mature (Colby et al. 1993).

Additional example are plentiful (Behrman et al. 2004; Colby et al. 1993).

4. Functional integration is not a necessary part of design. It does not exist in simple designs without multiple parts, for example. Since evolution always produces functional integration but design does not, functional integration is evidence more for evolution than for design.

Further Reading: Behrman, E.J., et al. 2004. Evidence from biochemical pathways in favor of unfinished evolution rather than intelligent design; Colby, C., et al. 1993. Evidence for jury-rigged design in nature. http://www.talkorigins.org/faqs/jury-rigged.html.

CI141: Similarities in DNA and anatomy are due to common design.

Similarities in anatomy and DNA sequences simply reflect the fact that the organisms had the same designer. (Sarfati 2002, chap. 6)

1. Different forms also (it is claimed) come from the same designer, so similar forms are not evidence of a common designer. Evidence for a designer must begin by specifying (before the fact) what is expected from the designer. When do we expect similar forms, and when do we expect different forms? Intelligent design theory will not answer that. Evolution theory has made that prediction, and the pattern of similarities and differences that we observe accords with what evolution predicts.

2. There are similarities that cannot rationally be attributed to design. For example, an endogenous retroviral element (ERV) is a retrovirus (a parasite) that has become part of the genome. There are several kinds of ERVs, and they can insert themselves at random locations. Humans and chimps have thousands of such ERVs in common—the same type of ERV at the same location in the genome (D.M. Taylor 2003).

3. The "form follows function" principle is the opposite from what we expect from known design (see CI141.1).

CI141.1: Similar structures for similar functions, different for different.

Similarity of structures in different life forms reflects the fact that they were created for similar purposes. Different structures reflect different functions. (H.M. Morris 1985, 70)

1. Actual known designers use similar forms for similar functions and different forms for different functions. In life, we often see different forms for similar functions (e.g., different designs for bird, bat, and pterodactyl wings) and similar forms for different functions (e.g., spider webs for trapping flying insects, reinforcing tunnels, protecting eggs, transferring sperm, ballooning, marking a trail, serving as a safety line after jumping to escape a threat, and detecting motion).

CI190: SETI researchers expect that they can detect design.

SETI researchers expect that they can recognize artificial signals, proving that there is an objective criterion for recognizing intelligent design. (Dembski 1998b)

1. SETI researchers do not expect to find recognizably designed messages in the signals they are looking for; in fact, they expect that the signal modulation would be smeared out and lost. They are looking for narrow-band signals, which are what people build and are not found in known natural radio signals (SETI Instutute n.d.). The objective criterion for recognizing intelligent design is to look for things that look like what people build.

   Further Reading: Petrich, L. 2003. Animal and extraterrestrial artifacts: Intelligently designed? http://www.secweb.org/asset.asp?AssetID=283.

CI191: Archaeologists and forensic scientists can detect design.

Just as sciences such as archaeology and forensics can detect design, so it is a valid scientific practice to detect intelligent design in nature. The success of those sciences shows that the methods of intelligent design work in practice. (Dembski 1999a, 2001a)

1. The methods of archaeology and forensics are unrelated to any methods proposed by ID advocates. Archaeologists and forensic scientists look for patterns that they know, from prior observation, are the sort of patterns that human designers make. The same goes for all other sciences that detect design. ID theorists have no prior observation of other designers to go by. Or, if they do use the methods of archaeologists and forensic scientists, they are implicitly assuming that the designers were human.

2. The only proposed ID method, Dembski's filter (see CI111), is eliminative; it tries to detect design only by eliminating other possibilities. The methods used by scientists are not eliminative. They consider many possibilities and choose the one that best fits the data. If none fit the data, the question is left unresolved.

FIRST CAUSE

CI200: There must have been a first cause.

Every event has a cause. The universe itself had a beginning, so it must have had a first cause, which must have been a creator God. (H. M. Morris 1985, 19–20)

1. The assumption that every event has a cause, although common in our experience, is not necessarily universal. The apparent lack of cause for some events, such as radioactive decay, suggests that there might be exceptions. There are also hypotheses, such as alternate dimensions of time or an eternally oscillating universe, that allow a universe without a first cause (see CE440).

2. By definition, a cause comes before an event. If time began with the universe, "before" does not even apply to it, and it is logically impossible that the universe be caused.

3. This claim raises the question of what caused God. If, as some claim, God does not need a cause, then by the same reasoning, neither does the universe.

ANTHROPIC PRINCIPLE

CI301: The cosmos is fine-tuned to permit human life.

The cosmos is fine-tuned to permit human life. If any of several fundamental constants were only slightly different, life would be impossible. (This claim is also known as the weak anthropic principle; H. Ross 1994)

1. The claim assumes life in its present form is a given; it applies not to life but to life only as we know it. The same outcome results if life is fine-tuned to the cosmos.

We do not know what fundamental conditions would rule out any possibility of any life. For all we know, there might be intelligent beings in another universe arguing that if fundamental constants were only slightly different, then the absence of free quarks and the extreme weakness of gravity would make life impossible.

Indeed, many examples of fine-tuning are evidence that life is fine-tuned to the cosmos, not vice versa. This is exactly what evolution proposes.

2. If the universe is fine-tuned for life, why is life such an extremely rare part of it?

3. Many fine-tuning claims are based on numbers being the "same order of magnitude," but this phrase gets stretched beyond its original meaning to buttress design arguments; sometimes numbers more than one thousandfold different are called the same order of magnitude (Klee 2002).

How fine is "fine" anyway? That question can only be answered by a human judgment call, which reduces or removes objective value from the anthropic principle argument.

4. The fine-tuning claim is weakened by the fact that some physical constants are dependent on others, so the anthropic principle may rest on only a very few initial conditions that are really fundamental (Kane et al. 2000). It is further weakened by the fact that different initial conditions sometimes lead to essentially the same outcomes, as with the initial mass of stars and their formation of heavy metals (Nakamura et al. 1997), or that the tuning may not be very fine, as with the resonance window for helium fusion within the sun (Livio et al. 1989).

5. If part of the universe were not suitable for life, we would not be here to think about it. There is nothing to rule out the possibility of multiple universes, most of which would be unsuitable for life. We happen to find ourselves in one where life is conveniently possible because we cannot very well be anywhere else.

6. The anthropic principle is an argument against an omnipotent creator. If God can do anything, he could create life in a universe whose conditions do not allow for it.

Further Reading: Goldsmith, D. 2004. The best of all possible worlds; Stenger, V. J. 1997. Intelligent design. http://www.talkorigins.org/faqs/cosmo.html; Stenger, V. J. 1999. The anthropic coincidences. http://www.stephenjaygould.org/ctrl/stenger_intel.html; Weinberg, S. 1999. A designer universe? http://www.physlink.com/Education/essay_weinberg.cfm.

CI302: The cosmos is fine-tuned to permit scientific discovery.

The conditions that enable life to exist also give the best overall setting for scientific discovery; habitability correlates with measurability. For example, the moon exists with the right size and distance so that a perfect total solar eclipse is observable, and the total solar eclipse of 1919 was crucial in testing general relativity. (Gonzalez and Richards 2002)

1. People tend to start with the easy stuff, and scientists are no different. The discoveries they have made would naturally tend to fall in the areas where discovery is easiest.

When airline traffic was halted for three days after the 9/11 attack, scientists took the opportunity to measure the effect of contrails (Travis et al. 2002). That does not mean the 9/11 attack was designed for scientific discovery. Likewise, the fact that scientists in general have taken what measurements they have does not mean the universe was designed to allow them to do so.

2. The argument boils down to the tautology "If things had been different, things would be different." There is no evidence that a different universe would be better or worse. Different conditions might make it harder to observe what is easy to observe now, but they would make other observations easier, probably leading to the discovery of things we do not yet know. The philosophy behind the claim was satirized by Voltaire (1759):

> "It is demonstrable," said he, "that things cannot be otherwise than as they are; for as all things have been created for some end, they must necessarily be created for the best end. Observe, for instance, the nose is formed for spectacles, therefore we wear spectacles. The legs are visibly designed for stockings, accordingly we wear stockings. Stones were made to be hewn and to construct castles, therefore My Lord has a magnificent castle; for the greatest baron in the province ought to be the best lodged. Swine were intended to be eaten, therefore we eat pork all the year round: and they, who assert that everything is right, do not express themselves correctly; they should say that everything is best."

3. In many ways, the configuration of the universe hinders scientific discovery. For example:

- Things in space are hard to get to. They are very far away; the intervening vacuum is hostile to life; our gravity well makes it costly to leave earth.

- The speed of light would hinder communication when and if we ever do explore the galaxy.

- Our life spans are too short to individually note many important changes in nature.

- Nobody knows of any easy way to find the structure of proteins.

- There appears to be no life on other planets with which to compare earth life.

- Underwater exploration is quite inconvenient for us.

- Our brains may be structured so that some things are difficult to understand. Many people, for example, cannot seem to comprehend that the question "What is outside the universe?" is meaningless. Quantum mechanics and relativity are also difficult to conceptualize. And the tendency to categorize things into distinct "kinds" may be biological. (It is helpful for language, but it hinders understanding of evolution.)

4. A universe whose laws are easy to discover implies a simple universe, but design theorists keep telling us how complex things are.

5. The importance of the 1919 eclipse in establishing the truth of relativity may be exaggerated (H. Collins 1998; see also M. Stanley 2003). General relativity, like most theories, was established by a variety of consistent observations.

Further Reading: Voltaire. 1759. *Candide.* http://www.literature.org/authors/voltaire/candide.

META-ARGUMENTS

CI401: The methodology of science rules out even considering design.

The methodology of science limits itself to considering only natural causes. This rules out the possibility of design as a cause, even though inferring design, consistent with scientific methods, is logically based on data observed in nature. (Calvert et al. 2001)

1. The claim is obviously false, because science can and does detect design in several contexts, such as archaeology and forensics. Design theorists themselves point to such examples as evidence that design can be detected (Dembski 1998b). Considering intelligent design besides human design, though, is ruled out by the fact that proponents say nothing positive about what such intelligent design implies.

2. Science does not limit itself to considering only natural causes. There have been numerous scientific investigations of phenomena that presumably do not have natural causes, such as the power of prayer (Astin et al. 2000; Cha et al. 2001), divination (Enright 1999; Randi 1982), and life after death (Schwartz et al. 2001). What matters to science is not that something be natural (whatever that means), but that observations can be objectively and reliably verified by others.

Further Reading: Isaak, M. 2002c. A philosophical premise of 'naturalism'? http://www.talk design.org/faqs/naturalism.html.

CI402: Evolutionists have blinded themselves to seeing design.

Due to their preconceptions and bias for materialism, evolutionists have blinded themselves to seeing design.

1. This claim is nothing more than an excuse that design theorists use to try to explain away their own failure to make their case. When someone proposes a new scientific theory, it is that person's responsibility to make a case for it. Scientific theories have, in the past, achieved wide acceptance despite strong cultural and scientific resistance. (Evolution itself is an example.) If there is substance to ID theory, its proponents must make it clear.

2. People who study evolution come from a variety of cultural backgrounds. Many of them are far from committed to materialism. Some students enter the field hoping to

challenge existing dogmas, and objectively detecting design in life would certainly accomplish that. If there were anything to ID theory, there should be more than enough biologists to help the design theorists make their case.

3. To all appearances, design theorists have blinded themselves to seeing flaws in their theories. Their religious motivation is obvious. Just as important, they do not follow the usual scientific procedure of testing their ideas.

A scientific theory is tested by subjecting it to a very real chance of falsification. Scientists make specific predictions based on the theory, look to see if the predictions pan out, and consider the theory false if the results cannot fit what was expected. ID theorists, unlike evolutionary scientists (see CA211), do not put their ideas to such risks. Apparently, they do not want their ideas at risk.

4. Design theory is older than Darwin's theory of evolution. Design theory has nothing but its own lack of worth to blame for its failure.

CI410: Design requires a designer.
Design requires a designer; contrivance requires a contriver. (W. Paley 1802, 11)

1. Design does not require an anthropomorphized designer. Designs appear in clouds, for example, with no more of a designer than uneven heating, evaporation, and other natural phenomena.

2. Evolution is a designer. Via variation and selection, it serves to favor reproduction and shape things according to environmental conditions.

3. If the designer does not need a designer to create it, why should other things?

OTHER CREATIONISM

VEDIC CREATIONISM

CJ001: Mankind has existed essentially unchanged for billions of years.
Modern humans have lived on earth for billions of years. (Cremo and Thompson 1993)

1. Cremo and Thompson supported their claim with Vedic scripture. They claimed scientific support for it, but the science they cited was old and discredited.

2. More seriously, Cremo and Thompson were selective in what they cited. The vast majority of evidence shows that hominids have developed more or less gradually over the past 6 or more million years and that modern humans are much younger (Tattersall 1995).

Further Reading: Brass, M. 2002. *The Antiquity of Man: Artifactual, Fossil, and Gene Records Explored*; Foley, J. 1996. NBC's "The Mysterious Origins of Man." http://www.talkorigins.org/faqs/mom.html; Groves, C. 1994. Creationism: The Hindu View. http://www.talkorigins.org/faqs/mom/groves.html.

NATIVE NORTH AMERICAN CREATIONISM

CJ311: The 9,300-year-old Kennewick Man was an Umatilla ancestor.
Kennewick Man is a 9,300-year-old human skeleton found along the Columbia River in Washington state. Oral tradition of the Umatilla tribe says that they had always inhabited that area,

making Kennewick Man their ancestor, and requiring, under the 1990 Native American Graves Protection and Repatriation Act, that his bones be returned to the Umatilla for reburial. (Minthorn, cited in Morell 1998)

1. Populations migrate and disperse. Very few family groups stay in one place for 9,000 years. It is unlikely that Kennewick Man was an ancestor of any of today's Umatilla.

2. Traditions from some other tribes in the area, such as the Quillayute, strongly imply that the Umatilla have not been native there forever (E. E. Clark 1953; Isaak 2002a). Even a pagan group has claimed Kennewick Man as an ancestor.

3. It is ironic to note that the only form of creationism to receive federal approval in United States in the last fifty years is Umatilla creationism (although a circuit court of appeal has overturned this decision; Chatters 2001, 266; Holden 2004; Paulson 2004).

Further Reading: Chatters, J. C. 2001. *Ancient Encounters*; Morell, V. 1998. Kennewick Man's trials continue.

ISLAMIC CREATIONISM

CJ530: The Qur'an's accuracy on scientific points shows overall accuracy.
The Qur'an's accuracy on various scientific and historical points shows its divine origin and indicates overall accuracy. (Yahya n.d.b)

1. Arguments about the scientific accuracy of the Bible apply exactly as well to the Qur'an (see CH130).

Further Reading: Answering Islam. n.d. Qur'an and science. http://www.answering-islam.org/Quran/Science/index.htm; Bannister, A. n.d. Can "modern science" be found in the Qur'an? http://www.answering-islam.org/Andy/fallacies.html.

BIBLIOGRAPHY

References that are recommended as further reading are marked with an asterisk. A few books that are particularly informative and well-written are marked with two asterisks.

All internet addresses were verified in May 2004. However, Web links are notoriously unstable. If a link does not work for you, it may still be possible to find the Web page elsewhere via a search engine. Many old Web pages can also be found at www.archive.org.

Abbreviations used: *PNAS* = *Proceedings of the National Academy of Sciences of the United States of America*; *RNCSE* = *Reports of the National Center for Science Education*.

*Aardsma, G. E. 1988a. Has the speed of light decayed recently? *Creation Research Society Quarterly* 25(1): 36f.

*Aardsma, G. E. 1988b. Has the speed of light decayed? *Impact* 179. El Cajon, CA: Institute for Creation Research.

Abelson, P. 1996. Chemical events on the primitive earth. *PNAS* 55: 1365–1372.

Ackerman, P. D. 1986. *It's a Young World After All*. Grand Rapids, MI: Baker Book House.

Acton, G. 1997. Behe and the blood clotting cascade. http://www.talkorigins.org/origins/postmonth/feb97.html.

*Adami, C., C. Ofria, and T. C. Collier. 2000. Evolution of biological complexity. *PNAS* 97: 4463–4468. (technical)

Adams, H. E., L. W. Wright Jr., and B. A. Lohr. 1996. Is homophobia associated with homosexual arousal? *Journal of Abnormal Psychology* 105(3): 440–445.

Agassiz, L. 1874. *Evolution and Permanence of Type*. Reprinted in Hull, D. L. 1973. *Darwin and His Critics*. Cambridge, MA: Harvard University Press 440.

Ahmad, Q. R., et al. (SNO Collaboration). 2002. Direct evidence for neutrino flavor transformation from neutral-current interactions in the Sudbury Neutrino Observatory. *Physical Review Letters* 89: 011301. http://xxx.lanl.gov/abs/nucl-ex/0204008.

AIG. 1990a. The amazing bombardier beetle. *Creation Ex Nihilo* 12(1): 29.

AIG. 1990b. Fossil pollen in Grand Canyon overturns plant evolution. *Creation Ex Nihilo* 12(1): 38–39. http://www.answersingenesis.org/home/area/Magazines/docs/v12n1_pollen.asp.

AIG. 1998. Skeptics choke on frog: Was Dawkins caught on the hop? http://www.answersingenesis .org/docs/3907.asp.

*AIG. n.d.a. Arguments we think creationists should NOT use. http://www.answersingenesis .org/home/area/faq/dont_use.asp.

AIG. n.d.b. Creation Education Center. http://www.answersingenesis.org/cec/docs/CvE_report.asp.

AIG. n.d.c. Statement of faith. http://www.answersingenesis.org/home/area/about/faith.asp.

*Aitken, M. J. 1990. *Science-Based Dating in Archaeology*. London: Longman.

*Alden, A. 2003. Tectonic plate motions, Eurasia/Africa. http://geology.about.com/library/bl/maps/ blplatemo_atlas.htm.

*Allen, J. A., M. Burns, and S. C. Sargent. 1986. *Cataclysms on the Columbia*. Portland, OR: Timber Press.

*Alley, R. B., and M. L. Bender. 1998. Greenland ice cores: Frozen in time. *Scientific American* 278(2): 80–85.

*Alters, B. J., and S. M. Alters. 2001. *Defending Evolution in the Classroom*. Boston: Jones and Bartlett.

Alves, M. J., M. M. Coelho, and M. J. Collares-Pereira. 2001. Evolution in action through hybridisation and polyploidy in an Iberian freshwater fish: A genetic review. *Genetica* 111: 375–385.

American Southwest. n.d. Mexican hat. http://www.americansouthwest.net/utah/mexican_hat/ index.html.

Aneshansley, D. J., and T. Eisner. 1969. Biochemistry at 100C: Explosive secretory discharge of bombardier beetles (Brachinus). *Science* 165: 61–63.

Aneshansley, D. J., et al. 1983. Thermal concomitants and biochemistry of the explosive discharge mechanism of some little known bombardier beetles. *Experientia* 39: 366–368.

Ankerberg, J., S. Austin, D. Gish, and K. Wise. 1990. The creation debate: Oxygen—the deathblow to life? http://www.johnankerberg.org/Articles/science/SC1202W3.htm.

Annweiler, E., W. Michaelis, and R. U. Meckenstock. 2002. Identical ring cleavage products during anaerobic degradation of naphthalene, 2-methylnaphthalene, and tetralin indicate a new metabolic pathway. *Applied and Environmental Microbiology* 68: 852–858.

Answering Islam. n.d. Qur'an and science. http://www.answering-islam.org/Quran/Science/index .htm.

Aranda-Espinoza, H., et al. 1999. Electrostatic repulsion of positively charged vesicles and negatively charged objects. *Science* 285: 394–397.

Arct, M. J. 1991. *Dendrochronology in the Fossil Forests of the Specimen Creek Area Yellowstone National Park*. PhD dissertation, Loma Linda University.

*Arens, N. C. 1998. Progymnosperms. http://www.ucmp.berkeley.edu/IB181/VPL/Osp/Osp2.html.

Armstrong, D. 1996. The New Testament canon. http://ic.net/~erasmus/RAZ45.HTM.

Associated Press. Feb. 23, 1933. Hitler aims blow at 'Godless' move, Lansing State Journal (Michigan). Reprinted at http://www.infidels.org/library/historical/unknown/hitler.html.

Astin, J. A., E. Harkness, and E. Ernst. 2000. The efficacy of "distant healing": A systematic review of randomized trials. *Annals of Internal Medicine* 132: 903–910.

*Atkins, P. W. 1984. *The Second Law*. New York: Scientific American Books.

*Attendorn, H.-G., and R.N.C. Bowen. 1997. *Radioactive and Stable Isotope Geology*. London: Chapman & Hall.

Augustine, St. 1982. *The Literal Meaning of Genesis* vol. 1. Translated by J. H. Taylor. New York: Newman Press. http://www.holycross.edu/departments/religiousstudies/alaffey/Augustine-Genesis.htm.

*Aulie, R. P. 1998. A reader's guide to *Of Pandas and People*. http://www.nabt.org/sub/evolution/ panda1.asp.

Austin, S. A. 1986. Mt. St. Helens and catastrophism. *Impact* 157 (July). http://www.icr.org/ pubs/imp/imp-157.htm.

Austin, S. A. 1988. Grand Canyon lava flows: A survey of isotope dating methods. *Impact* 178 (Apr.). http://www.icr.org/pubs/imp/imp-178.htm.

Austin, S.A. 1992. Excessively old "ages" for Grand Canyon lava flows. *Impact* 224 (Feb.). http://www.icr.org/pubs/imp/imp-224.htm.

Austin, S.A. 1995. *Grand Canyon: Monument to Catastrophe.* Santee, CA: Institute for Creation Research.

Austin, S.A. 1996. Excess argon within mineral concentrates from the New Dacite Lava Dome at Mount St. Helens volcano. *Creation Ex Nihilo Technical Journal* 10(3): 335–343. http://www.icr.org/research/sa/sa-r01.htm.

Austin, S.A. 2000. *Archaeoraptor*: Feathered dinosaur from National Geographic doesn't fly. *Impact* 321 (Mar).

Austin, S.A., et al. 1994. Catastrophic plate tectonics: A global flood model of earth history. *Proceedings of the Third International Conference on Creationism.* Pittsburgh, PA: Creation Science Fellowship, pp. 609–621.

Avarello, R., et al. 1992. Evidence for an ancestral alphoid domain on the long arm of human chromosome 2. *Human Genetics* 89: 247–249.

Awadalla, P., A. Eyre-Walker, and J. Maynard Smith. 1999. Linkage disequilibrium and recombination in hominid mitochondrial DNA. *Science* 286: 2524–2525.

Awramik, S.M. 1992. The oldest records of photosynthesis. *Photosynthesis Research* 33: 75–89.

Babinski, E.T. 1995. *Leaving the Fold: Testimonies of Former Fundamentalists.* Amherst, NY: Prometheus Books.

*Babinski, E.T. 2003. Cetacean evolution (whales, dolphins, porpoises). http://www.edwardtbabinski.us/babinski/whale_evolution.html.

*Babinski, E.T. n.d. An old, out of context quotation. http://www.talkorigins.org/faqs/ce/3/part8.html.

Bagemihl, B. 1999. *Biological Exuberance: Animal Homosexuality and Natural Diversity.* New York: St. Martin Press.

*Bahar, S. 2002. Evolution of the eye: Lessons from freshman physics and Richard Dawkins. *The Biological Physicist* 2(2): 2–5. http://www.aps.org/units/dbp/newsletter/jun02.pdf.

**Bailey, L.R. 1989. *Noah: The Person and the Story in History and Tradition.* Columbia: University of South Carolina Press.

Baker, V.R. 1978. The Spokane flood controversy and the Martian outflow channels. *Science* 202: 1249–1256.

Baldwin, J.M. 1896. A new factor in evolution. *American Naturalist* 30: 441–451, 536–553. http://www.santafe.edu/sfi/publications/Bookinforev/baldwin.html.

Baliunas, S.L., et al. 1995. Chromospheric variations in main-sequence stars. *Astrophysical Journal* 438: 269–287.

Ball, P. 2001. Missing links made simple. *Nature Science Update.* http://www.nature.com/nsu/010308/010308-5.html.

*Ball, P. 2003. Lab tests tenets' limits. *Nature Science Update.* http://www.nature.com/nsu/030428/030428-20.html.

*Bannister, A. n.d. Can "modern science" be found in the Qur'an? http://www.answering-islam.org/Andy/fallacies.html.

*Bannister, R.C. 1979. *Social Darwinism: Science and Myth in Anglo-American Social Thought.* Philadelphia: Temple University Press.

Banta, J. 2001. Whale transitional fossil evidence. http://fp.bio.utk.edu/darwin/1997/whale.html.

Bard, E., B. Hamelin, R.G. Fairbanks, and A. Zindler. 1990. Calibration of the ^{14}C timescale over the past 30,000 years using mass spectrometric U-Th ages from Barbados corals. *Nature* 345: 405–410.

Barger, R.N. 2000. A summary of Lawrence Kohlberg's stages of moral development. http://www.nd.edu/~rbarger/kohlberg.html.

Barker, D. 1990. Leave no stone unturned. *Freethought Today* (Mar.). http://www.ffrf.org/lfif/stone.html.

Barnes, F. A. 1975. The case of the bones in stone. *Desert Magazine* 38(2): 36–39.

Barnes, T. G. 1973. *Origin and destiny of Earth's magnetic field*. ICR Technical Monograph No. 4. El Cajon, CA: ICR.

Barnes, T. G. 1982. Young age for the moon and earth. *Impact* 110 (Aug.). http://www.icr .org/pubs/imp/imp-110.htm.

Barnhart, C. L. (ed.). 1948. *The American College Dictionary*. New York: Random House.

Barton, N. H., and B. Charlesworth. 1998. Why sex and recombination? *Science* 281: 1986–1990.

Bateman, R. M., et al. 1998. Early evolution of land plants: Phylogeny, physiology, and ecology of the primary terrestrial radiation. *Annual Review of Ecology and Systematics* 29: 263–292.

Batten, D. 2002. It's not science. http://www.answersingenesis.org/docs2002/0228not_science.asp.

Batten, R. P. 1976. *Living Trophies*. New York: Thomas Y. Crowell.

Baumgardner, J. R. 1990a. Changes accompanying Noah's Flood. *Proceedings of the Second International Conference on Creationism*, vol. 2. Pittsburgh, PA: Creation Science Fellowship, pp. 35–45.

Baumgardner, J. R. 1990b. The imperative of non-stationary natural law in relation to Noah's Flood. *Creation Research Society Quarterly* 27(3): 98–100.

Baumgardner, J. R. 2003. Carbon dating undercuts evolution's long ages. *Impact* 364 (Oct.). http://www.icr.org/newsletters/impact/impactoct03.html.

Baumgardner, J. R., and D. W. Barnette. 1994. Patterns of ocean circulation over the continents during Noah's Flood. *Proceedings of the Third International Conference on Creationism*. Pittsburgh, PA: Creation Science Fellowship, pp. 77–86.

Baymann, F., M. Brugna, U. Mühlenhoff, and W. Nitschke. 2001. Daddy, where did (PS)I come from? *Biochimica et Biophysica Acta* 1507: 291–310.

Becker, B., and B. Kromer. 1993. The continental tree-ring record—absolute chronology, ^{14}C calibration and climatic change at 11 ka. *Palaeogeography, Palaeoclimatology, Palaeoecology* 103: 67–71.

Becker, B., B. Kromer, and P. Trimborn. 1991. A stable-isotope tree-ring timescale of the Late Glacial/Holocene boundary. *Nature* 353: 647–649.

*Becker, L. 2002. Repeated blows. *Scientific American* 286(3): 76–83.

Becker, L., R. J. Poreda, and T. E. Bunch. 2000. Fullerenes: An extraterrestrial carbon carrier phase for noble gases. *PNAS* 97: 2979–2983.

Behe, M. J. 1996a. *Darwin's Black Box*. New York: Free Press.

Behe, M. J. 1996b. Darwin under the microscope. *New York Times*, Oct. 29, 1996, sec. A, p. 25. http://www.arn.org/docs/behe/mb_dm11496.htm.

Behe, M. J. 2003. A functional pseudogene?: An open letter to Nature. http://www.arn.org/ docs2/news/behepseudogene052003.htm.

Beheregaray, L. B., and P. Sunnucks. 2001. Fine-scale genetic structure, estuarine colonization and incipient speciation in the marine silverside fish *Odontesthes argentinensis*. *Molecular Ecology* 10(12): 2849–2866.

*Behrman, E. J., G. A. Marzluf, and R. Bentley. 2004. Evidence from biochemical pathways in favor of unfinished evolution rather than intelligent design. *Journal of Chemical Education* 81: 1051–1052.

Benner, S. A., M. D. Caraco, J. M. Thomson, and E. A. Gaucher. 2002. Planetary biology—paleontological, geological, and molecular histories of life. *Science* 296: 864–868.

*Benton, M. J. 1991. *The Rise of the Mammals*. New York: Crescent Books.

Benzing, D. H. 1990. *Vascular Epiphytes*. Cambridge: Cambridge University Press.

Bergman, J. 2000. Why abiogenesis is impossible. *Creation Research Society Quarterly* 36(4). http://www.creationresearch.org/crsq/articles/36/36_4/abiogenesis.html.

Berlinski, D. 1996. The deniable Darwin. *Commentary* 101(6). http://www.rae.org/dendar.html.

Berlitz, C. 1984. *Atlantis, the Eighth Continent*. New York: Putnam.

Bermúdez de Castro, J. M., et al. 1997. A hominid from the Lower Pleistocene of Atapuerca, Spain: Possible ancestor to Neandertals and modern humans. *Science* 276: 1392–1395.

*Bernstein, M. P., S. A. Sandford, and L. J. Allamandola. 1999a. Life's far-flung raw materials. *Scientific American* 281(1): 42–49.

Bernstein, M. P., et al. 1999b. UV irradiation of polycyclic aromatic hydrocarbons in ices: Production of alcohols, quinones, and ethers. *Science* 283: 1135–1138. See also pp. 1123–1124.

Berthault, G. 2000. Experiments in stratification. *Impact* 328 (Oct.). http://www.icr.org/pubs/imp/imp-328.htm.

*Beus, S. S., and M. Morales (eds.). 2002. *Grand Canyon Geology*. 2nd ed. New York: Oxford University Press. (technical)

Big-Bang-Theory. 2002. http://www.big-bang-theory.com.

*Birkeland, B. 2004. Fossil soils (paleosols) at Joggins. http://www.evcforum.net/ubb/Forum7/HTML/000116.html#7.

Bishop, A. C. 1981. The development of the concept of continental drift. In *The Evolving Earth*, ed. L.R.M. Cocks, 155–164. London: British Museum.

Bishop, G. 2002. Majority of Ohio science professors and public agree: "Intelligent design" mostly about religion. http://www.ncseweb.org/resources/articles/733_ohio_scientists39_intellige_10_15_2002.asp.

Bize, S., et al. 2003. Testing the stability of fundamental constants with the $^{199}Hg^+$ single-ion optical clock. *Physical Review Letters* 90: 150802.

Blankenship, R. E. 1992. Origin and early evolution of photosynthesis. *Photosynthesis Research* 33: 91–111.

Blankenship, R. E., and H. Hartman. 1998. The origin and evolution of oxygenic photosynthesis. *Trends in Biochemical Sciences* 23: 94–97.

Blocker, A., K. Komoriya, and S.-I. Aizawa. 2003. Type III secretion systems and bacterial flagella: Insights into their function from structural similarities. *PNAS* 100(6): 3027–3030.

Boardman, D. R., II, and P. H. Heckel. 1989. Glacial-eustatic sea-level curve for early Late Pennsylvanian sequence in north-central Texas and biostratigraphic correlation with curve for mid-continent North America. *Geology* 17: 802–805.

*Boggs, S. 1995. *Principles of Sedimentology and Stratigraphy*. New York: Freeman.

Böhler, C., P. E. Nielsen, and L. E. Orgel. 1995. Template switching between PNA and RNA oligonucleotides. *Nature* 376: 578–581. See also pp. 548–549.

*Bonner, J. T. 2000. *First Signals: The evolution of multicellular development*. Princeton, NJ: Princeton University Press.

Boraas, M. E. 1983. Predator induced evolution in chemostat culture. *Eos* 64: 1102.

Boraas, M. E., D. B. Seale, and J. E. Boxhorn. 1998. Phagotrophy by a flagellate selects for colonial prey: A possible origin of multicellularity. *Evolutionary Ecology* 12: 153–164.

*Bowler, P. J. 1983. *The Eclipse of Darwinism: Anti-Darwinian Evolution Theories in the Decades Around 1900*. Baltimore: Johns Hopkins University Press.

*Bowler, P. J. 1993. Biology and social thought, 1850–1914. *Berkeley Papers in History of Science*, 15. Berkeley: Office for History of Science and Technology, University of California at Berkeley, 95.

*Bowman, S. 1990. *Radiocarbon Dating*. Berkeley: University of California Press.

Boyden, A. M., et al. 2002. High bone density due to a mutation in LDL-receptor-related protein 5. *New England Journal of Medicine* 346: 1513–1521.

*Boyer, P. 2001. *Religion Explained*. New York: Basic Books.

Branch, G. 2002. Evolving banners at the Discovery Institute. *RNCSE* 22(5): 12. http://www.ncseweb.org/resources/articles/4116_evolving_banners_at_the_discov_8_29_2002.asp.

*Branch, G., and E. Scott. 2003. The antievolution law that wasn't. *The American Biology Teacher* 65(3): 165–166.

Brand, L. R. 1978. Footprints in the Grand Canyon. *Origins* 5(2): 64–82. http://www.grisda.org/origins/05064.htm.

Brand, L. R. 1996. Variations in salamander trackways resulting from substrate differences. *Journal of Paleontology* 70(6): 1004–1011.

Brand, L. R., and T. Tang. 1991. Fossil vertebrate footprints in the Coconino Sandstone (Permian) of northern Arizona: Evidence for underwater origin. *Geology* 19(12): 1201–1204.

*Brass, M. 2002. *The Antiquity of Man: Artifactual, Fossil, and Gene Records Explored*. Baltimore: Publish America.

*Brawley, J. 1992. Evolution's tiny violences: The Po-halo mystery. http://www.talkorigins.org/faqs/po-halos/violences.html.

Bretz, J. H. 1969. The Lake Missoula floods and the Channeled Scabland. *Journal of Geology* 77: 505–543.

Brewster, E. T. 1927. *Creation: A History of Non-Evolutionary Theories*. Indianapolis, IN: Bobbs-Merrill.

Briggs, D.E.G., E.N.K. Clarkson, and R. J. Aldridge. 1983. The conodont animal. *Lethaia* 16: 1–14.

*Bringas, E. 1996. *Going by the Book*. Charlottesville, VA: Hampton Roads Publishing.

*Brinkman, M. 1995. Ice core dating. http://www.talkorigins.org/faqs/icecores.html.

*Britian, T. n.d. Darwin on race and slavery. http://home.att.net/~troybritain/articles/darwin_on_race.htm.

Brocks, J. J., G. A. Logan, R. Buick, and R. E. Summons. 1999. Archean molecular fossils and the early rise of eukaryotes. *Science* 285: 1033–1036. See also Knoll, A. H. 1999. A new molecular window on early life. *Science* 285: 1025–1026. http://www.sciencemag.org/cgi/content/full/285/5430/1025.

*Brodsky, A. K. 1994. *The Evolution of Insect Flight*. Oxford: Oxford University Press.

Bronowski, J. 1973. *The Ascent of Man*. Boston: Little, Brown.

Brooks, C., D. E. James, and S. R. Hart. 1976. Ancient lithosphere: Its role in young continental volcanism. *Science* 193: 1086–1094.

Brooks, D. R., and E. O. Wiley. 1988. *Evolution as Entropy*. Chicago: University of Chicago Press.

Brown, C. J., K. M. Todd, and R. F. Rosenzweig. 1998. Multiple duplications of yeast hexose transport genes in response to selection in a glucose-limited environment. *Molecular Biology and Evolution* 15(8): 931–942. http://mbe.oupjournals.org/cgi/reprint/15/8/931.pdf.

Brown, C. W. n.d. *Ensatina eschscholtzi* speciation in progress: A classic example of Darwinian evolution. http://www.santarosa.edu/lifesciences/ensatina.htm.

Brown, K. S. 1999. Deep Green rewrites evolutionary history of plants. *Science* 285: 990–991.

Brown, W. 1995. *In the Beginning: Compelling Evidence for Creation and the Flood*. 6th ed. Phoenix, AZ: Center for Scientific Creation.

Brown, W. 2001. *In the Beginning: Compelling Evidence for Creation and the Flood*. 7th ed. Phoenix, AZ: Center for Scientific Creation. http://www.creationscience.com.

*Brunvand, J. H. 2000. *The Truth Never Stands in the Way of a Good Story*. Urbana: University of Illinois Press.

*Brush, S. G. 1983. Ghosts from the nineteenth century: Creationist arguments for a young earth. In *Scientists Confront Creationism*, ed. L. R. Godfrey, 49–84. New York: W. W. Norton.

Bull, J. J., and H. A. Wichman. 2001. Applied evolution. *Annual Review of Ecology and Systematics* 32: 183–217.

Burdick, C. L. 1966. Microflora of the Grand Canyon. *Creation Research Society Quarterly* 3: 38–50.

Burdick, C. L. 1972. Progress report on Grand Canyon palynology. *Creation Research Society Quarterly* 9: 25–30.

Burke, R. B. n.d. Some aspects of salt dissolution in the Williston Basin of North Dakota. *NDGS Newsletter* 28(1): 1–5. http://www.state.nd.us/ndgs/Newsletter/NL01S/PDF/salts01.pdf.

*Burkert, W. 1996. *Creation of the Sacred: Tracks of Biology in Early Religions*. Cambridge, MA: Harvard University Press.

Burr, C. 1997. The geophysics of God. *US News and World Report* 122 (June 16): 55–58.

*Burton, J. D., and D. Wright. 1981. Sea water and its evolution. In *The Evolving Earth*, ed. L.R.M. Cocks, 89–101. London: British Museum.

Byrne, K., and R. A. Nichols. 1999. *Culex pipiens* in London Underground tunnels: Differentiation between surface and subterranean populations. *Heredity* 82: 7–15.

Cairn-Smith, A. G. 1985. *Seven Clues to the Origin of Life*. Cambridge: Cambridge University Press.

Caldwell, M. W., and M.S.Y. Lee. 1997. A snake with legs from the marine Cretaceous of the Middle East. *Nature* 386: 705–709.

*Callaghan, C. A. 1987. Instances of observed speciation. *American Biology Teacher* 49: 34–36.

Calvert, J. H., W. S. Harris, and J. F. Sjogren. 2001. Science standards (letter to Kansas State Board of Education, Feb. 8, 2001). http://www.intelligentdesignnetwork.org/Feb8letterKSBE.htm.

Campagna, T. 2002. Bob Bakker on creationism. http://www.prehistoricplanet.com/features/articles/bakker/index.htm.

Canfield, D. E., and A. Teske. 1996. Late Proterozoic rise in atmospheric oxygen concentration inferred from phylogenetic and sulphur-isotope studies. *Nature* 382: 127–132. See also pp. 111–112.

Cann, R. L., M. Stoneking, and A. C. Wilson. 1987. Mitochondrial DNA and human evolution. *Nature* 325: 31–36. See also p. 13.

Carlberg, R., et al. 1999. Ask the experts: Astronomy: What process creates and maintains the beautiful spiral arms around spiral galaxies? http://www.sciam.com/askexpert_question.cfm?articleID=0008A68A-8C7F-1C72-9EB7809EC588F2D7.

*Carrier, R. 1998. Does the Christian theism advocated by J. P. Moreland provide a better reason to be moral than secular humanism? http://www.infidels.org/library/modern/richard_carrier/moreland.html.

*Carroll, R. L. 1997. *Patterns and Processes of Vertebrate Evolution*. Cambridge: Cambridge University Press. (technical)

Carter, R. 2000. Marx of respect. http://www.gruts.com/darwin/articles/2000/marx/index.htm.

*Catalano, J. (ed.). 1998. Publish or perish: Some published works on biochemical evolution. http://www.talkorigins.org/faqs/behe/publish.html.

Cavalier-Smith, T. 1987. The origin of eukaryote and archaebacterial cells. *Annals of the New York Academy of Sciences* 503: 17–54.

*Cavalier-Smith, T. 1997. The blind biochemist. *Trends in Ecology and Evolution* 12: 162–163.

Cavalier-Smith T. 2001. Obcells as proto-organisms: membrane heredity, lithophosphorylation, and the origins of the genetic code, the first cells, and photosynthesis. *Journal of Molecular Evolution* 53: 555–595.

Cavalier-Smith, T. 2002. The phagotrophic origin of eukaryotes and phylogenetic classification of Protozoa. *International Journal of Systematic and Evolutionary Microbiology* 52: 297–354.

Ceragioli, R. C. 1996. Solving the puzzle of "red" Sirius. *Journal for the History of Astronomy* 27: 93–128.

Cha, K. Y., D. P. Wirth, and R. A. Lobo. 2001. Does prayer influence the success of in vitro fertilization–embryo transfer? Report of a masked, randomized trial. *Journal of Reproductive Medicine* 46: 781–787.

Chadwick, A. V. 1973. Grand Canyon palynology—a reply. *Creation Research Society Quarterly* 9: 238.

Chadwick, A. V. 1981. Precambrian pollen in the Grand Canyon—a re-examination. *Origins* 8(1): 7–12.

Chand, H., R. Srianand, P. Petitjean, and B. Aracil. 2004. Probing the cosmological variation of the fine-structure constant: Results based on VLT-UVES sample. *Astronomy and Astrophysics* 417: 853. http://arxiv.org/abs/astro-ph/0401094.

Chang, S., et al. 1983. Prebiotic organic syntheses and the origin of life. In *Earth's Earliest Biosphere: Its Origin and Evolution*, ed. J. W. Schopf, 53–92. Princeton, NJ: Princeton University Press.

*Charles, R. H. (ed.). 1913. The Apocrypha and Pseudepigrapha of the Old Testament in English. Oxford: Clarendon Press. http://www.ccel.org/c/charles/pseudepigrapha.

*Chase, S. 1999. Is Haeckel's law of recapitulation a problem? http://www.talkorigins.org/origins/postmonth/feb99.html.

*Chatters, J. C. 2001. *Ancient Encounters: Kennewick Man and the First Americans.* New York: Simon & Schuster.

Chellapilla, K., and D. B. Fogel. 2001. Evolving an expert checkers playing program without using human expertise. *IEEE Transactions on Evolutionary Computation* 5: 422–428. http://www.natural-selection.com/Library/2001/IEEE-TEVC.pdf.

Chen, J.-Y., D.-Y. Huang, and C.-W. Li. 1999. An early Cambrian craniate-like chordate. *Nature* 402: 518–522.

Chen, J.-Y., et al. 2000. Precambrian animal diversity: Putative phosphatized embryos from the Doushantuo Formation of China. *PNAS* 97: 4457–4462.

Chen, J.-Y., et al. 2004. Small bilaterian fossils from 40 to 55 million years before the Cambrian. *Science* 305: 218–222. See also Stokstad, E. 2004. Controversial fossil could shed light on early animals' blueprint. *Science* 304: 1425.

Chen, P., Z. Dong, and S. Zhen. 1998. An exceptionally well-preserved theropod dinosaur from the Yixian Formation of China. *Nature* 391: 147–152.

Chiappe, L. M. 2002a. Basal bird phylogeny. In *Mesozoic Birds: Above the Heads of Dinosaurs*, eds. L. M. Chiappe and L. M. Witmer, 448–472. Berkeley: University of California Press.

Chiappe, L. M. 2002b. Osteology of the flightless *Patagopteryx deferrariisi* from the Late Cretaceous of Patagonia (Argentina). In *Mesozoic Birds: Above the Heads of Dinosaurs*, eds. L. M. Chiappe and L. M. Witmer, 281–316. Berkeley: University of California Press.

*Chiappe, L. M., and G. J. Dyke. 2002. The Mesozoic radiation of birds. *Annual Review of Ecology and Systematics* 33: 91–124. (technical)

Chiappe, L. M., S. Ji, Q. Ji, and M. A. Norell. 1999. Anatomy and systematics of the Confuciusornithidae (Theropoda: Aves) from the Late Mesozoic of northeastern China. *Bulletin of the American Museum of Natural History* 242: 1–89.

Chiappe, L. M., M. A. Norell, and J. M. Clark. 2001. A new skull of *Gobipteryx minuta* (Aves: Enantiornithes) from the Cretaceous of the Gobi Desert. *American Museum Novitates* 3346: 1–15.

Chiappe, L. M., M. A. Norell, and J. M. Clark. 2002. The Cretaceous, short-armed Alvarezsauridae. In *Mesozoic Birds: Above the Heads of Dinosaurs*, eds. L. M. Chiappe and L. M. Witmer, 87–120. Berkeley: University of California Press.

Chiappe, L. M., and L. M. Witmer (eds.). 2002. *Mesozoic Birds: Above the Heads of Dinosaurs.* Berkeley: University of California Press.

*Chronic, H. 1983. *Roadside Geology of Arizona.* Missoula, MT: Mountain Press Publishing.

*Clack, J. A. 2002. *Gaining Ground: The Origin and Early Evolution of Tetrapods.* Bloomington, IN: Indiana University Press.

*Claridge, M., H. Dawah, and M. Wilson (eds.). 1997. *Species: The Units of Biodiversity.* London: Chapman & Hall. (technical)

Clark, E. E. 1953. *Indian Legends of the Pacific Northwest.* Berkeley: University of California Press.

Clark, J. M., M. A. Norell, and L. M. Chiappe. 1999. An oviraptorid skeleton from the Late Cretaceous of Ukhaa Tolgod, Mongolia, preserved in an avianlike brooding position over an oviraptorid nest. *American Museum Novitates* 3265: 1–36.

*Clark, R. W. 1984. *The Survival of Charles Darwin: A Biography of a Man and an Idea.* New York: Random House.

*Clark, S. 1999. Polarized starlight and the handedness of life. *American Scientist* 87(4): 336–343.

Clarke, J. A., and M. A. Norell. 2002. The morphology and phylogenetic position of *Apsaravis ukhaana* from the late Cretaceous of Mongolia. *American Museum Novitates* 3387: 1–46.

Clementson, S. P. 1970. A critical examination of radioactive dating of rocks. *Creation Research Society Quarterly* 7: 137–141; citing Cherdyntsev, V. V., et al. Geological Institute Academy of Sciences, USSR, Earth Science Section, 172, p. 178. Cited in Morris, H. M. 1974 *Scientific Creationism.* Green Forest, AR: Master Books, 143.

Coates, M. I. 1996. The Devonian tetrapod *Acanthostega gunnari* Jarvik: Postcranial anatomy, basal tetrapod interrelationships and patterns of skeletal evolution. *Transactions of the Royal Society of Edinburgh: Earth Sciences* 87: 363–421.

Coates, M. I., and J. A. Clack. 1990. Polydactyly in the earliest known tetrapod limbs. *Nature* 347: 66–69.

Coates, M. I., and J. A. Clack. 1991. Fish-like gills and breathing in the earliest known tetrapod. *Nature* 352: 234–236.

*Cocks, L.R.M. (ed.). 1981. *The Evolving Earth*. London: British Museum.

Cody, G. D., et al. 2000. Primordial carbonylated iron-sulfur compounds and the synthesis of pyruvate. *Science* 289: 1337–1340.

Coffin, H. G. 1983. Erect floating stumps in Spirit Lake, Washington. *Geology* 11: 298–299.

Coghlan, A. 1998. A sexual revolution. *New Scientist* 160 (Nov. 21): 4. See also Maxygen, 1998. When genes "breed" in the lab, a surprising number of their offspring are supergenes. (Press release, Nov. 18, 1998). http://www.eurekalert.org/pub_releases/1998-11/NS-WGIT-181198.php.

*Cohen, P. 1996. Let there be life. *New Scientist* 151 (July 6): 22–27. http://www.newscientist.com/hottopics/astrobiology/letthere.jsp.

*Cohn, M. J., and C. Tickle. 1999. Developmental basis of limblessness and axial patterning in snakes. *Nature* 399: 474–479. (technical)

*Colby, C. 1993. Evidence for evolution: An eclectic survey. http://www.talkorigins.org/faqs/evolution-research.html.

*Colby, C., et al. 1993. Evidence for jury-rigged design in nature. http://www.talkorigins.org/faqs/jury-rigged.html.

Cole, A. A. 2000. The distance to the Large Magellanic Cloud. *Science* 289: 1149–1150.

*Cole, J. R. 1985. If I had a hammer. *Creation/Evolution* 5(1): 46–47.

*Cole, J. R., and L. R. Godfrey (eds.). 1985. The Paluxy River footprint mystery—solved. *Creation/Evolution* 5(1).

Collins, A. G. 1994. Metazoa: Fossil record. http://www.ucmp.berkeley.edu/phyla/metazoafr.html.

Collins, H. 1998. Hit or myth? *New Scientist* 159 (Sept. 12): 36–39.

*Collins, L. G. 1997. Polonium halos and myrmekite in pegmatite and granite. http://www.csun.edu/~vcgeo005/revised8.htm.

*Collins, L. G., and D. F. Fasold. 1996. Bogus "Noah's Ark" from Turkey exposed as a common geologic structure. *Journal of Geoscience Education* 44(4): 439–444. http://www.csun.edu/~vcgeo005/bogus.html.

*Colp, R., Jr. 1982. The myth of the Darwin–Marx letter. *History of Political Economy* 14: 461–482.

*Conner, S. R., and H. Ross. 1999. The unraveling of starlight and time. http://www.reasons.org/resources/apologetics/unravelling.shtml?main.

Conner, S. R., and D. N. Page. 1998. Starlight and time is the Big Bang. *Creation Ex Nihilo* 12(2): 174–194. See also letters in *Creation Ex Nihilo* 13(1), 1999, 49–52.

Conrad, E. C. 1981. Tripping over a trilobite: A study of the Meister tracks. *Creation/Evolution* 2(4): 30–33.

Conrad, E. C. 1982. Are there human fossils in the "wrong place" for evolution? *Creation/Evolution* 3(2): 14–22.

*Conway Morris, S. 1998. *The Crucible of Creation*. Oxford: Oxford University Press.

*Conway Morris, S. 2000. The Cambrian "explosion": Slow-fuse or megatonnage? *PNAS* 97: 4426–4429. (technical)

Cook, L. M. 2003. The rise and fall of the *carbonaria* form of the peppered moth. *Quarterly Review of Biology* 78: 399–417.

Cook, P. J., and J. H. Shergold (eds.). 1986. *Phosphate Deposits of the World, Volume 1. Proterozoic and Cambrian Phosphorites*. Cambridge: Cambridge University Press.

*Coon, M. 1998. Is the complement system irreducibly complex? http://www.talkorigins.org/faqs/behe/icsic.html.

Cooper, G., et al. 2001. Carbonaceous meteorites as a source of sugar-related organic compounds for the early Earth. *Nature* 414: 879–883. See also pp. 857–858.

Copely, J. 2001. The story of O. *Nature* 410: 862–864.

Corliss, W. R. 1988. Why do spiral galaxies stay that way? Or do they? *Science Frontiers Online* 55. http://www.science-frontiers.com/sf055/sf055p07.htm.

*Cosmides, L., and J. Tooby. 1997. Evolutionary psychology: A primer. http://www.psych.ucsb.edu/research/cep/primer.html.

Coulmas, F. 1989. *The Writing Systems of the World*. Malden, MA: B. Blackwell.

Courlander, H. 1996. *A Treasury of African Folklore*. New York: Marlowe & Co.

Cowling, T. G. 1981. The present status of dynamo theory. *Annual Review of Astronomy and Astrophysics* 19: 115–135.

*Cracraft, J. 1987. Species concepts and the ontology of evolution. *Biology and Philosophy* 2: 329–346.

Creed, E. R., D. R. Lees, and M. G. Bulmer. 1980. Pre-adult viability differences of melanic *Biston betularia* (L.) (Lepidoptera). *Biological Journal of the Linnean Society* 13: 251–262.

Cremo, M., and R. Thompson. 1993. *Forbidden Archaeology: The Hidden History of the Human Race*. Los Angeles: Bhaktivedanta.

Cronin, J. R., and S. Pizzarello. 1999. Amino acid enantiomer excesses in meteorites: Origin and significance. *Advances in Space Research* 23(2): 293–299.

Cronin, T. M. 1985. Speciation and stasis in marine ostracoda: Climatic modulation of evolution. *Science* 227: 60–63.

CRSC. 1998. The wedge strategy. http://www.antievolution.org/features/wedge.html.

Csete, M. E., and J. C. Doyle. 2002. Reverse engineering of biological complexity. *Science* 295: 1664–1669.

*Cuffey, C. A. 2001. The fossil record: Evolution or "scientific creation." http://www.gcssepm.org/special/cuffey_00.htm.

Cullen, B. 1998. Parasite ecology and the evolution of religion. In *The Evolution of Complexity*, ed. F. Heylighen, http://pespmc1.vub.ac.be/Papers/Cullen.html.

Cuny, H. 1965. *Louis Pasteur: The Man and His Theories*. Translated by P. Evans. London: The Scientific Book Club.

Currie, P. J., and P. Chen. 2001. Anatomy of *Sinosauropteryx prima* from Liaoning, northeastern China. *Canadian Journal of Earth Sciences* 38: 1705–1727.

Cuzzi, J. N., and P. R. Estrada. 1998. Compositional evolution of Saturn's rings due to meteoroid bombardment. *Icarus* 132: 1–35.

Daeschler, E. B., and N. Shubin. 1998. Fish with fingers? *Nature* 391: 133.

Dalley, S. 1989. *Myths from Mesopotamia*. Oxford: Oxford University Press.

Dalrymple, G. B. 1969. ^{40}Ar/^{36}Ar analyses of historic lava flows. *Earth and Planetary Science Letters* 6: 47–55.

*Dalrymple, G. B. 1991. *The Age of the Earth*. Stanford, CA: Stanford University Press.

Dalrymple, G. B. 2000. Radiometric dating does work! *RNCSE* 20(3): 14–17. http://www.ncseweb.org/resources/rncse_content/vol20/6061_radiometeric_dating_does_work_12_30_1899.asp.

Dalrymple, G. B., G. A. Izett, L. W. Snee, and J. D. Obradovich. 1993. ^{40}Ar/^{39}Ar age spectra and total-fusion ages of tektites from Cretaceous–Tertiary boundary sedimentary rocks in the Beloc formation, Haiti. *United States Geological Survey Bulletin* no. 2065.

Dalton, R. 2000. Feathers fly over Chinese fossil bird's legality and authenticity. *Nature* 403: 689–690.

*Darwin, C. 1859. *The Origin of Species*. 1st ed. London: Senate. http://www.talkorigins.org/faqs/origin.html.

Darwin, C. 1868. *Variation of Animals and Plants under Domestication*. Vol. 2, chap. 20. London: John Murray. http://pages.britishlibrary.net/charles.darwin/texts/variation/variation20.html.

Darwin, C. 1872. *The Origin of Species*. 6th ed. London: Senate. http://www.literature.org/authors/darwin-charles/the-origin-of-species-6th-edition/index.html.

Darwin, C. 1881. Letter to W. Graham. In *The Life and Letters of Charles Darwin*, ed. F. Darwin. New York: D. Appleton & Co., 1905. http://pages.britishlibrary.net/charles.darwin/texts/letters/letters1_08.html.

Darwin, C. 1913. *Voyage Round the World of H.M.S. Beagle.* 11th ed. London: John Murray. http://pages.britishlibrary.net/charles.darwin/texts/beagle_voyage/beagle21.html.

Daubin, V., N. A. Moran, and H. Ochman. 2003. Phylogenetics and the cohesion of bacterial genomes. *Science* 301: 829–832. See also pp. 745–746.

Davidson, C. 1997. Creatures from primordial silicon. *New Scientist* 156 (Nov. 15): 30–34. http://www.newscientist.com/hottopics/ai/primordial.jsp.

*Davidson, J. P., W. E. Reed, and P. M. Davis. 1997. *Exploring Earth: An Introduction to Physical Geology.* Upper Saddle River, NJ: Prentice Hall.

Davies, E. K., A. D. Peters, and P. D. Keightley. 1999. High frequency of cryptic deleterious mutations in *Caenorhabditis elegans. Science* 285: 1748–1751.

Davies, K. 1994. Distribution of supernova remnants in the galaxy. *Proceedings of the Third International Conference on Creationism.* Pittsburgh, PA: Creation Science Fellowship. http://www.creationdiscovery.org/cdp/articles/snrart.html.

Davies, K. 1996. Evidences for a young sun. *Impact* 276 (June). http://www.icr.org/pubs/imp/imp-276.htm.

Davis, P., and D. H. Kenyon. 1989. *Of Pandas and People: The Central Question of Biological Origins.* 2nd ed. Dallas, TX: Haughton.

Davis, P. G., and D.E.G. Briggs. 1995. Fossilization of feathers. *Geology* 23: 783–786.

*Dawkins, R. 1986. *The Blind Watchmaker: Why the Evidence of Evolution Reveals a Universe without Design.* New York: Norton.

*Dawkins, R. 1995. *River out of Eden.* New York: Basic Books.

*Dawkins, R. 1996. *Climbing Mount Improbable.* New York: W. W. Norton.

*Dawkins, R. 2000. There's more to books than titles. http://archive.workersliberty.org/wlmags/wl61/dawkins.htm.

*Dawkins, R. 2003. The information challenge. In *A Devil's Chaplain,* R. Dawkins. Boston: Houghton Mifflin. http://www.skeptics.com.au/journal/dawkins1.htm.

Day, J. 1992. Leviathan. In *The Anchor Bible Dictionary,* Vol. 4, ed. David Noel Freedman, 295–296. New York: Doubleday.

Day, R.P.J. 1990. An account of a debate with a creationist. http://www.talkorigins.org/faqs/debate-rob-day.html.

*Deacon, T. W. 1998. *The Symbolic Species: The Co-Evolution of Language and the Brain.* New York: W. W. Norton.

*Deamer, D. W., and J. Ferris. 1999. The origins and early evolution of life. http://www.chemistry.ucsc.edu/~deamer/home.html.

Dean, M., et al. 1996. Genetic restriction of HIV-1 infection and progression to AIDS by a deletion allele of the CKR5 structural gene. *Science* 273: 1856–1862.

Decker, R., and B. Decker. 1998. *Volcanoes.* 3rd ed. New York: Freeman.

Decu, V., M. Gruia, S. L. Keffer, and S. M. Sarbu. 1994. Stygobiotic waterscorpion, *Nepa anophthalma,* n. sp. (Heteroptera: Nepidae), from a sulfurous cave in Romania. *Annals of the Entomological Society of America* 87: 755–761.

*de Duve, C. 1995a. The beginnings of life on earth. *American Scientist* 83: 428–437. http://www.americanscientist.org/template/AssetDetail/assetid/21438?fulltext=true.

*de Duve, C., 1995b. *Vital Dust: Life as a Cosmic Imperative.* New York: Basic Books.

Delano, J. W. 2001. Redox history of the Earth's interior since ~3900 Ma: Implications for prebiotic molecules. *Origins of Life and Evolution of the Biosphere* 31: 311–341.

Dembski, W. A. 1996. What every theologian should know about creation, evolution and design. http://www.arn.org/docs/dembski/wd_theologn.htm.

Dembski, W. A. 1998a. *The Design Inference: Eliminating Chance through Small Probabilities.* Cambridge: Cambridge University Press.

Dembski, W. A. 1998b. Science and Design. *First Things* 86 (Oct.): 21–27. http://www.firstthings.com/ftissues/ft9810; shdembski.html.

Dembski, W. A. 1999a. *Intelligent Design: The Bridge between Science and Theology*. Downers Grove, IL: InterVarsity Press.

Dembski, W. A. 1999b. Signs of intelligence: A primer on the discernment of intelligent design. *Touchstone* 12(4): 76–84.

Dembski, W. A. 1999c. Why evolutionary algorithms cannot generate specified complexity. *Metaviews* 152 (Nov. 1). (www.meta-list.org). http://www.leaderu.com/offices/dembski/docs/bd-algorithms.html.

Dembski, W. A. 2001a. Is intelligent design a form of natural theology? http://www.designinference.com/documents/2001.03.ID_as_nat_theol.htm.

Dembski, W. A. 2001b. Is intelligent design testable? http://www.arn.org/docs/dembski/wd_isidtestable.htm.

Dembski, W. A. 2001c. Teaching intelligent design—What happened when? A response to Eugenie Scott. http://www.arn.org/docs/dembski/wd_teachingid0201.htm.

Dembski, W. A. 2002a (Nov. 6). ARN Board: What genetic algorithms can do. http://www.arn.org/ubb/ultimatebb.php?ubb=get_topic;f=13;t=000428;p=1.

Dembski, W. A. 2002b. *No Free Lunch*. Lanham, MD: Rowman & Littlefield.

Dembski, W. A., and J. W. Richards. 2001. *Unapologetic Apologetics*. Downers Grove, IL: InterVarsity Press. http://www.gospelcom.net/ivpress/title/exc/1563-1.pdf.

Demetrius, L. 2000. Theromodynamics and evolution. *Journal of Theoretical Biology* 206: 1–16.

Dennett, D. C. 1991. *Consciousness Explained*. Boston: Little, Brown.

Densmore, R., and J. Zasada. 1983. Seed dispersal and dormancy patterns in northern willows: Ecological and evolutionary significance. *Canadian Journal of Botany* 61: 3207–3216.

Denton, M. 1986. *Evolution: A Theory in Crisis*. Bethesda, MD: Adler & Adler.

Derbyshire, J. 2003. Pseudoscience vs. snobbery: A Doonesbury lesson. *National Review Online* (Apr. 22). http://www.nationalreview.com/derbyshire/derbyshire042203.asp.

de Wet, J.M.J. 1971. Polyploidy and evolution in plants. *Taxon* 20: 29–35.

Dewey, D. 1996. Introduction to the Mandelbrot set. http://www.ddewey.net/mandelbrot.

Dewey, J. 1910. The influence of Darwinism on philosophy. In *The Influence of Darwin on Philosophy and Other Essays in Contemporary Thought*. New York: Holt. Reprinted in *Classic American Philosophers*, ed. M. H. Fisch. New York: Appleton-Century-Crofts, 1951.

DeWitt, D. A. 2002. The dark side of evolution. http://www.answersingenesis.org/docs2002/0510eugenics.asp.

Dial, K. P. 2003. Wing-assisted incline running and the evolution of flight. *Science* 299: 402–404. See also p. 329.

*Dickin, A. P. 1995. *Radiogenic Isotope Geology*. Cambridge: Cambridge University Press.

Dieckmann, U., and M. Doebeli. 1999. On the origin of species by sympatric speciation. *Nature* 400: 354–357.

Dilcher, D. 2000. Toward a new synthesis: major evolutionary trends in the angiosperm fossil record. *PNAS* 97: 7030–7036.

*Dingus, L., and T. Rowe. 1997. *The Mistaken Extinction: Dinosaur Evolution and the Origin of Birds*. New York: Freeman.

*Di Peso, C. C. 1953. The clay figurines of Acambaro, Guanajuato, Mexico. *American Antiquity* 18: 388–389.

Discovery Institute. 2001. A scientific dissent from Darwinism. http://www.discovery.org/articleFiles/PDFs/100ScientistsAd.pdf.

Discovery Institute. 2003. A preliminary analysis of the treatment of evolution in biology textbooks currently being considered for adoption by the Texas State Board of Education. http://www.discovery.org/articleFiles/PDFs/TexasPrelim.pdf.

*Dobzhansky, T. 1973. Nothing in biology makes sense except in the light of evolution. *The American Biology Teacher* 35: 125–129. http://www.pbs.org/wgbh/evolution/library/10/2/text_pop/l_102_01.html.

Dolphin, L. n.d. Table 1: Master Set of 193 Values of c. http://www.ldolphin.org/cdata.txt.

Domning, D. P. 2001a. The earliest known fully quadupedal sirenian. *Nature* 413: 625–627.

Domning, D. P. 2001b. New "intermediate form" ties seacows firmly to land. *RNCSE* 21(5–6): 38–42.

Doolan, R. 1993. Are dinosaurs alive today? Where Jurassic Park went wrong. *Creation Ex Nihilo* 15(4): 12–15. http://www.answersingenesis.org/docs/1302.asp.

*Doolittle, R. F. 1997. A delicate balance. *Boston Review* (Feb./Mar.). http://bostonreview.net/BR22.1/doolittle.html.

*Doolittle, W. F. 2000. Uprooting the tree of life. *Scientific American* 282 (Feb.): 90–95.

Dorale, J. A., et al. 1998. Climate and vegetation history of the midcontinent from 75 to 25 ka: A speleothem record from Crevice Cave, Missouri, USA. *Science* 282: 1871–1874.

Dornhaus, A., and L. Chittka. 1999. Evolutionary origins of bee dances. *Nature* 401: 38.

*Drake, J. W., B. Charlesworth, D. Charlesworth, and J. F. Crow. 1998. Rates of spontaneous mutation. *Genetics* 148: 1667–1686. (technical)

*Drange, T. M. 1998. Why be moral? http://www.infidels.org/library/modern/theodore_drange/whymoral.html.

*Drews, C. 2002. Transitional fossils of hominid skulls. http://www.theistic-evolution.com/transitional.html.

*Drummond, H. 1904. *The Lowell Lectures on the Ascent of Man.* Glasgow, Scotland: Robert Maclehose. http://www.ccel.org/d/drummond/ascent/ascent01.htm.

Duff, P.M.D., A. Hallam, and E. K. Walton. 1967. *Cyclic Sedimentation.* New York: Elsevier.

*Dunbar, R. 2003. Evolution: Five big questions: 5. What's God got to do with it? *New Scientist* 178 (June 14): 38–39.

*Dundes, A. (ed.). 1988. *The Flood Myth.* Berkeley: University of California Press.

*Dunkelberg, P. 2003. Irreducible complexity demystified. http://www.talkdesign.org/faqs/icdmyst/ICDmyst.html.

DYG. 2000. Evolution and creationism in public education: An in-depth reading of public opinion. http://www.pfaw.org/pfaw/dfiles/file_36.pdf.

*Earth Impact Database. 2003. http://www.unb.ca/passc/ImpactDatabase.

Easterbrook, G. 2003. NFL teams should hire psychics, and why, why, why are you punting?! http://www.nfl.com/nflnetwork/story/6908185.

Eberhard, W. G. 2000. Spider manipulation by a wasp larva. *Nature* 406: 255–256.

Edwards, R. L., et al. 1993. A large drop in atmospheric $^{14}C/^{12}C$ and reduced melting in the Younger Dryas, documented with ^{230}Th ages of corals. *Science* 260: 962–968.

Edwards v. Aguillard. 1986. U.S. Supreme Court *amicus curiae* brief of 72 Nobel laureates (and others). (Case 482 U.S. 578, 1987). http://www.talkorigins.org/faqs/edwards-v-aguillard/amicus1.html.

*Edwords, F. 1981. Why creationism should not be taught as science; Part 2: The educational issues. *Creation/Evolution* 2(1): 6–36. http://www.ncseweb.org/resources/articles/3955_issue_03_volume_2_number_1__2_21_2003.asp.

Eicher, D. L. 1976. *Geologic Time.* Englewood Cliffs, NJ: Prentice–Hall.

Eigen, M., and P. Schuster. 1977. The hypercycle. A principle of natural self-organization. Part A: Emergence of the hypercycle. *Naturwissenschaften* 64: 541–565.

Eisen, J., and M. Wu. 2002. Phylogenetic analysis and gene functional predictions: Phylogenomics in action. *Theoretical Population Biology* 61: 481–487.

Eisner, T. 1958. The protective role of the spray mechanism of the bombardier beetle, *Brachynus ballistarius* Lec. *Journal of Insect Physiology* 2: 215–220.

Eisner, T., and D. J. Aneshansley. 1982. Spray aiming in bombardier beetles: Jet deflection by the Coanda effect. *Science* 215: 83–85.

Eisner, T., et al. 1989. Chemical defense of an Ozaenine bombardier beetle from New Guinea. *Psyche* 96: 153–160.

Eisner, T., D. J. Aneshansley, M. Eisner, A. B. Attygalle, D. W. Alsop, and J. Meinwald. 2000. Spray mechanism of the most primitive bombardier beetle (*Metrius contractus*). *Journal of Experimental Biology* 203: 1265–1275.

*Elders, W. A. 1998. Bibliolatry in the Grand Canyon. *RNCSE* 18(4): 8–15.

Eldredge, N. 1972. Systematics and evolution of *Phacops rana* (Green, 1832) and *Phacops iowensis* Delo, 1935 (Trilobita) from the Middle Devonian of North America. *Bulletin of the American Museum of Natural History* 147: 45–114.

Eldredge, N. 1974. Stability, diversity, and speciation in Paleozoic epeiric seas. *Journal of Paleontology* 48: 540–548.

Eldredge, N., and S. J. Gould. 1972. Punctuated equilibria: An alternative to phyletic gradualism. In *Models in Paleobiology*, ed. T.J.M. Schopf, 82–115. San Francisco: Freeman, Cooper & Co.

Elena, S. F., V. S. Cooper, and R. E. Lenski. 1996. Punctuated evolution caused by selection of rare beneficial mutations. *Science* 272: 1802–1804.

*Ellington, A. D., and M. Levy. 2003. Gas, discharge, and the Discovery Institute. *RNCSE* 23(3–4): 39–40.

*Elsberry, W. R. 1995. Transitional fossil challenge. http://www.rtis.com/nat/user/elsberry/evobio/evc/argresp/tranform.html.

*Elsberry, W. R. 1996. Punctuated equilibria. http://www.talkorigins.org/faqs/punc-eq.html.

*Elsberry, W. R. 1997. Enterprising science needs naturalism. http://www.utexas.edu/cola/depts/philosophy/faculty/koons/ntse/papers/Elsberry.html.

*Elsberry, W. R. 1998. Population size and time of creation or Flood. http://www.rtis.com/nat/user/elsberry/evobio/evc/argresp/populate.html.

*Elsberry, W. R. 2000. The anti-evolutionists: William A. Dembski. http://www.antievolution.org/people/dembski_wa/sc.html.

Elsberry, W. R. 2002. Commentary on William A. Dembski's "No Free Lunch: Why specified complexity cannot be purchased without intelligence." http://www.antievolution.org/people/dembski_wa/rev_nfl_wre_bn.html.

Elsberry, W. R. n.d. What does "intelligent agency by proxy" do for the design inference? http://www.talkreason.org/articles/wre_id_proxy.cfm.

Elsberry, W. R., and M. Perakh. 2004. How Intelligent Design advocates turn the sordid lessons from Soviet and Nazi history upside down. http://www.talkreason.org/articles/eandp.cfm.

*Elsberry, W., and J. Shallit. 2003. Information theory, evolutionary computation, and Dembski's "complex specified information." http://www.talkreason.org/articles/eandsdembski.pdf.

Elzanowski, A. 2002. Archaeopterygidae (Upper Jurassic of Germany). In *Mesozoic Birds: Above the Heads of Dinosaurs*, eds. L. M. Chiappe and L. M. Witmer, 129–159. Berkeley: University of California Press.

Emery, G. T. 1972. Perturbation of nuclear decay rates. *Annual Review of Nuclear Science* 22: 165–202.

Encyclopaedia Britannica. 1984. Population. Vol. 14: 816.

Engel, M. S., and D. A. Grimaldi. 2004. New light shed on the oldest insect. *Nature* 427: 627–630.

Engels, J. C. 1971. Effects of sample purity on discordant mineral ages found in K-Ar dating. *Journal of Geology* 79: 609–616.

Engwer, J. n.d. Catholicism and the Canon. http://members.aol.com/jasonte2/canon.htm.

Enoch, H. 1916. Darwin's final recantation. *Bombay Guardian*, March 25, 1916, quoted at http://www.forerunner.com/forerunner/X0724_Darwins_Final_Recant.html.

Enright, J. T. 1999. Testing dowsing: The failure of the Munich experiments. *Skeptical Inquirer* 23(1): 39–46.

Erdoes, R., and A. Ortiz. 1984. *American Indian Myths and Legends*. New York: Pantheon.

Ertem, G., and J. P. Ferris. 1996. Synthesis of RNA oligomers on heterogeneous templates. *Nature* 379: 238–240.

Erwin, T. L. 1967. Bombardier beetles (Coleoptera, Carabidae) of North America: Part II. Biology and behavior of *Brachinus pallidus* Erwin in California. *Coleopterists' Bulletin* 21: 41–55.

Erwin, T. L. 1970. A reclassification of bombardier beetles and a taxonomic revision of the North and Middle American species (Carabidae: Brachinida). *Quaestiones Entomologicae* 6: 4–215.

Esterhuysen, A., and J. Smith. 1998. Evolution: 'the forbidden word'? *South African Archaeological Bulletin* 53: 135–137. Quoted in Stear, J. 2004. It's official! Racism is an integral part of creationist dogma. http://home.austarnet.com.au/stear/aig_and_racism_response.htm.

Evans, S. 2003a. Unlocking the mystery of Illustra Media. http://www.ncseweb.org/resources/articles/6786_unlocking_the_mystery_of_illus_7_1_2003.asp.

Evans, S. 2003b. Who promotes *Unlocking the Mystery of Life?* http://www.ncseweb.org/resources/articles/6304_who_promotes_iunlocking_the__7_3_2003.asp.

*EvoWiki. 2004. Junk DNA. http://www.evowiki.org/wiki.phtml?title=Junk_DNA.

Eyre-Walker, A., N. H. Smith, and J. Maynard Smith. 1999. How clonal are human mitochondria? *Proceedings of the Royal Society, Series B* 266: 477–483. http://www.biols.susx.ac.uk/CSE/members/aeyrewalker/pdfs/EWPRS99a.pdf.

FAO/IAEA. 1977. *Manual on Mutation Breeding.* 2nd ed. Vienna: International Atomic Energy Agency.

Farquhar, J., H. Bao, and M. Thiemens. 2000. Atmospheric influence of earth's earliest sulfur cycle. *Science* 289: 756–758.

Farrand, W. R. 1961. Frozen mammoths and modern geology. *Science* 133: 729–735.

*Farrar, P., and B. Hyde. n.d. The vapor canopy hypothesis holds no water. http://www.talkorigins.org/faqs/canopy.html.

Faulkner, D. 1998. The young faint sun paradox and the age of the solar system. *Impact* 300 (June). http://www.icr.org/pubs/imp/imp-300.htm.

*Faure, G. 1986. *Principles of Isotope Geology.* 2nd ed. New York: Wiley.

*Faure, G. 1998. *Principles and Applications of Geochemistry.* 2nd ed. Upper Saddle River, NJ: Prentice-Hall.

Fehr, E., and S. Gächter. 2002. Altruistic punishment in humans. *Nature* 415: 137–140.

Ferris, J. P., et al. 1984. The investigation of the HCN derivative diiminosuccinonitrile as a prebiotic condensing agent. The formation of phosphate esters. *Origins of Life and Evolution of the Biosphere* 15: 29–43.

Ferris, J. P., A. R. Hill Jr., R. Liu, and L. E. Orgel. 1996. Synthesis of long prebiotic oligomers on mineral surfaces. *Nature* 381: 59–61.

**Ferris, T. 1997. *The Whole Shebang.* New York: Simon & Schuster.

*Festinger, L., H. W. Riecken, and S. Schachter. 1956. *When Prophecy Fails.* New York: Harper & Row.

*Fichter, L. S. 1999. The Wilson cycle. http://csmres.jmu.edu/geollab/Fichter/Wilson/Wilson.html.

Filchak, K. E., J. B. Roethele, and J. L. Feder. 2000. Natural selection and sympatric divergence in the apple maggot *Rhagoletis pomonella. Nature* 407: 739–742.

*Finkelstein, I., and N. A. Silberman. 2001. *The Bible Unearthed.* New York: Free Press.

Fischer, M., et al. 2004. New limits on the drift of fundamental constants from laboratory measurements. *Physical Review Letters* 92: 230802.

*Fitelson, B., C. Stephens, and E. Sober. 1999. How not to detect design. *Philosophy of Science* 66: 472–488. http://philosophy.wisc.edu/sober/dembski.pdf.

Flank, L. 1995a. Does science discriminate against creationists? http://www.geocities.com/Cape Canaveral/Hangar/2437/discrim.htm.

*Flank, L. 1995b. The therapsid–mammal transitional series. http://www.geocities.com/Cape Canaveral/Hangar/2437/therapsd.htm.

Flank, L. n.d. "Dr." Hovind, "created kinds", and his $250,000 "reward." http://www.geocities.com/CapeCanaveral/Hangar/2437/hovind.htm.

Fleay, D. 1958. Flight of the platypus. *National Geographic* 114(4) 512–525.

Flint, R. F. 1971. *Glacial and Quaternary Geology.* New York: Wiley.

*FLMNH. n.d. Fossil horses in hyperspace. http://www.flmnh.ufl.edu/natsci/vertpaleo/fhc/fhc.htm.

**Foley, J. 1996–2004. Fossil hominids: The evidence for human evolution. http://www.talkorigins .org/faqs/homs.

*Foley, J. 1996. NBC's "The Mysterious Origins of Man." http://www.talkorigins.org/faqs/mom .html.

*Foley, J. 1997. Creationist arguments: Australopithecines. http://www.talkorigins.org/faqs/homs/ a_piths.html.

Foley, J. 2002. Comparison of all skulls. http://www.talkorigins.org/faqs/homs/compare.html.

*Foley, J. 2003. Creationist arguments: The monkey quote. http://www.talkorigins.org/faqs/ homs/monkeyquote.html.

*Foley, J. 2004a. Creationist arguments: Anomalous fossils. http://www.talkorigins.org/faqs/homs/ a_anomaly.html.

*Foley, J. 2004b. Prominent hominid fossils. http://www.talkorigins.org/faqs/homs/specimen.html.

Ford, D.C., and C.A. Hill. 1999. Dating of speleothems in Kartchner Caverns, Arizona. *Journal of Cave and Karst Studies* 61(2): 84–88. http://www.caves.org/pub/journal/PDF/V61/v61n2-Ford.pdf.

*Forey, P.L. 1998. *History of the Coelacanth Fishes.* London: Chapman & Hall.

*Forrest, B. 2002. The wedge at work: How intelligent design creationism is wedging its way into the cultural and academic mainstream. In *Intelligent Design Creationism and Its Critics*, ed. R.T. Pennock, 5–53. Cambridge, MA: MIT Press.

*Forrest, B., and P.R. Gross. 2004. *Creationism's Trojan Horse: The Wedge of Intelligent Design.* Oxford: Oxford University Press.

Forster, C.A., S.D. Sampson, L.M. Chiappe, and D.W. Krause. 1998. The theropod ancestry of birds: New evidence from the Late Cretaceous of Madagascar. *Science* 279: 1915–1919.

Forsyth, D.J. 1970. The structure of the defence glands of the Cicindelidae, Amphizoidae, and Hygrobiidae (Insecta: Coleoptera). *Journal of Zoology, London* 160: 51–69.

Fox, S.W. 1960. How did life begin? *Science* 132: 200–208.

Fox, S.W. 1984. Creationism and evolutionary protobiogenesis. In *Science and Creationism*, ed. A. Montagu, 194–239. Oxford: Oxford University Press.

*Fox, S.W. 1988. *The Emergence of Life: Darwinian Evolution from the Inside.* New York: Basic Books.

Fox, S.W., and K. Dose. 1977. *Molecular Evolution and the Origin of Life.* Revised ed. New York: Marcel Dekker.

Fox, S.W., et al. 1995. Experimental retracement of the origins of a protocell: It was also a protoneuron. In *Chemical Evolution: Structure and Model of the First Cell*, eds. C. Ponnamperuma and J. Chela-Flores, 17–36. Dordrecht, Netherlands: Kluwer Academic Publishers.

*Frack, D. 1999. Peppered moths, round 2. http://www.calvin.edu/archive/evolution/199904/ 0100.html, . . . /0103.html, . . . /0200.html and . . . /0201.html.

Francis, J.E., and P.E. Hansche. 1972. Directed evolution of metabolic pathways in microbial populations. I. Modification of the acid phosphatase pH optimum in *S. cerevisiae*. *Genetics* 70: 59–73.

Francis, J.E., and P.E. Hansche. 1973. Directed evolution of metabolic pathways in microbial populations. II. A repeatable adaptation in *Saccharaomyces cerevisiae*. *Genetics* 74: 259–265.

*Frey, R.W. 1982. Sedimentology photo. *Journal of Sedimentary Petrology* 52: 614.

Frey, E., H-D. Sues, and W. Munk. 1997. Gliding mechanism in the Late Permian reptile *Coelurosauravus*. *Science* 275: 1450–1452.

Friedman, M., R.H.H. Hugman III, and J. Handin. 1980. Experimental folding of rocks under confining pressure, part VIII—Forced folding of unconsolidated sand and of lubricated layers of limestone and sandstone. *Geological Society of America Bulletin* 91: 307–312.

**Friedman, R.E. 1987. *Who Wrote the Bible?* New York: Summit Books.

Fritz, W.J. 1980. Reinterpretation of the depositional environment of the Yellowstone "fossil forests." *Geology* 8: 309–313.

Fritz, W.J. 1983. Comment and reply on "Erect floating stumps in Spirit Lake, Washington." *Geology* 11: 733–734.

Fritz, W. J. 1984. Comment and reply on "Yellowstone fossil forests: New evidence for burial in place." *Geology* 12: 638–639.

*Fry, I. 2000. *The Emergence of Life on Earth: A Historical and Scientific Overview*. New Brunswick, NJ: Rutgers University Press.

Fujii, Y., et al. 2000. The nuclear interaction at Oklo 2 billion years ago. *Nuclear Physics B* 573: 377–401. http://arxiv.org/abs/hep-ph/9809549.

Funkhouser, J. G., and J. J. Naughton. 1968. Radiogenic helium and argon in ultramafic inclusions from Hawaii. *Journal of Geophysical Research* 73: 4601–4607.

Furnes, H., et al. 2004. Early life recorded in Archean pillow lavas. *Science* 304: 578–581.

Gandolfo, M. A., et al. 1998. Oldest known fossils of monocotyledons. *Nature* 394: 532–533.

Garner, P. 1997. Green River blues. *Creation* 19(3): 18–19.

Garwood, N. C. 1989. Tropical soil seed banks: A review. In *Ecology of Soil Seed Banks*, eds. M. A. Leck, V. T. Parker, and R. L. Simpson, 149–209. San Diego: Academic Press.

Gaunt, M. W., and M. A. Miles. 2002. An insect molecular clock dates the origin of the insects and accords with palaeontological and biogeographic landmarks. *Molecular Biology and Evolution* 19: 748–761.

Gee, J. S., et al. 2000. Geomagnetic intensity variations over the past 780 kyr obtained from near-seafloor magnetic anomalies. *Nature* 408: 827–832.

Gentry, R. V. 1986. *Creation's Tiny Mystery*. Knoxville, TN: Earth Science Associates.

Gibbons, A. 1998. Calibrating the mitochondrial clock. *Science* 279: 28–29.

*Gilbert, S. F. 1988. *Developmental Biology*. 2nd ed. Sunderland MA: Sinauer.

Gilchrist, G. W. 1997. The elusive scientific basis of intelligent design theory. *RNCSE* 17(3): 14–15. http://www.ncseweb.org/resources/articles/2083_the_elusive_scientific_basis_o_3_16_2001 .asp.

*Gillette, D. D., and M. G. Lockley (eds). 1989. *Dinosaur Tracks and Traces*. Cambridge: Cambridge University Press. (technical)

*Gilovich, T. 1991. *How We Know What Isn't So*. New York: Free Press.

Gingerich, P. D. 1976. Paleontology and phylogeny: Patterns of evolution of the species level in early Tertiary mammals. *American Journal of Science* 276(1): 1–28.

Gingerich, P. D. 1980. Evolutionary patterns in early Cenozoic mammals. *Annual Review of Earth and Planetary Sciences* 8: 407–424.

Gingerich, P. D. 1983. Evidence for evolution from the vertebrate fossil record. *Journal of Geological Education* 31: 140–144.

Gingerich, P. D., B. H. Smith, and E. L. Simons. 1990. Hind limb of Eocene *Basilosaurus*: Evidence of feet in whales. *Science* 249: 154–157.

Gingerich, P. D., et al. 1983. Origin of whales in epicontinental remnant seas: New evidence from the Early Eocene of Pakistan. *Science* 220: 403–406.

Gingerich, P. D., et al. 1994. New whale from the Eocene of Pakistan and the origin of cetacean swimming. *Nature* 368: 844–847.

Gish, D. T. 1977. *Dinosaurs—Those Terrible Lizards*. El Cajon, CA: Master Books.

Gish, D. T. 1979. *Evolution: The Fossils Say No!* 3rd ed. San Diego: Creation-Life Publishers.

Gish, D. T. 1985. *Evolution: The Challenge of the Fossil Record*. El Cajon, CA: Creation-Life Publishers.

Gish, D. T. 1994. When is a whale a whale? *Impact* 250 (Apr.). http://www.icr.org/pubs/imp/imp-250.htm.

Gish, D. T. 1997. Gish responds to critique. *The Skeptic* 5(2): 37–41. http://mypage.direct.ca/w/writer/gish-response.html.

Gish, D. T., R. B. Bliss, and W. R. Bird. 1981. Summary of scientific evidence for creation. *Impact* 95–96 (May/Jun.). http://www.icr.org/pubs/imp/imp-095.htm.

*Gishlick, A. D. n.d. Icons of evolution? http://www.ncseweb.org/icons.

Gishlick, A., N. Matzke, and W. R. Elsberry. 2004. Meyer's hopeless monster. http://www.pandasthumb.org/pt-archives/000430.html.

Gislén, A., et al. 2003. Superior underwater vision in a human population of sea gypsies. *Current Biology* 13: 833–836. See also Pilcher, H. R. 2003. How to see shells on the sea floor. http://www.nature.com/nsu/030512/030512-14.html.

Gitt, W. 1998. What about the big bang? *Creation* 20(3): 42–44. http://www.answersingenesis .org/creation/v20/i3/big_bang.asp.

Gitt, W., and C. Wieland. 1998. Weasel words. *Creation Ex Nihilo* 20(4): 20–21. http://www .answersingenesis.org/docs/3746.asp.

Glatzmaier, G. A., and P. H. Roberts. 1995. A three-dimensional self-consistent computer simulation of a geomagnetic field reversal. *Nature* 377: 203–209.

Gliboff, S. 2000. Paley's design argument as an inference to the best explanation, or, Dawkins' dilemma. *Studies in History and Philosophy of Biological and Biomedical Sciences* 31(4): 579–597.

Gödel, K. 1949. An example of a new type of cosmological solutions of Einstein's field equations of gravitation. *Reviews of Modern Physics* 21(3): 447–450.

Godfrey, L. R. 1985. Foot notes of an anatomist. *Creation/Evolution* 5(1): 16–36.

*Godfrey, L. R. (ed.). 1983. *Scientists Confront Creationism*. New York: W. W. Norton.

Goff, F., and J. N. Gardner. 1994. Evolution of a mineralized geothermal system, Valles Caldera, New Mexico. *Economic Geology* 89: 1803–1832.

*Goldsmith, D. 2004. The best of all possible worlds. *Natural History* 113(6): 44–49.

Gonzalez, G., and J. W. Richards. *The Privileged Planet*. Washington, DC: Regnery.

Gosse, P. H. 1857. *Omphalos*. Reprint, Woodbridge, CT: Ox Bow Press, 1999.

Gottsch, J. D. 2001. Mutation, selection, and vertical transmission of theistic memes in religious canons. *Journal of Memetics* 5. http://jom-emit.cfpm.org/2001/vol5/gottsch_jd.html.

*Gould, J. L., and C. G. Gould. 1995. *The Honey Bee*. New York: Scientific American Library.

*Gould, S. J. 1976. Darwin's untimely burial. In *Philosophy of Biology*, ed. M. Ruse, 93–98. New York: Prometheus Books, 1998. http://www.stephenjaygould.org/ctrl/gould_tautology.html.

Gould, S. J. 1977a. The child as man's real father. In *Ever Since Darwin*, 63–69. New York: W. W. Norton.

*Gould, S. J. 1977b. *Ontogeny and Phylogeny*. Cambridge, MA: Belknap Press.

*Gould, S. J. 1977c. The problem of perfection, or How can a clam mount a fish on its rear end? In *Ever Since Darwin*, 103–110. New York: W. W. Norton. Excerpted at http://www.indiana .edu~ensiweb/lessons/contriv.pdf.

*Gould, S. J. 1983. Evolution as fact and theory. In *Hen's Teeth and Horse's Toes*, 253–262. New York: W. W. Norton. http://www.stephenjaygould.org/library/gould_fact-and-theory.html.

*Gould, S. J. 1991a. An essay on a pig roast. In *Bully for Brontosaurus*, 432–447. New York: W. W. Norton.

*Gould, S. J. 1991b. Life's little joke. In *Bully for Brontosaurus*, 168–181. New York: W. W. Norton.

*Gould, S. J. 1995. Hooking leviathan by its past. In *Dinosaur in a Haystack*, 395–376. New York: Harmony Books.

*Gould, S. J. 1998a. The tallest tale. In *Leonardo's Mountain of Clams and the Diet of Worms*, 301–318. New York: Three Rivers Press.

Gould, S. J. 1998b. The upwardly mobile fossils of Leonardo's living earth. In *Leonardo's Mountain of Clams and the Diet of Worms*, 17–44. New York: Three Rivers Press.

*Gould, S. J. 2002. What does the dreaded 'E' word mean anyway? In *I Have Landed*, 241–256. New York: Harmony Books.

Graham, A. L. 1981. Plate tectonics. In *The Evolving Earth*, ed. L.R.M. Cocks, 165–178. London: British Museum.

Grant, B. 2000. Letter: Charges of fraud misleading. *Pratt Tribune*, Dec. 13, 2000. Reprinted at http://www.indiana.edu/~ensiweb/lessons/icon.cr.html.

Grant, B. R., and P. R. Grant. 2003. What Darwin's finches can teach us about the evolutionary origin and regulation of biodiversity. *BioScience* 53: 965–975.

*Grant, B. S. 1999. Fine tuning the peppered moth paradigm. *Evolution* 53: 980–984. http://mason.gmu.edu~jlawrey/biol471/melanism.pdf.

Grant, H. J., A. W. Ivins, and C. W. Nibley. n.d. Mormon view of evolution. http://www.lightplanet .com/mormons/basic/gospel/creation/evolution_eom.htm.

*Gray, T. M. 1999. Complexity—yes! Irreducible—maybe! Unexplainable—no! A creationist criticism of irreducible complexity. http://tallship.chm.colostate.edu/evolution/irred_compl.html.

*Greene, T. S. 2000, 2002. Impact craters on Earth. http://www.geocities.com/Athens/Thebes/ 7755/ancientproof/impactcraters.html.

Greene, T. S. n.d. My motivation. http://www.geocities.com/Athens/Thebes/7755/motivation.html.

Greenlees, P. 2000. Theory of alpha decay. http://www.phys.jyu.fi/research/gamma/publications/ ptgthesis/node26.html.

*Greig, R. 1996. Did Darwin recant? *Creation* 18(1): 36–37. http://www.answersingenesis .org/docs/1315.asp.

Grieve, R.A.F. 1997. Extraterrestrial impact events: The record in the rocks and the stratigraphic column. *Palaeogeography, Paleoclimatology, Paleoecology* 132: 5–23.

Grimaldi, J. V., and K. Barker. 2003. Spelman's style polarizing zoo employees. *Washington Post*, April 28, 2003, A01. http://www.washingtonpost.com/wp-dyn/articles/A46238-2003Apr27. html.

Grimm, J. and W. Grimm. 1944. *The Complete Grimm's Fairy Tales*. New York: Pantheon Books.

Grinnell, G. B. 1961. *Pawnee Hero Stories and Folk-Tales*. Lincoln: University of Nebraska Press; reprinted from Forest and Stream Publishing Company, New York, 1889.

*Groves, C. 1994. Creationism: The Hindu view. *The Skeptic* 14(3): 43–45. http://www.talkorigins .org/faqs/mom/groves.html.

*Groves, C. 1999. Getting desperate. http://www.talkorigins.org/faqs/homs/desperate_cg.html.

*Grünbaum, A. 1995. The poverty of theistic morality. In *Science, Mind and Art*, eds. K. Gavroglu, J. Stachel, and M. W. Wartofsky, 203–242. Dordrecht, Netherlands: Kluwer Academic Publishers. http://www.infidels.org/library/modern/adolf_grunbaum/poverty.html.

**Guth, A. H. 1997. *The Inflationary Universe*. Reading, MA: Addison-Wesley.

*Guthrie, R. D. 1990. *Frozen Fauna of the Mammoth Steppe*. Chicago: University of Chicago Press.

Hajek, A. E., and R. J. St. Leger. 1994. Interactions between fungal pathogens and insect hosts. *Annual Review of Entomology* 39: 293–322.

Haldane, J.B.S. 1957. The cost of natural selection. *Journal of Genetics* 55: 511–524.

Hall, B. G. 1981. Changes in the substrate specificities of an enzyme during directed evolution of new functions. *Biochemistry* 20: 4042–4049.

Hall, B. G., and T. Zuzel. 1980. Evolution of a new enzymatic function by recombination within a gene. *PNAS* 77: 3529–3533.

Hallam, A. 1968. Morphology, palaeoecology and evolution of the genus *Gryphaea* in the British Lias. *Philosophical Transactions of the Royal Society of London B* 254: 91–128.

Ham, K. 1989. Were you there? *Back to Genesis* 10a (Oct.). http://www.icr.org/pubs/btg-a/btg-010a.htm.

Ham, K., J. Sarfati, and C. Wieland. 2000. *The Revised and Expanded Answers Book*. Green Forest, AR: Master Books.

Ham, K., and C. Wieland. 1997. Your appendix: It's there for a reason. *Creation Ex Nihilo* 20(1): 41–43. http://www.answersingenesis.org/docs/357.asp.

Han, J., and H. G. Craighead. 2000. Separation of long DNA molecules in a microfabricated entropic trap array. *Science* 288: 1026–1029.

*Handprint Media. 1999. Human evolution. http://www.handprint.com/LS/ANC/evol.html#chart.

Hansche, P. E. 1975. Gene duplication as a mechanism of genetic adaptation in *Saccharomyces cerevisiae*. *Genetics* 79: 661–674.

Hardin, D. M., Jr. 1999. Acute appendicitis: review and update. *American Family Physician* 60: 2027–2034.

*Harding, K. 1999. What would we expect to find if the world had flooded? http://www .creationism.ws/what_if_flood.htm.

Harland, W. B., et al. 1990. *A Geologic Time Scale 1989*. Cambridge: Cambridge University Press.

*harlequin2. 2001a. Ar-Ar dating assumes there is no excess argon? http://members.cox.net/ardipi thecus/evol/lies/lie024.html.

*harlequin2. 2001b. 200 year old lava dated 2.96 billion years old? http://members.cox.net/ardipi thecus/evol/lies/lie023.html.

*Harris, A. N. 2000. An observed example of morphological evolution. http://www.talkorigins .org/origins/postmonth/jul00.html.

Harris, D. V., and E. P. Kiver. 1985. *The Geologic Story of the National Parks and Monuments*. New York: Wiley.

*Harter, R. 1996. Piltdown Man: The bogus bones caper. http://www.talkorigins.org/faqs/ piltdown.html.

*Harter, R. 1999. Are mutations harmful? http://www.talkorigins.org/faqs/mutations.html.

Hartley, B. S. 1984. Experimental evolution of ribitol dehydrogenase. In *Microorganisms as Model Systems for Studying Evolution*, ed. R. P. Mortlock, 23–54. New York: Plenum.

Hastings, R. J. 1987. New observations on Paluxy Tracks confirm their dinosaurian origin. *Journal of Geological Education* 35: 4–15.

*Hastings, R. J. 1988. Rise and fall of the Paluxy mantracks. *Perspectives on Science and Christian Faith* 40: 144–155.

Hauert, C., S. De Monte, J. Hofbauer, and K. Sigmund. 2002. Volunteering as Red Queen mechanism for cooperation in public goods games. *Science* 296: 1129–1132.

Hawaii Christians Online. n.d. Missing day. http://www.hawaiichristiansonline.com/the_missing_ day.html.

Hawkes, N. 2000. Even early man was late for first date. *Times* (UK), Oct. 31, 2000. http://groups.yahoo.com/group/TNUKdigest/message/708.

*Hawking, S. 1988. *A Brief History of Time*. Toronto: Bantam.

*Hawking, S. 2001. *The Universe in a Nutshell*. New York: Bantam.

Hayashi, Y., et al. 2003. Can an arbitrary sequence evolve towards acquiring a biological function? *Journal of Molecular Evolution* 56: 162–168.

Haynes, J. D. 1995. A critique of the possibility of genetic inheritance of homosexual orientation. *Journal of Homosexuality* 28: 91–113.

Heckel, P. H. 1986. Sea-level curve for Pennsylvanian eustatic marine transgressive–regressive depositional cycles along midcontinent outcrop belt, North America. *Geology* 14: 330– 334.

*Heinrich, P. 1996. The mysterious origins of man: Atlantis, mammoths, and crustal shift. http://www.talkorigins.org/faqs/mom/atlantis.html.

Heitkötter, J., and D. Beasley (eds.). 2000. The Hitch-hiker's guide to evolutionary computation (FAQ for comp.ai.genetic), Issue 8.2. http://ai.ia.ac.cn/alife.santafe.edu/~joke/encore/www/ default.htm.

Henahan, S. 1996. From primordial soup to the prebiotic beach: An interview with exobiology pioneer, Dr. Stanley L. Miller. http://www.accessexcellence.org/WN/NM/miller.html.

*Henke, K. R. n.d.a. Berthault's "stratigraphy": Rediscovering what geologists already know and strawperson misrepresentations of modern applications of Steno's principles. http://home.aus tarnet.com.au/stear/henke_steno.htm. See also Berthault, G. 2003. Guy Berthault's response to Kevin Henke's article. http://home.austarnet.com.au/stear/guy_response_henke.htm. See also Henke, K. R. 2004. Some Questions for Dr. Berthault. http://home.austarnet.com.au/stear/ questions_berthault_k_henke.htm.

Henke, K. R. n.d.b. How can Woodmorappe sell us a bill of goods if he doesn't know the costs? http://home.austarnet.com.au/stear/woodmorappe_bill_of_goods_henke.htm.

Henke, K. R. n.d.c. Young-earth creationist 'dating' of a Mt. St. Helens dacite: The failure of Austin and Swenson to recognize obviously ancient minerals. http://home.austarnet.com.au/stear/mt_ st_helens_dacite_kh.htm.

Herrick, S., Jr. 1935. Tables for the reduction of radial velocities to the Sun. *Lick Observatory Bulletin* 470: 85–90.

*Higham, T. 1999. Radiocarbon WEB-Info. http://www.c14dating.com/.

Higham, T. n.d. Corrections to radiocarbon dates. http://www.c14dating.com/corr.html.

*Hildeman, E. J. 2004. *Creationism: The Bible Says No!* Bloomington, IN: Author House.

Hilgen, F. J., W. Krijgsman, C. G. Langereis, and L. J. Lourens. 1997. Breakthrough made in dating of the geological record. *Eos* 78(28): 285, 288–289. http://www.agu.org/sci_soc/eos96336.html.

*Hillis, D. M., et al. 1992. Experimental phylogenetics: Generation of a known phylogeny. *Science* 255: 589–592. (technical)

Hitching, F. 1982. *The Neck of the Giraffe.* New York: Meridian.

Hitler, A. *Mein Kampf.* Translated by R. Manheim. Boston: Houghton Mifflin.

Hoffman, P. F., et al. 1998. A Neoproterozoic snowball earth. *Science* 281: 1342–1346. See also pp. 1259, 1261.

*Hofstadter, R. 1944. *Social Darwinism in American Thought.* Philadelphia: University of Pennsylvania Press.

Holden, C. 2004. Scientists hope ruling will lead them to bones. *Science* 303: 943.

*Holland, J. H. 1975. *Adaptation in Natural and Artificial Systems.* Ann Arbor: University of Michigan Press. (technical)

*Holloway, R. n.d. Evolution of a creationist quote. http://www.ntanet.net/quote.html.

Hooper, J. 2002. *Of Moths and Men: An Evolutionary Tale.* New York: W. W. Norton.

Hooper, S. D., and O. G. Berg. 2003. On the nature of gene innovation: Duplication patterns in microbial genomes. *Molecular Biology and Evolution* 20: 945–954.

*Hopkins, M. 2002. Quotations and misquotations: Why what antievolutionists quote is not valid evidence against evolution. http://www.talkorigins.org/faqs/quotes.

Hoppe, R. B. 2004. Introduction to multiple designers theory. http://www.pandasthumb.org/pt-archives/000509.html.

*Ho-Stuart, C. 2003. Muller and mutations. http://www.talkorigins.org/faqs/quotes/muller.html.

Hovind, K. 1998. Dr. Hovind's "Creation Seminar" part 3b: Dinosaurs alive today. http://www.algonet.se/~tourtel/hovind_seminar/seminar_part3b.html.

Hovind, K. 2003. Introduction to Dr. Hovind's "Creation Seminar." http://www.algonet.se~tourtel/hovind_seminar/seminar_introduction.html.

Hovind, K. n.d.a. Doesn't carbon dating or potassium argon dating prove the Earth is millions of years old? http://www.drdino.com/QandA/index.jsp?varFolder=CreationEvolution&varPage=CarbonPotassiumargondating.jsp.

Hovind, K. n.d.b. Dr. Hovind's $250,000 offer. http://www.drdino.com/Ministry/250k/index.jsp.

Hoy, R. R., and D. Robert. 1996. Tympanal hearing in insects. *Annual Review of Entomology* 41: 433–450.

Hoyle, F. 1975. *Astronomy and Cosmology: A Modern Course.* San Francisco: Freeman.

Hoyle, F. 1983. *The Intelligent Universe.* New York: Holt, Rinehart and Winston.

Hueck, C. J. 1998. Type III protein secretion systems in bacterial pathogens of animals and plants. *Microbiology and Molecular Biology Reviews* 62: 379–433.

Hughes, A. L., and R. Friedman. 2003. Parallel evolution by gene duplication in the genomes of two unicellular fungi. *Genome Research* 13: 794–799.

*Hull, D. L. 1997. The ideal species concept—and why we can't get it. In *Species: The Units of Biodiversity*, eds. M. Claridge, H. Dawah and M. Wilson, 357–380. London: Chapman & Hall.

Humber, P. G. 1987a. The ascent of racism. *Impact* 164 (Feb.). http://www.icr.org/pubs/imp/imp-164.htm.

Humber, P. G. 1987b. Stalin's brutal faith. *Impact* 172 (Oct.). http://www.icr.org/pubs/imp/imp-172.htm.

*Hume, D. 1947. *Dialogues Concerning Natural Religion.* Edited by N. K. Smith. Indianapolis IN: Bobbs-Merrill. http://www.anselm.edu/homepage/dbanach/dnr.htm.

Humphreys, D. R. 1993. The Earth's magnetic field is young. *Impact* 242 (Aug.). http://www.icr.org/pubs/imp/imp-242.htm.

Humphreys, D. R. 1994. *Starlight and Time.* Green Forest, AR: Master Books.

Humphreys, D. R. 1999. Evidence for a young world. http://www.answersingenesis.org/docs/4005.asp.

Humphreys, D. R. 2002. Seven years of Starlight and Time. http://www.icr.org/starlightandtime/starlightandtime.html.

Humphreys, D. R. 2003. Light from creation illuminates cosmic axis. *Acts and Facts* 32(6): 4.

Humphreys, D. R., S. A. Austin, J. R. Baumgardner, and A. A. Snelling. 2003. Helium diffusion rates support accelerated nuclear decay. http://www.icr.org/research/icc03/pdf/Helium_ICC_7-22-03.pdf.

Humphreys, D. R., S. A. Austin, J. R. Baumgardner, and A. A. Snelling. 2004. Helium diffusion age of 6,000 years supports accelerated nuclear decay. *Creation Research Society Quarterly* 41(1): 1–16. http://www.creationresearch.org/crsq/articles/41/41_1/Helium.htm.

Humphreys, K. 2003. Nazareth—The town that theology built. http://www.jesusneverexisted.com/nazareth.html.

*Hunt, K. 1995. Horse evolution. http://www.talkorigins.org/faqs/horses/horse_evol.html.

*Hunt, K. 1997. Transitional vertebrate fossils FAQ. http://www.talkorigins.org/faqs/faq-transitional.html.

*Hunt, K. 2002. Carbon-14 in coal deposits. http://www.talkorigins.org/faqs/c14.html.

Hunter, R. E. 1977. Basic types of stratification in small eolian dunes. *Sedimentology* 24: 361–387.

*Hurd, G. S. 2004. Dino-blood and the young earth. http://www.talkorigins.org/faqs/dinosaur/blood.html. or http://home.austarnet.com.au/stear/YEC_and_dinoblood.html.

Huse, S. 1983. *The Collapse of Evolution*. Grand Rapids, MI: Baker Book House.

*Huxley, T.H.H. 1893. Evolution and ethics. http://aleph0.clarku.edu/huxley/CE9/index.html.

Hwang, S. H., M. A. Norell, Q. Ji, and K. Gao. 2002. New specimens of *Microraptor zhaoianus* (Theropoda: Dromaeosauridae) from northeastern China. *American Museum Novitates* 3381: 1–44.

Hyatt, A. 1866. On the paralellism between the different stages of life in the individual and those in the entire group of the molluscous order Tetrabranchiata. *Memoirs Read Before the Boston Society of Natural History* 1: 193–209.

ICR. 1980. The ICR scientists. *Impact* 86 (Aug.). http://www.icr.org/pubs/imp/imp-086.htm.

ICR. 2000. ICR tenets of creationism. http://www.icr.org/abouticr/tenets.htm.

IJdo, J. W., A. Baldini, D. C. Ward, S. T. Reeders, and R. A. Wells. 1991. Origin of human chromosome 2: An ancestral telomere-telomere fusion. *PNAS* 88: 9051–9055.

Ingman, M., H. Kaessmann, S. Pääbo, and U. Gyllensten. 2000. Mitochondrial genome variation and the origin of modern humans. *Nature* 408: 708–713. See also pp. 552–553. See also Thomson, J. 2000. Humans did come out of Africa, says DNA. *Nature Science Update*. http://www.nature.com/nsu/001207/001207-8.html.

*Inlay, M. 2002. Evolving immunity. http://www.talkdesign.org/faqs/Evolving_Immunity.html.

International Council on Biblical Inerrancy. 1978. The Chicago statement on Biblical inerrancy. http://www.jpusa.org/jpusa/documents/biblical.htm.

Irwin, D. E., S. Bensch, and T. D. Price. 2001. Speciation in a ring. *Nature* 409: 333–337.

*Isaak, M. 1995. Five major misconceptions about evolution. http://www.talkorigins.org/faqs/faq-misconceptions.html.

*Isaak, M. 1997. Bombardier beetles and the argument of design. http://www.talkorigins.org/faqs/bombardier.html.

*Isaak, M. 1998. Problems with a global flood. 2nd ed. http://www.talkorigins.org/faqs/faq-noahs-ark.html.

*Isaak, M. 2000. What is creationism? http://www.talkorigins.org/faqs/wic.html.

Isaak, M. 2002a. Flood stories from around the world. http://www.talkorigins.org/faqs/flood-myths.html.

*Isaak, M. 2002b. Is that so? The art of evaluating information. http://home.earthlink.net~misaak/claims.html.

*Isaak, M. 2002c. A philosophical premise of "naturalism"? http://www.talkdesign.org/faqs/naturalism.html.

*Isaak, M. 2003. What design looks like. *RNCSE* 23(5–6): 25–26, 31–35.

IVP Online. n.d. About us. http://www.gospelcom.net/ivpress/info/aboutus.

Jarvik, E. 1996. The Devonian tetrapod Ichthyostega. *Fossils and Strata* 40: 1–213. (technical)

Jeffares, D. C., A. M. Poole, and D. Penny. 1998. Relics from the RNA world. *Journal of Molecular Evolution* 46: 18–36.

Jeffrey, G. R. 1996. *The Signature of God.* Toronto: Frontier Research Publications.

Jenkins, R. O., et al. 2003. Bacterial degradation of arsenobetaine via dimethylarsinoylacetate. *Archives of Microbiology* 180: 142–150.

Jennings, S., and G. R. Thompson. 1986. Diagensis of Plio-Pleistocene sediments of the Colorado River Delta, southern California. *Journal of Sedimentary Petrology* 56(1): 89–98.

*Jessey, D. n.d. Isotope geochemistry. http://geology.csupomona.edu/drjessey/class/gsc300/isotope1.pdf.

*Jewitt, D. n.d. Kuiper Belt. http://www.ifa.hawaii.edu/~jewitt/kb.html.

Ji, Q., P. J. Currie, M. A. Norell, and S. Ji. 1998. Two feathered dinosaurs from northeastern China. *Nature* 393: 753–761.

*Johanson, D. C., and B. Edgar. 1996. *From Lucy to Language.* New York: Simon & Schuster.

Johanson, D. C., and M. A. Edey. 1981. *Lucy: The Beginnings of Humankind.* New York: Simon & Schuster.

*Johansson, S. 1998. The Solar FAQ. http://www.talkorigins.org/faqs/faq-solar.html.

*Johansson, S. 2002. The evolution of human language capacity. Master's Thesis, University of Lund. http://home.hj.se/~lsj/langevod.pdf.

*Johnson, B. 1993. How to change nuclear decay rates. http://math.ucr.edu/home/baez/physics/ParticleAndNuclear/decay_rates.html.

Johnson, G. R., R. K. Jain, and J. C. Spain. 2002. Origins of the 2,4-dinitrotoluene pathway. *Journal of Bacteriology.* 184: 4219–4232. (Erratum on p. 6084.)

Johnson, P. E. 1990. Evolution as dogma: The establishment of naturalism. *First Things* (Oct.): 15–22. http://www.arn.org/docs/johnson/pjdogma1.htm.

Johnson, P. E. 1999. The church of Darwin. *Wall Street Journal,* Aug. 16, 1999. http://www.arn.org/docs/johnson/chofdarwin.htm.

Johnson, P. E. 2003. American Family Radio, Jan. 10.

*Jones, D. E. 2000. *An Instinct for Dragons.* New York: Routledge.

Joyce, G. F. 2004. Directed evolution of nucleic acid enzymes. *Annual Review of Biochemistry* 73: 791–836.

**Judson, O. 2002. *Dr. Tatiana's Sex Advice to All Creation.* New York: Metropolitan Books.

Kaessmann, H., F. Heissig, A. von Haeseler, and S. Pääbo. 1999. DNA sequence variation in a non-coding region of low recombination on the human X chromosome. *Nature Genetics* 22: 78–81.

Kane, G. L., M. J. Perry, and A. N. Zytkow. 2000. The beginning of the end of the anthropic principle. *New Astronomy* 7: 45–53. http://xxx.lanl.gov/abs/astro-ph/0001197.

Kang, C. H., and E. R. Nelson. 1979. *The Discovery of Genesis.* St. Louis, MO: Concordia Publishing House.

Kapan, D. D. 2001. Three-butterfly system provides a field test of mullerian mimicry. *Nature* 409: 338–340.

Kaplan, H., K. Hill, J. Lancaster, and A. M. Hurtado 2000. A theory of human life history evolution: Diet, intelligence, and longevity. *Evolutionary Anthropology* 9: 156–185. http://www.soc.upenn.edu/courses/2003/spring/soc621_iliana/readings/kapl00d.pdf.

Kasahara, M., et al. 1997. Chromosomal duplication and the emergence of the adaptive immune system. *Trends in Genetics* 13: 90–92.

Kashiwaya, K., S. Ochiai, H. Sakai, and T. Kawai. 2001. Orbit-related long-term climate cycles revealed in a 12-Myr continental record from Lake Baikal. *Nature* 410: 71–74.

Kasting, J. F. 1993. Earth's early atmosphere. *Science* 259: 920–926.

*Kauffman, S. A. 1993. *The Origins of Order.* New York: Oxford University Press. (technical)

Kawamoto, K., and M. Akaboshi. 1982. Study on the chemical evolution of low molecular weight compounds in a highly oxidized atmosphere using electric discharges. *Origins of Life* 12(2): 133–141.

Keefe, A. D., and J. W. Szostak. 2001. Functional proteins from a random-sequence library. *Nature* 410: 715–718.

Keegan, L. P., A. Gallo, and M. A. O'Connell. 2000. Survival is impossible without an editor. *Science* 290: 1707–1709.

Keith, M. L., and G. M. Anderson. 1963. Radiocarbon dating: Fictitious results with mollusk shells. *Science* 141: 634–637.

Kendler, K. S., et al. 2000. Sexual orientation in a U.S. national sample of twin and nontwin sibling pairs. *American Journal of Psychiatry* 157: 1843–1846.

Kenrick, P., and P. R. Crane. 1997. The origin and early evolution of plants on land. *Nature* 389: 33–39.

Kepler, C. n.d. Another false link discovered in the theory of evolution. http://www.canadianheritagealliance.com/channels/articles/kepler/evolution.html.

Kermack, K. A., F. Mussett, and H. W. Rigney. 1981. The skull of *Morganucodon. Zoological Journal of the Linnean Society* 71: 1–158.

Kerr, R. A. 2000. An appealing snowball earth that's still hard to swallow. *Science* 287: 1734–1736.

Kestenbaum, D. 1998. Gentle force of entropy bridges disciplines. *Science* 279: 1849.

*Kevles, D. 1995. *In the Name of Eugenics: Genetics and the Uses of Human Heredity.* New York: Knopf.

Keyes, A. 2001. Survival of the fittest? *WorldNetDaily.* http://www.worldnetdaily.com/news/article.asp?ARTICLE_ID=23533.

*Kidwell, S. M., and S. M. Holland. 2002. The quality of the fossil record: Implications for evolutionary analyses. *Annual Review of Ecology and Systematics* 33: 561–588. (technical)

Kim, J., and B. A. Salisbury. 2001. A tree obscured by vines: Horizontal gene transfer and the median tree method of estimating species phylogeny. *Pacific Symposium on Biocomputing* 6: 571–582.

*Kimball, J. W. 2003. Speciation. http://users.rcn.com/jkimball.ma.ultranet/BiologyPages/S/Speciation.html.

Kimura, M. 1983. *The Neutral Theory of Molecular Evolution.* Cambridge: Cambridge University Press.

Kirk, K. M., J. M. Bailey, M. P. Dunne, and N. G. Martin. 2000. Measurement models for sexual orientation in a community twin sample. *Behavior Genetics* 30(4): 345–356.

*Kirkpatrick, R. C. 2000. The evolution of human homosexual behavior. *Current Anthropology* 41: 385–413. (technical)

Kitagawa, H., and J. van der Plicht. 1998. Atmospheric radiocarbon calibration to 45,000 yr B.P.: Late glacial fluctuations and cosmogenic isotope production. *Science* 279: 1187–1190. See also Kitagawa, H., and J. van der Plicht. 2000. PE-04. A 45,000 year varve chronology from Japan. http://www.cio.phys.rug.nl/HTML-docs/Verslag/97/PE-04.htm.

Kivisild, T., et al. 2000. Questioning evidence for recombination in human mitochondrial DNA. *Science* 288: 1931. www.sciencemag.org/cgi/content/full/288/5473/1931a.

Klee, R. 2002. The revenge of Pythagoras: How a mathematical sharp practice undermines the contemporary design argument in astrophysical cosmology. *British Journal for the Philosophy of Science* 53: 331–354.

*Knight, J. 2002. Evolutionary genetics: All genomes great and small. *Nature* 417: 374–376. http://www.nature.com/cgi-taf/DynaPage.taf?file=/nature/journal/v417/n6887/full/417374a_r.html.

Knödlseder, J. 2000. Constraints on stellar yields and Sne from gamma-ray line observations. *New Astronomy Reviews* 44: 315–320. http://xxx.lanl.gov/abs/astro-ph/9912131.

Knox, J.R., P.C. Moews, and J.-M. Frere. 1996. Molecular evolution of bacterial beta-lactam resistance. *Chemistry and Biology* 3: 937–947.

Kofahl, R.E. 1977. *Handy Dandy Evolution Refuter*. Chap. 5. San Diego: Beta Books. http://www.parentcompany.com/handy_dandy/hder5.htm.

Kofahl, R.E. 1981. The bombardier beetle shoots back. *Creation/Evolution* 2(3): 12–14.

Kolosick, J. n.d. Kent Hovind's "$250,000 award to prove evolution!" http://home.austarnet.com .au/stear/kent_hovind's_phony_challenge.htm.

Kondrashov, A.S. 1997. Evolutionary genetics of life cycles. *Annual Review of Ecology and Systematics* 28: 391–435.

Kondrashov, A.S., and F.A. Kondrashov. 1999. Interactions among quantitative traits in the course of sympatric speciation. *Nature* 400: 351–354.

*Konner, M. 2002. *The Tangled Wing: Biological Constraints on the Human Spirit*. New York: Holt.

Korol, A., et al. 2000. Nonrandom mating in *Drosophila melanogaster* laboratory populations derived from closely adjacent ecologically contrasting slopes at "Evolution Canyon." *PNAS* 97: 12637–12642. See also Schneider, C.J. 2000. Natural selection and speciation. *PNAS* 97: 12398–12399.

*Kossy, D. 2001. *Strange Creations: Aberrant Ideas of Human Origins from Ancient Astronauts to Aquatic Apes*. Los Angeles: Feral House.

Koza, J.R., M.A. Keane, and M.J. Streeter. 2003. Evolving inventions. *Scientific American* 288(2): 52–59.

Krane, K.S. 1987. *Introductory Nuclear Physics*. New York: Wiley.

Kreimer, G. 1999. Reflective properties of different eyespot types in dinoflagellates. *Protist* 150: 311–323. http://www.urbanfischer.de/journals/protist/content/issue3/Pro0021.pdf.

Krings, M., et al. 2000. A view of Neandertal genetic diversity. *Nature Genetics* 26: 144–146.

Krumenaker, L. 1995. Rhythm section. *The Sciences* (Nov/Dec.): 14–17.

Kuban, G.J. 1989. Color distinctions and other curious features of dinosaur tracks near Glen Rose, Texas. In *Dinosaur Tracks and Traces*, eds. D.D. Gillette and M.G. Lockley, 427–440. Cambridge: Cambridge University Press.

*Kuban, G.J. 1996a. A review of NBC's "The Mysterious Origins of Man." http://www.talkorigins .org/faqs/paluxy/nbc.html.

*Kuban, G.J. 1996b. The Texas dinosaur/"man track" controversy. http://www.talkorigins.org/faqs/ paluxy.html.

*Kuban, G.J. 1997. Sea monster or shark? An analysis of a supposed plesiosaur carcass netted in 1977. *RNCSE* 17(3): 16–28.

*Kuban, G.J. 1998a. The life and death of Malachite Man. http://members.aol.com/gkuban/moab .htm.

*Kuban, G.J. 1998b. The "Meister Print": An alleged human sandal print from Utah. http://www.talkorigins.org/faqs/paluxy/meister.html.

*Kuban, G.J. 1999. The London Hammer: An alleged out-of-place artifact. http://paleo.cc/paluxy/hammer.htm.

*Kuechmann, F.C. 2000. Creationist comedy. http://home.austarnet.com.au/stear/kuechmann_ cretin_comedy.htm.

Kukla, G., and Z. An. 1989. Loess stratigraphy in Central China. *Palaeogeography, Palaeoclimatology, Palaeoecology* 72: 203–225.

*Kunchithapadam, K. 1995, 2000. What, if anything, is a Mitochondrial Eve? http://www .talkorigins.org/faqs/homs/mitoeve.html.

Kurland, C.G., B. Canback, and O.G. Berg. 2003. Horizontal gene transfer: A critical view. *PNAS* 100: 9658–9662.

*Kurtén, B. 1986. *How to Deep-Freeze a Mammoth*. New York: Columbia University Press.

Kuwajima, G. 1988. Construction of a minimum-size functional flagellin of *Escherichia coli*. *Journal of Bacteriology* 170: 3305–3309.

Kuzicheva, E. A., and N. B. Gontareva. 1999. The possibility of nucleotide abiogenetic synthesis in conditions of 'KOSMOS-2044' satellite space flight. *Advances in Space Research* 23(2): 393–396.

Kuzicheva, E. A., and N. B. Gontareva. 2002. Prebiotic synthesis of nucleotides at the Earth orbit in presence of Lunar soil. *Advances in Space Research* 30(6): 1525–1531.

Kyte, F. T., and J. T. Wasson. 1986. Accretion rate of extraterrestrial matter: Iridium deposited 33 to 67 million years ago. *Science* 232: 1225–1229.

Labandeira, C. C., B. S. Beall, and F. M. Hueber. 1988. Early insect diversification: Evidence from a Lower Devonian bristletail from Quebec. *Science* 242: 913–916.

*Lacey, J. C., N. S. Wickramasinghe, and G. W. Cook. 1992. Experimental studies on the origin of the genetic code and the process of protein synthesis: A review update. *Origins of Life and Evolution of the Biosphere* 22: 243–275. (technical)

Lague, M. R., and W. L. Jungers. 1996. Morphometric variation in Plio-Pleistocene hominid distal humeri. *American Journal of Physical Anthropology* 101: 401–427.

LaHaye, T., and J. Morris. 1976. *The Ark on Ararat*. Nashville, TN: Thomas Nelson.

*Lambert, F. L. 2003. The second law of thermodynamics. http://www.secondlaw.com.

*Land, M. F., and D.-E. Nilsson. 2001. *Animal Eyes*. Oxford: Oxford University Press.

Lane, L. 2003. Intelligent design in the scientific literature. http://www.geocities.com/lclane2/idlit.html.

Lang, D., et al. 2000. Structural evidence for evolution of the β/α barrel scaffold by gene duplication and fusion. *Science* 289: 1546–1550. See also p. 1490.

*LANL. n.d. The Los Alamos built spectrometers. http://lunar.lanl.gov/pages/spectros.html.

*Larson, G. 1998. *There's a Hair in My Dirt!: A Worm's Story*. New York: HarperCollins.

*Lazare, D. 2002. False testament: Archaeology refutes the Bible's claim to history. *Harper's* 304 (March): 39–47. http://www.findarticles.com/cf_0/m1111/1822_304/83553507/p1/article.jhtml?term=bible++!2Barchaeology.

Lederberg, J., and E. M. Lederberg. 1952. Replica plating and indirect selection of bacterial mutants. *Journal of Bacteriology* 63: 399–406.

Lee, D. H., et al. 1996. A self-replicating peptide. *Nature* 382: 525–528.

Lee, M.S.Y., G. L. Bell Jr., and M. W. Caldwell. 1999. The origin of snake feeding. *Nature* 400: 655–659.

Lee, P. N., P. Callaerts, H. G. de Couet, and M. Q. Martindale. 2003. Cephalopod *Hox* genes and the origin of morphological novelties. *Nature* 424: 1061–1065.

Lee, R. E. 1981. Radiocarbon: Ages in error. *Anthropological Journal of Canada* 19(3): 9–29. Reprinted in *Creation Research Society Quarterly* 19(2): 117–127 (1982).

*Leeming, D., and M. Leeming. 1994. *A Dictionary of Creation Myths*. New York: Oxford University Press.

Lefalophodon. n.d. Alpheus Hyatt (1838–1902). http://www.nceas.ucsb.edu~alroy/lefa/Hyatt.html.

Lehmann, T., et al. 2003. Population structure of *Anopheles gambiae* in Africa. *Journal of Heredity* 94: 133–147.

Leipe, D. D., L. Aravind, and E. V. Koonin. 1999. Did DNA replication evolve twice independently? *Nucleic Acids Research* 27: 3389–3401.

Lenner, G., M. Rajock, and J. Browning. 1995. Evidence #4. http://emporium.turnpike.net/C/cs/evid4.htm.

Lenski, R. E. 1995. Evolution in experimental populations of bacteria. In *Population Genetics of Bacteria, Society for General Microbiology, Symposium 52*, eds. S. Baumberg et al., 193–215. Cambridge: Cambridge University Press.

Lenski, R. E., C. Ofria, T. C. Collier, and C. Adami. 1999. Genome complexity, robustness and genetic interactions in digital organisms. *Nature* 400: 661–664.

Lenski, R. E., C. Ofria, R. T. Pennock, and C. Adami. 2003. The evolutionary origin of complex features. *Nature* 423: 139–144. http://myxo.css.msu.edu/papers/nature2003. See also National

Science Foundation. 2003. Artificial life experiments show how complex functions can evolve. http://www.sciencedaily.com/releases/2003/05/030508075843.htm.

Lenski, R. E., M. R. Rose, S. C. Simpson, and S. C. Tadler. 1991. Long-term experimental evolution in *Escherichia coli*. I. Adaptation and divergence during 2,000 generations. *American Naturalist* 138: 1315–1341.

Levy, M., and A. D. Ellington. 2003. Exponential growth by cross-catalytic cleavage of deoxyribozymogens. *PNAS* 100: 6416–6421.

Levy, M., and S. L. Miller. 1998. The stability of the RNA bases: Implications for the origin of life. *PNAS* 95: 7933–7938.

Lewin, R. 1981. No gap here in the fossil record. *Science* 214: 645–646.

Lewis, H. 1993. Clarkia. In *The Jepson Manual: Higher Plants of California*, ed. J. C. Hickman, 786–793. Berkeley: University of California Press.

*Lewis, R. 1997. Scientists debate RNA's role at beginning of life on earth. *The Scientist* 11(7): 11. http://www.mhhe.com/biosci/genbio/life/articles/article28.mhtml.

*Li, W.-H. 1997. *Molecular Evolution*. Sunderland, MA: Sinauer.

Lie-Svendsen, Ø., and M. H. Rees. 1996. Helium escape from the terrestrial atmosphere—the ion outflow mechanism. *Journal of Geophysical Research* 101: 2435–2443.

Lin, E. C. C., and T. T. Wu. 1984. Functional divergence of the L-Fucose system in mutants of *Escherichia coli*. In *Microorganisms as Model Systems for Studying Evolution*, ed. R. P. Mortlock, 135–164. New York: Plenum Press.

*Lindsay, D. 1996. Review: "Darwin's Black Box, the Biochemical Challenge to Evolution" by Michael Behe. http://www.don-lindsay-archive.org/creation/behe.html.

Lindsay, D. 1997a. Astronomical cycles. http://www.don-lindsay-archive.org/creation/astro_cycles.html.

*Lindsay, D. 1997b. Is "survival of the fittest" a tautology? http://www.don-lindsay-archive.org/creation/tautology.html.

Lindsay, D. 1997c. A smooth fossil transition: Orbulina, a foram. http://www.don-lindsay-archive.org/creation/orbulina.html.

*Lindsay, D. 1998a. How long would the fish eye take to evolve? http://www.don-lindsay-archive.org/creation/eye_time.html.

*Lindsay, D. 1998b. Scenarios and "just so" stories http://www.don-lindsay-archive.org/creation/stories.html.

Lindsay, D. 1999a. Are radioactive dating methods consistent with each other? http://www.don-lindsay-archive.org/creation/crater_chain.html.

Lindsay, D. 1999b. How could the immune system evolve? http://www.don-lindsay-archive.org/creation/evolve_immune.html.

Lindsay, D. 2000a. Are radioactive dating methods consistent with the deeper-is-older rule? http://www.don-lindsay-archive.org/creation/confirm.html.

*Lindsay, D. 2000b. Fresh lava dated as 22 million years old. http://www.don-lindsay-archive.org/creation/hawaii.html.

*Lindsay, D. 2000c. Living fossils like the coelacanth. http://www.don-lindsay-archive.org/creation/coelacanth.html.

*Lindsay, D. 2004. Famous quotes found in books. http://www.don-lindsay-archive.org/creation/quotes.html.

Lines, D. 1995. The fossilized human finger. http://www.creationevidence.org/museum_tour/finger/finger.html.

Linne, C. 1760. *Disquisitio de sexu plantarum ab Academia Imperiali Scientiarum Petropolitana praemio ornata*. *Amoenitates Academicae* 10: 100–131. Quoted in Rieseberg, L. H. 1997. Hybrid origins of plant species. *Annual Review of Ecology and Systematics* 28: 359–389.

*Lippard, J. 1999. Lucy's knee joint: A case study in creationists' willingness to admit their errors. http://www.talkorigins.org/faqs/knee-joint.html.

Lipps, J. H., and P. W. Signor (eds.). 1992. *Origin and Early Evolution of the Metazoa.* New York: Plenum Press.

Lipson, H., and J. B. Pollack. 2000. Automatic design and manufacture of robotic lifeforms. *Nature* 406: 974–978.

*Littleton, K. 2002. Fish fossils. http://www.talkorigins.org/origins/postmonth/sep02.html.

Liu, T., et al. 1985. Loess–paleosol sequence in China and climatic history. *Episodes* 8: 21–28.

Living Word Bible Church. n.d. The result of believing evolution. http://www.lwbc.co.uk/Genesis/results%20of%20believing%20evolution.htm.

*Livingstone, D. N. 1984. *Darwin's Forgotten Defenders: The Encounter Between Evangelical Theology and Evolutionary Thought.* Vancouver, BC: Regent College Publishing.

Livio, M., D. Hollowell, A. Weiss, and J. Truran. 1989. The anthropic significance of the existence of an excited state of 12_C. *Nature* 340: 281–284.

Lockley, M., and A. P. Hunt. 1995. *Dinosaur Tracks and Other Fossil Footprints of the Western United States.* New York: Columbia University Press.

Lockley, M. G. 1992. Comment and reply on "Fossil vertebrate footprints in the Coconino Sandstone (Permian) of northern Arizona: Evidence for underwater origin." *Geology* 20: 666–667.

Loewe, L., and S. Scherer. 1997. Mitochondrial Eve: The plot thickens. *Trends in Ecology and Evolution* 12: 422–423.

Logan, G. A., J. M. Hayes, G. B. Hieshima, and R. E. Summons. 1995. Terminal Proterozoic reorganization of biogeochemical cycles. *Nature* 376: 53–56. See also pp. 16–17.

Long, P. 1994. A town with a golden gene. *Health* 8(1): 60–66.

Loope, D. B. 1992. Comment and reply on "Fossil vertebrate footprints in the Coconino Sandstone (Permian) of northern Arizona: Evidence for underwater origin." *Geology* 20: 667–668.

Lorenz, K. 1966. *On Aggression.* New York: Bantam.

Lowell, P. (as "A Mathematician"). 1907. The evidence of life on Mars: A simple mathematical proof that the canals are not due to natural causes. *Scientific American* (Oct. 26). Reprinted in Becker, Barbara J. 2003. History 135C: Exploring the Cosmos. http://eee.uci.edu/clients/bjbecker/ExploringtheCosmos/week10b.html.

Lubenow, M. L. 1992. *Bones of Contention: A Creationist Assessment of Human Fossils.* Grand Rapids, MI: Baker Books.

Lumsden, R. Quoted in Alters, B. J. 1995. A content analysis of the Institute for Creation Research's Institute on Scientific Creationism. *Creation/Evolution* 15(2): 1–15.

Luo, Z.-X., A. W. Crompton, and A.-L. Sun. 2001. A new mammaliaform from the Early Jurassic and evolution of mammalian characteristics. *Science* 292: 1535–1546.

*Lyell, C. 1830. *Principles of Geology.* London: John Murray. http://www.esp.org/books/lyell/principles/facsimile/title3.html.

Lynch, M., and J. S. Conery. 2000. The evolutionary fate and consequences of duplicate genes. *Science* 290: 1151–1155. See also pp. 1065–1066.

*MacAndrew, A. n.d. Misconceptions around Mitochondrial Eve. http://www.evolutionpages.com/Mitochondrial%20Eve.htm.

Macbeth, N. 1971. *Darwin Retried.* Boston: Delta.

Mace, S. R., B. A. Sims, and T. C. Wood. 2003. Fellowship, creation, and schistosomes. *Impact* 357 (Mar.). http://www.icr.org/newsletters/impact/impactmar03.html.

Macks, S. 2000. Madd Macks Debate Archives. http://tccsa.tc/m_debate_archives.html.

Macnair, M. R. 1989. A new species of *Mimulus* endemic to copper mines in California. *Botanical Journal of the Linnean Society* 100: 1–14.

*MacRae, A. 1994a. Could coal deposits be explained by a global flood? http://www.talkorigins.org/faqs/polystrate/coal.html.

*MacRae, A. 1994b. "Polystrate" tree fossils. http://www.talkorigins.org/faqs/polystrate/trees.html.

*MacRae, A. 1994c. Yellowstone National Park (U.S.) fossil forests. http://www.talkorigins.org/faqs/polystrate/yellowstone.html.

*MacRae, A. 1998. Radiometric dating and the geological time scale: Circular reasoning or reliable tools? http://www.talkorigins.org/faqs/dating.html.

*Majerus, M.E.N. 1998. *Melanism: Evolution in Action.* Oxford: Oxford University Press.

Majerus, M.E.N. 1999. (Letter). Quoted in Frack, D. 1999. Peppered moths, round 2, part 2. http://www.calvin.edu/archive/evolution/199904/0103.html.

Majerus, M.E.N., C.F.A. Brunton, and J. Stalker. 2000. A bird's eye view of the peppered moth. *Journal of Evolutionary Biology* 13: 155–159.

Makovicky, P.J., M.A. Norell, J.M. Clark, and T. Rowe. 2003. Osteology and relationships of *Byronosaurus jaffei* (Theropoda: Troodontidae). *American Museum Novitates* 3402, 1–32.

Mallatt, J., and J.-Y. Chen. 2003. Fossil sister group of craniates: Predicted and found. *Journal of Morphology* 258: 1–31.

Malmgren, B.A., W.A. Berggren, and G.P. Lohmann. 1984. Species formation through punctuated gradualism in planktonic foraminifera. *Science* 225: 317–319.

Mani, G.S. 1982. A theoretical analysis of the morph frequency variation in the peppered moth over England and Wales. *Biological Journal of the Linnean Society* 17: 259–267.

Mani, G.S. 1990. Theoretical models of melanism in *Biston betularia*—a review. *Biological Journal of the Linnean Society* 39: 355–371.

Marden, J.H., and M.G. Kramer. 1995. Locomotor performance of insects with rudimentary wings. *Nature* 377: 332–334.

Margulis, L. 1981. *Symbiosis in Cell Evolution.* San Francisco: Freeman.

*Margulis, L., and D. Sagan. 1990. *The Origins of Sex: Three Billion Years of Genetic Recombination.* New Haven, CT: Yale University Press.

Marion, H., et al. 2003. Search for variations of fundamental constants using atomic fountain clocks. *Physical Review Letters* 90: 150801.

Martin, G. 1999. And baby makes 6 billion. *San Francisco Chronicle,* Oct. 11, 1999, A1.

Martin, M.W., et al. 2000. Age of Neoproterozoic bilatarian body and trace fossils, White Sea, Russia: Implications for metazoan evolution. *Science* 288: 841–845. See also p. 789.

Martin, W., and M.J. Russell. 2003. On the origins of cells: A hypothesis for the evolutionary transitions from abiotic geochemistry to chemoautotrophic prokaryotes, and from prokaryotes to nucleated cells. *Philosophical Transactions, Biological Sciences* 358: 59–85.

Maryanska, T., H. Osmólska, and M. Wolsan. 2002. Avialan status for oviraptorosauria. *Acta Palaeontologica Polonica* 47(1): 97–116. http://app.pan.pl/acta47/app47-097.pdf.

Mastropaolo, J. 1998. Re: The evolutionist: Liar, believer in miracles, king of criminals. http://www.asa3.org/archive/evolution/199811/0040.html.

*Mathews, D. 2000. Domesticated dinosaurs? http://www.geocities.com/athens/agora/3958/weekly/weekly56.htm.

*Matson, D.E. 1994. How good are those young-earth arguments? http://www.talkorigins.org/faqs/hovind/howgood.html.

*Matsumura, M. 1997. Miracles in, creationism out: "The geophysics of God." *RNCSE* 17(3): 29–32. http://www.ncseweb.org/resources/rncse_content/vol17/4787_miracles_in_creationism_out_12_30_1899.asp.

Matthews, M. 2003. A century of fraud. http://www.answersingenesis.org/docs2003/1118piltdown.asp.

*Matzke, N.J. 2003. Evolution in (brownian) space: A model for the origin of the bacterial flagellum. http://www.talkdesign.org/faqs/flagellum.html.

*Max, E.E. 1999. The evolution of improved fitness by random mutation plus selection. http://www.talkorigins.org/faqs/fitness.

Max, E.E. 2003. Plagiarized errors and molecular genetics. http://www.talkorigins.org/faqs/molgen.

*Mayden, R.L. 1997. A hierarchy of species concepts: The denoument in the saga of the species problem. In *Species: The Units of Biodiversity,* eds. M.F. Claridge, H.A. Dawah, and M.R. Wilson, 381–424. London: Chapman & Hall.

Maynard, S. 2001. Life's intelligent design. *Tacoma News Tribune*, May 7, 2001. http://www.arn.org/docs/news/lifesintelligentdesign050701.htm.

Mayor, A. 2000. *The First Fossil Hunters*. Princeton, NJ: Princeton University Press.

Mayr, E. 1942. *Systematics and the Origin of Species*. New York: Columbia University Press.

Mayr, E. 1963. *Animal Species and Evolution*. Cambridge, MA: Belknap Press.

*Mayr, E. 2000. Darwin's influence on modern thought. *Scientific American* 283(1): 78–83.

McCarthy, M. D., J. I. Hedges, and R. Benner. 1998. Major bacterial contribution to marine dissolved organic nitrogen. *Science* 281: 231–234.

*McClendon, J. H. 1999. The origin of life. *Earth-Science Reviews* 47: 71–93. (technical)

*McCourt, R. M., R. L. Chapman, M. A. Buchheim, and B. D. Mishler. 1996. Green plants. http://tolweb.org/tree?group=Green_plants&contgroup=Eukaryotes.

McKee, E. D. 1979. A study of global sand seas: Ancient sandstones considered to be eolian. U.S. Geological Survey Professional Paper 1052.

McLean v. Arkansas Board of Education. 1982. http://www.talkorigins.org/faqs/mclean-v-arkansas. html.

McMullen, E. T. 1998. The death of the dinosaurs, superfloods and other megacatastrophes: Catastrophes and scientific change. http://www.georgiasouthern.edu~etmcmull/DINO.htm.

**McPhee, J. 1998. *Annals of the Former World*. New York: Farrar, Straus & Giroux. (This collects four previous books by McPhee—*Basin and Range*, *In Suspect Terrain*, *Rising from the Plains*, and *Assembling California*—which may also be obtained separately.)

McShea, D. W. 1998. Possible largest-scale trends in organismal evolution: Eight live hypotheses. *Annual Review of Ecology and Systematics* 29: 293–318.

Meert, J. 2000. Consistent radiometric dates. http://gondwanaresearch.com/radiomet.htm.

Meert, J. 2002. Were Adam and Eve toast? http://gondwanaresearch.com/hp/adam.htm.

*Meert, J. 2003a. Andrew Snelling and the iron concretion? http://gondwanaresearch.com/hp/crefaqs.htm#who.

Meert, J. 2003b. Walt Brown's pseudochallenge. http://gondwanaresearch.com/hp/walt_brown.htm.

Meléndez-Hevia, E., T. G. Waddell, and M. Cascante. 1996. The puzzle of the Krebs citric acid cycle: Assembling the pieces of chemically feasible reactions, and opportunism in the design of metabolic pathways during evolution. *Journal of Molecular Evolution* 43(3): 293–303.

*Mellett, J. S. and J. Wolf. 1985. The role of "Nebraska man" in the creation–evolution debate. *Creation/Evolution* 5(2): 31–43. http://www.talkorigins.org/faqs/homs/wolfmellett.html.

*Merling, D. n.d. Has Noah's Ark been found? http://www.tentmaker.org/WAR/HasNoahs ArkBeenFound1.html.

Merriam-Webster's Dictionary. 1974. New York: Simon & Schuster.

Meyer, R. 1997. *Paleoalterites and Paleosols: Imprints of Terrestrial Processes in Sedimentary Rocks*. Rotterdam: A. A. Balkema.

Meyer, S. C. 2002. Teach the controversy on origins. *Cincinnati Enquirer*, March 30, 2002.

Meyer, S. C. 2004. The origin of biological information and the higher taxonomic categories. *Proceedings of the Biological Society of Washington* 117: 213–239.

Meyer-Berthaud, B., S. E. Scheckler, and J. Wendt. 1999. *Archaeopteris* is the earliest known modern tree. *Nature* 398: 700–701.

*Meyers, S. 2000. The Signature of God by Grant R. Jeffrey (review). http://members.aol.com/ibss2/ gospel.html.

Meyers, S., and R. Doolan. 1987. Rapid stalactites? *Creation Magazine* 9(4): 6–8. http://answersin genesis.org/home/area/magazines/docs/cen_v9n4_stalactites.asp.

Michigan House Bill 4946. 2003 (July 2). House introduced bill. http://www.michiganlegislature .org/mileg.asp?page=getObject&Name=2003-HB-4946.

Middleton, K. M. 2002. Evolution of the perching foot in theropods. *Journal of Vertebrate Paleontology* 22: 88A.

*Mikkelson, B., and D. P. Mikkelson. 2000. The lost day. http://www.snopes.com/religion/ lostday.htm.

Milgrom, M. 2002. Does dark matter really exist? *Scientific American* 287(2): 42–52.

Miller, A. I. 1997. Dissecting global diversity patterns: Examples from the Ordovician radiation. *Annual Review of Ecology and Systematics* 28: 85–104.

*Miller, H. 1857. *The Testimony of the Rocks; or, Geology in Its Bearings on the Two Theologies, Natural and Revealed.* Edinburgh: Shepherd & Elliot. See also MacRae, A. n.d. Hugh Miller—19th-century creationist geologist. http://home.tiac.net~cri/1998/miller.html.

*Miller, K. B. n.d. Taxonomy, transitional forms, and the fossil record. http://www.asa3.org/ASA/resources/Miller.html.

*Miller, K. R. 1999. *Finding Darwin's God.* New York: HarperCollins.

*Miller, K. R. 2002. The truth about the "Santorum Amendment" language on evolution. http://www.millerandlevine.com/km/evol/santorum.html.

Miller, K. R. 2003. Answering the biochemical argument from design. In *God and Design: The Teleological Argument and Modern Science*, ed. N. Manson, 292–307. London: Routledge. http://www.millerandlevine.com/km/evol/design1/article.html.

Miller, K. R. 2004. The flagellum unspun. In *Debating Design: From Darwin to DNA*, eds. W. Dembski and M. Ruse. New York: Cambridge University Press, 81–97. http://www.millerandlevine.com/km/evol/design2/article.html.

*Miller, K. R. n.d. Of pandas and people: A brief critique. http://www.kcfs.org/pandas.html.

*Miller, L. 2001. King David was a nebbish. http://dir.salon.com/books/feature/2001/02/07/solomon/index.html.

Miller, S. L. 1987. Which organic compounds could have occurred on the prebiotic earth? *Cold Spring Harbor Symposia on Quantitative Biology* 52: 17–27.

Minsky, M. 1985. *The Society of Mind.* New York: Simon & Schuster.

Mitchell, S. 1987. *The Book of Job.* San Francisco: North Point Press. Cited in Pennock, R. 1999. *Tower of Babel.* Cambridge, MA: MIT Press, p. 217.

*MNSU. n.d. Radio-carbon dating. http://emuseum.mnsu.edu/archaeology/dating/radio_carbon.html.

Moffat, A. S. 2000. Transposons help sculpt a dynamic genome. *Science* 289: 1455–1457.

Mohr, R. E. 1975. Measured periodicities of the Biwabik (Precambrian) stromatolites and their geophysical significance. In *Growth Rhythms and the History of the Earth's Rotation*, eds. G. D. Rosenberg and S. K. Runcorn, 43–56. New York: Wiley.

Moiseeva, O. V., et al. 2002. A new modified *ortho* cleavage pathway of 3-chlorocatechol degradation by *Rhodococcus opacus* 1CP: Genetic and biochemical evidence. *Journal of Bacteriology.* 184: 5282–5292.

Møller, A. P. 1993. A fungus infecting domestic flies manipulates sexual behaviour of its host. *Behavioral Ecology and Sociobiology* 33: 403–407.

*Monastersky, R. 1999. Waking up to the dawn of vertebrates. *Science News* 156: 292. http://www.sciencenews.org/sn_arc99/11_6_99/fob1.htm.

*Montanari, A., and C. Koeberl. 2000. *Impact Stratigraphy: The Italian Record.* Heidelberg: Springer Verlag. (technical)

Mooney, C. 2003. John Zogby's creative polls. *The American Prospect*, Feb. 1, 2003. http://www.prospect.org/print/V14/1/mooney-c.html.

Moon, S. M. 1976. The central figure and the transitional period. http://www.unification.net/1976/760307.html.

Moon, S. M. 1990. Parents day and I. http://www.unification.net/1990/900327.html.

*Moore, D. 2001. Supernovae, supernova remnants and young earth creationism FAQ. http://www.talkorigins.org/faqs/supernova.

*Moorey, P.R.S. 1991. *A Century of Biblical Archaeology.* Louisville, KY: Westminster/John Knox Press.

*Moran, L. 1993. Evolution is a fact and a theory. http://www.talkorigins.org/faqs/evolution-fact.html.

*Morell, V. 1998. Kennewick Man's trials continue. *Science* 280: 190–192.

Morris, H. M., and M. E. Clark. 1987. Why did God create carnivorous animals if there was to be no death in the world as first created? FAQ #46. From *The Bible Has the Answer*. 3rd ed. Green Forest, AR: Master Books. http://www.icr.org/bible/bhta46.html.

Morris, H. M. 1976. *The Genesis Record: A Scientific and Devotional Commentary on the Book of Beginnings*. San Diego: Creation-Life Publishers.

Morris, H. M. 1982. Bible-believing scientists of the past. *Impact* 103 (Jan.). http://www.icr.org/pubs/imp/imp-103.htm.

Morris, H. M. 1983. Creation is the foundation. *Impact* 126 (Dec.). http://www.icr.org/pubs/imp/imp-126.htm.

Morris, H. M. 1984. Recent creation is a vital doctrine. *Impact* 132 (June). http://www.icr.org/pubs/imp/imp-132.htm.

Morris, H. M. 1985. *Scientific Creationism*. Green Forest, AR: Master Books.

Morris, H. M. 1986. The vanishing case for evolution. *Impact* 156 (June). http://www.icr.org/pubs/imp/imp-156.htm.

Morris, H. M. 1998a. Bigotry in science. *Back to Genesis* 114a (June). http://www.icr.org/pubs/btg-a/btg-114a.htm.

Morris, H. M. 1998b. The Fall, the curse, and evolution. *Back to Genesis* 112 (Apr.). http://www.icr.org/pubs/btg-a/btg-112a.htm.

Morris, H. M. 1998c. The literal week of creation. *Back to Genesis* 113a (May). http://www.icr.org/pubs/btg-a/btg-113a.htm.

Morris, H. M. 2000a. Evil-ution. *Back to Genesis* 140 (Aug.). http://www.icr.org/pubs/btg-a/btg-140a.htm.

Morris, H. M. 2000b. The vital importance of believing in recent creation. *Back to Genesis* 138 (June). http://www.icr.org/pubs/btg-a/btg-138a.htm.

Morris, H. M. 2003a. The bounds of the dominion mandate. *Back to Genesis* 175 (July). http://www.icr.org/pubs/btg-a/btg-175a.htm.

Morris, H. M. 2003b. The mathematical impossibility of evolution. *Back to Genesis* 179 (Nov.). http://www.icr.org/pubs/btg-a/btg-179a.htm.

Morris, J. D. 1989. Was 'Lucy' an ape-man? *Back to Genesis* 11b (Nov.). http://www.icr.org/pubs/btg-b/btg-011b.htm.

Morris, J. D. 1995. The Yellowstone petrified forests. *Impact* 268 (Oct.). http://www.icr.org/pubs/imp/imp-268.htm.

Morris, J. D. 1996. ICR, for such a time as this. *Back to Genesis* 87a (Mar.). http://www.icr.org/pubs/btg-a/btg-087a.htm.

Morris, J. D. 1999a. "Forum." KQED radio, Sep. 15.

Morris, J. D. 1999b. Open letter included with mailing of April 1999 *Acts and Facts*.

Morris, J. D. 2002. Does salt come from evaporated sea water? *Back to Genesis* 167d (Nov.). http://www.icr.org/newsletters/drjohn/drjohnnov02.html.

Morris, J. D. 2003. Dating Niagara Falls. *Impact* 359 (May), iv. http://www.icr.org/pubs/imp/imp-359.htm.

Morris, S. 1983. The hollow earth: A maddening theory that can't be disproved. *Omni* 5(11): 128–129.

Morton, G. R. 1979. Can the canopy hold water? *Creation Research Society Quarterly* 16: 164–169.

Morton, G. R. 1996. Salt in the sea. http://www.asa3.org/archive/evolution/199606/0051.html.

*Morton, G. R. 1997a. Fish to amphibian transition. http://home.entouch.net/dmd/transit.htm.

*Morton, G. R. 1997b. Precambrian pollen. http://www.asa3.org/archive/asa/199709/0101.html.

Morton, G. R. 1998a. Fish cause problems for the global flood. http://home.entouch.net/dmd/fish.htm.

Morton, G. R. 1998b. The global flood produces acidic flood waters. http://home.entouch.net/dmd/acid.htm.

Morton, G. R. 1998c. The letter the *Creation Research Society Quarterly* didn't want you to read. http://home.entouch.net/dmd/letter.htm.

Morton, G. R. 1998d. Young-earth arguments: A second look. http://home.entouch.net/dmd/age.htm.

*Morton, G. R. 2000a. The demise and fall of the water vapor canopy: A fallen creationist idea. http://home.entouch.net/dmd/canopy.htm.

*Morton, G. R. 2000b. Phylum level evolution. http://home.entouch.net/dmd/cambevol.htm.

Morton, G. R. 2000c. The transformation of a young-earth creationist. *Perspectives on Science and Christian Faith* 52(2): 81–83. http://home.entouch.net/dmd/transform.htm.

Morton, G. R. 2001a. The geologic column and its implications for the flood. http://www.talkorigins.org/faqs/geocolumn.

Morton, G. R. 2001b. Nineteenth century design arguments. *RNCSE* 21(3–4): 21–22, 27. http://home.entouch.net/dmd/design.htm.

*Morton, G. R. 2002a. The imminent demise of evolution: The longest running falsehood in creationism. http://home.entouch.net/dmd/moreandmore.htm.

*Morton, G. R. 2002b. Morton's Demon. http://www.talkorigins.org/origins/postmonth/feb02.html.

Morton, G. R. 2002c. Pollen order presents problems for the flood. http://home.entouch.net/dmd/pollen.htm.

Morton, G. R. n.d. Personal stories of the creation/evolution struggle. http://home.entouch.net/dmd/person.htm.

Morton, G. R., H. S. Slusher, R. C. Bartman, and T. G. Barnes. 1983. Comments on the velocity of light. *Creation Research Society Quarterly* 20: 63–65.

Mulkidjanian, A. Y., D. A. Cherepanov, and M. Y. Galperin. 2003. Survival of the fittest before the beginning of life: Selection of the first oligonucleotide-like polymers by UV light. *BMC Evolutionary Biology* 3: 12. http://www.biomedcentral.com/1471-2148/3/12/abstract.

Mullen, L. 2003. Shining light on life's origin. *Astrobiology Magazine.* http://www.astrobio.net/news/article492.html.

Muller, H. J. 1939. Reversibility in evolution considered from the standpoint of genetics. *Biological Reviews* 14: 261–280.

*Murphy, D. C. 2002. Devonian times: Ichthyostega stensioei. http://www.mdgekko.com/devonian/Order/re-ichthyostega.html.

*Musgrave, I. 1998a. Lies, damned lies, statistics, and probability of abiogenesis calculations. http://www.talkorigins.org/faqs/abioprob/abioprob.html.

*Musgrave, I. 1998b. Re: Abiogenesis. http://www.talkorigins.org/origins/postmonth/apr98.html.

Musgrave, I. 1999. Weasels, ReMine, and Haldane's dilemma. http://www.talkorigins.org/origins/postmonth/sep99.html.

*Musgrave, I. 2000. Evolution of the bacterial flagella. http://www.health.adelaide.edu.au/Pharm/Musgrave/essays/flagella.htm.

*Musgrave, I. 2001. The *Period* gene of Drosophila. http://www.talkorigins.org/origins/postmonth/apr01.html.

*Musgrave, I., S. Pirie-Shepherd, and D. Theobald. 2003a. Apolipoprotein AI mutations and information. http://www.talkorigins.org/faqs/information/apolipoprotein.html.

*Musgrave, I., et al. 2003b. Information theory and creationism. http://home.mira.net~reynella/debate/informat.htm.

Musser, G. 1998. Inconstant constants. *Scientific American* 279(5): 24, 28. http://members.tripod.com/unifier2/inconstantconstants.html.

*Musser, G. 2002. Been there, done that. *Scientific American* 286(3): 25–26. http://www.sciam.com/article.cfm?articleID=000D59C8-5512-1CC6-B4A8809EC588EEDF.

*Myers, P. Z. 2003. Wells and Haeckel's embryos: A review of chapter 5 of Icons of Evolution. http://www.talkorigins.org/faqs/wells/haeckel.html.

Nachman, M. W., and S. L. Crowell. 2000. Estimate of the mutation rate per nucleotide in humans. *Genetics* 156(1): 297–304.

Nachman, M. W., S. N. Boyer, J. B. Searle, and C. F. Aquadro. 1994. Mitochondrial DNA variation and the evolution of Robertsonian chromosomal races of house mice, *Mus domesticus. Genetics* 136(3): 1105–1120.

Nakamura, T., H. Uehara, and T. Chiba. 1997. The minimum mass of the first stars and the anthropic principle. *Progress of Theoretical Physics* 97: 169–171. http://xxx.lanl.gov/abs/astro-ph/9612113.

Nanayama, F., et al. 2003. Unusually large earthquakes inferred from tsunami deposits along the Kuril trench. *Nature* 424: 660–663.

*NAS. 1999. Science and creationism. http://www.nap.edu/html/creationism.

NASA Quest. n.d. Mars Team online photo gallery. http://quest.arc.nasa.gov/mars/photos/pathfinder.html; see especially http://quest.arc.nasa.gov/mars/photos/images/marspfsite.gif.

*National Science Foundation. 2003. Artificial life experiments show how complex functions can evolve. http://www.sciencedaily.com/releases/2003/05/030508075843.htm.

*NCSE. 2002a. Analysis of the Discovery Institute's bibliography. http://www.ncseweb.org/resources/articles/3878_analysis_of_the_discovery_inst_4_5_2002.asp.

*NCSE. 2002b. Is there a federal mandate to teach intelligent design creationism? http://www.ncseweb.org/resources/articles/ID-activists-guide-v1.pdf.

NCSE. 2003. Project Steve. http://www.ncseweb.org/article.asp?category=18.

*NCSE. n.d. Voices for evolution. http://www.ncseweb.org/article.asp?category=2.

*Nedin, C. 1999. All about *Archaeopteryx*. http://www.talkorigins.org/faqs/archaeopteryx/info.html.

Negoro, S., et al. 1994. The nylon oligomer biodegradation system of *Flavobacterium* and *Pseudomonas*. *Biodegradation* 5: 185–194.

Nelson, C. 2001. Purposelessness personified. http://www.creationequation.com/archives/MotherNatureMasquerade.htm.

Nelson, K.E., M. Levy, and S.L. Miller. 2000. Peptide nucleic acids rather than RNA may have been the first genetic molecule. *PNAS* 97: 3868–3871.

*Nesse, R.M., and G.C. Williams. 1994. *Why We Get Sick*. New York: Times Books.

Nesvorný, D., W.F. Bottke Jr., L. Dones, and H.F. Levison. 2002. The recent breakup of an asteroid in the main-belt region. *Nature* 417: 720–722.

*Netting, J. 2000. Model of good (and bad) behaviour. *Nature Science Update*. http://www.nature.com/nsu/001026/001026-2.html.

Nevins, S.E. 1976. Continental drift, plate tectonics, and the Bible. *Impact* 32 (Feb.). http://icr.org/pubs/imp/imp-032.htm.

Nevo, E. 1999. *Mosaic Evolution of Subterranean Mammals: Regression, Progression and Global Convergence*. Oxford: Oxford University Press.

Newcomb, R.D., et al. 1997. A single amino acid substitution converts a carboxylesterase to an organophosporus hydrolase and confers insecticide resistance on a blowfly. *PNAS* 94: 7464–7468.

Newton, W.C.F., and C. Pellew. 1929. *Primula kewensis* and its derivatives. *Journal of Genetics* 20: 405–467.

Niemi, T.M., Z. Ben-Avraham, and J.R. Gat (eds.). 1997. *The Dead Sea: The Lake and Its Setting*. Oxford Monographs on Geology and Geophysics, No. 36.

Nilsson, D.-E., and S. Pelger. 1994. A pessimistic estimate of the time required for an eye to evolve. *Proceedings of the Royal Society, Biological Sciences* 256: 53–58.

NIST Time and Frequency Division. n.d. Frequently asked questions. http://www.boulder.nist.gov/timefreq/general/leaps.htm.

*NIST. Updated monthly. NIST time scale data archive. http://www.boulder.nist.gov/timefreq/pubs/bulletin/leapsecond.htm.

NMSR. 2003. Sandia National Laboratories says that the Intelligent Design Network (IDNet-NM/Zogby) "Lab Poll" is BOGUS! http://www.nmsr.org/id-poll.htm.

Nobel Foundation. 1977. The Nobel Prize in chemistry 1977. http://nobelprize.org/chemistry/laureates/1977.

Nóbrega, M. A. et al. 2004. Megabase deletions of gene deserts result in viable mice. *Nature* 431: 988–993.

Nomoto, K., et al. 1997a. Nucleosynthesis in type 1A supernovae. http://xxx.lanl.gov/abs/astro-ph/9706025.

Nomoto, K., et al. 1997b. Nucleosynthesis in type II supernovae. http://xxx.lanl.gov/abs/astro-ph/9706024.

Norell, M. A., and J. A. Clarke. 2001. Fossil that fills a critical gap in avian evolution. *Nature* 409: 181–184.

Norman, H. 1990. *Northern Tales, Traditional Stories of Eskimo and Indian Peoples*. New York: Pantheon Books.

Norman, T. G., and B. Setterfield. 1987. *The Atomic Constants, Light, and Time*. Flinders University of South Australia, School of Mathematical Sciences, Technical Report. http://www.ldolphin.org/setterfield/report.

Nowak, M. A., K. M. Page, and K. Sigmund. 2000. Fairness versus reason in the ultimatum game. *Science* 289: 1773–1775.

Numbers, R. L. 1992. *The Creationists*. New York: Knopf.

Nuttall, N. 1998. Stand clear of the Tube's 100-year-old super-bug. *Times* (London), Aug. 26, 1998, 1. http://www.gene.ch/gentech/1998/Jul–Sep/msg00188.html.

Oard, M. J. 2003. Are polar ice sheets only 4,500 years old? *Impact* 361 (July).

O'Connell, P. 1969. *Science of Today and the Problems of Genesis*. 2nd ed. Hawthorne, CA: Christian Book Club of America. Cited in Foley, J. 2003. Creationist arguments: The monkey quote. http://www.talkorigins.org/faqs/homs/monkeyquote.html.

O'Hanlon, R. 1997. *No Mercy: A Journey to the Heart of the Congo*. New York: Knopf. See also: Book reviews. *Smithsonian Magazine*. 1998 (Aug.). http://www.smithsonianmag.com/smithsonian/issues98/aug98/bookreview_aug98.html.

Ohta, T. 1992. The nearly neutral theory of molecular evolution. *Annual Review of Ecology and Systematics* 23: 263–286.

Ohta, T. 2003. Evolution by gene duplication revisited: Differentiation of regulatory elements versus proteins. *Genetica* 118: 209–216.

Oliver, A., et al. 2000. High frequency of hypermutable *Pseudomonas aeruginosa* in cystic fibrosis lung infection. *Science* 288: 1251–1253. See also pp. 1186–1187. See also LeClerc, J. E., and T. A. Cebula. 2000. *Pseudomonas* survival strategies in cystic fibrosis [letter]. 2000. *Science* 289: 391–392.

Open University Team. 1989. *The Ocean Basins: Their Structure and Evolution*. Oxford: Pergamon Press.

*Orgel, L. E. 1994. The origin of life on the earth. *Scientific American* 271(4): 76–83.

Orgel, L. E. 1998. Polymerization on the rocks: Theoretical introduction. *Origins of Life and Evolution of the Biosphere* 28: 227–234.

*Otte, D., and J. A. Endler (eds.). 1989. *Speciation and Its Consequences*. Sunderland, MA: Sinauer.

Overn, W. n.d. Isochron rock dating is fatally flawed. http://www.tccsa.tc/articles/isochrons2.html.

Ovid, 1958. *The Metamorphoses*. Translated by Horace Gregory. New York: Viking.

Padgett, A. G. 2000. Creation by design. *Books and Culture* 6(4): 30. Available from http://www.christianitytoday.com/bc/2000/004/13.30.html.

Paley, R. 2000. The eye. http://objective.jesussave.us/eye.html.

Paley, W. 1802. *Natural Theology: or, Evidences of the Existence and Attributes of the Deity*. London: J. Faulder.

*Pannella, G. 1972. Paleontological evidence on the Earth's rotational history since the early Precambrian. *Astrophysics and Space Science* 16: 212–237. (technical)

Pannella, G. 1976. Tidal growth patterns in recent and fossil mollusc bivalve shells: A tool for the reconstruction of paleotides. *Naturwissenschaften* 63: 539–543.

Pannella, G., C. MacClintock, and M. Thompson. 1968. Paleontological evidence of variation in length of synodic month since Late Cambrian. *Science* 162: 792–796.

Pappelis, A., and S. W. Fox. 1995. Domain protolife: Protocells and metaprotocells within thermal protein matrices. In *Chemical Evolution: Structure and Model of the First Cell*, eds. C. Ponnamperuma and J. Chela-Flores, 129–132. Dordrecht, Netherlands: Kluwer Academic Publishers.

Parfit, M. 1995. The floods that carved the West. *Smithsonian* 26: 48–59.

Park, I. S., C.-H. Lin, and C. T. Walsh. 1996. Gain of D-alanyl-D-lactate or D-lactyl-D-alanine synthetase activities in three active-site mutants of the *Escherichia coli* D-alanyl-D-alanine ligase B. *Biochemistry* 35: 10464–10471.

Parker, E. N. 1965. Dynamical theory of the solar wind. *Space Science Reviews* 4: 666–708.

Parsons, T. J., et al. 1997. A high observed substitution rate in the human mitochondrial control region. *Nature Genetics* 15: 363–368.

Pathlights. n.d.a. The age of the earth—1. http://www.pathlights.com/ce_encyclopedia/ 05agee2.htm.

Pathlights. n.d.b. The age of the earth—2. http://www.pathlights.com/ce_encyclopedia/ 05agee3.htm.

Pathlights. n.d.c. Chromosome comparisons. http://www.pathlights.com/ce_encyclopedia/15sim03 .htm.

*Patterson, B. 2002. Transitional fossil species and modes of speciation. http://www.origins .tv/darwin/transitionals.htm.

Patton, D. n.d.a. Fossilized hammer. http://www.bible.ca/tracks/fossilized-hammer.htm.

Patton, D. n.d.b. Official world site Malachite Man. http://www.bible.ca/tracks/malachite-man.htm.

*Paul, G. S. 2002. *Dinosaurs of the Air*. Baltimore: Johns Hopkins University Press.

Paulson, T. 2004. Scientists win Kennewick Man ruling. *Seattle Post-Intelligencer*, Feb. 5, 2004. http://seattlepi.nwsource.com/local/159408_kennewickman05.html.

Pearson, P. N., N. J. Shackleton, and M. A. Hall. 1997. Stable isotopic evidence for the sympatric divergence of *Globigerinoides trilobus* and *Orbulina universa* (planktonic foraminifera). *Journal of the Geological Society, London* 154: 295–302.

*Peck, J. R., and A. Eyre-Walker. 1997. The muddle about mutations. *Nature* 387: 135–136.

*Pellmyr, O., J. N. Thompson, J. M. Brown, and R. G. Harrison. 1996. Evolution of pollination and mutualism in the yucca moth lineage. *The American Naturalist* 148: 827–847. (technical)

Pennisi, E. 2001. Genome duplications: The stuff of evolution? *Science* 294: 2458–2460.

*Pennock, R. T. 1996. Naturalism, evidence, and creationism: The case of Phillip Johnson. In *Intelligent Design Creationism and Its Critics*, ed. R. T. Pennock, 77–97. Cambridge, MA: MIT Press, 2002. Orig. pub. in *Biology and Philosophy* 11: 543–549.

*Pennock, R. T. 1999. *Tower of Babel: The Evidence against the New Creationism*. Cambridge, MA: MIT Press.

*Pennock, R. T. 2003. Creationism and intelligent design. *Annual Review of Genomics and Human Genetics* 4: 143–163.

*Perakh, M. 2003a. The No Free Lunch theorems and their application to evolutionary algorithms. http://www.talkreason.org/articles/orr.cfm.

*Perakh, M. 2003b. *Unintelligent Design*. Amherst, NY: Prometheus Books.

Perlmutter, S., et al. 1998. Discovery of a supernova explosion at half the age of the universe and its cosmological implications. *Nature* 391: 51–54. http://xxx.lanl.gov/abs/astro-ph/9712212.

*Petrich, L. 2003. Animal and extraterrestrial artifacts: Intelligently designed? http://www.secweb .org/asset.asp?AssetID=283.

Péwé, T. L. 1975. Quaternary stratigraphic nomenclature in unglaciated Central Alaska. U.S. Geological Survey Professional Paper 862.

*Pieret, J. 2002. Kent Hovind's $250,000 offer. http://talkorigins.org/faqs/hovind.html.

*Pieret, J. (ed.). 2003. The quote mine project. http://www.talkorigins.org/faqs/quotes/mine/ project.html.

*Pigliucci, M. 1999. Where do we come from? A humbling look at the biology of life's origin. *Skeptical Inquirer* 23(5): 21–27.

*Pinker, S. 1994. *The Language Instinct*. New York: Morrow.

Pizzarello, S., and A.L. Weber. 2004. Prebiotic amino acids as asymmetric catalysts. *Science* 303: 1151.

Plantinga, A. 1991a. An evolutionary argument against naturalism. *Logos* 12: 27–49. http://hisdefense.org/audio/ap_audio.html.

Plantinga, A. 1991b. When faith and reason clash: Evolution and the Bible. In *Intelligent Design Creationism and Its Critics*, ed. R.T. Pennock, 113–145. Cambridge, MA: MIT Press, 2002. Orig. pub. in *Christian Scholar's Review* 21: 8–32.

*Poindexter, B. 2003. The horse's mouth. http://home.kc.rr.com/bnpndxtr/download/HorsesMouth-BP007.pdf.

Poirier, J., and K.B. Cumming. 1993. Design features of the monarch butterfly life cycle. *Impact* 237 (Mar.). http://www.icr.org/pubs/imp/imp-237.htm.

*Pojeta, J., Jr., and D.A. Springer. 2001. *Evolution and the Fossil Record*. Alexandria, VA: American Geological Institute. http://www.agiweb.org/news/evolution.pdf; see also http://www.agiweb .org/news/spot_06apr01_evolutionbk.htm.

Poldervaart, A. 1955. Chemistry of the earth's crust. In *Crust of the Earth*, ed. A. Poldervaart, 119–144. Geological Society of America Special Paper 62. Baltimore, MD: Waverly Press.

Pole, A., I.J. Gordon, and M.L. Gorman. 2003. African wild dogs test the 'survival of the fittest' paradigm. *Proceedings of the Royal Society, Biological Sciences* 270: S57.

Polidoro, M. 2002. Ica Stones: Yabba-dabba do! *Skeptical Inquirer* 26(5): 24. http://www.csicop.org/si/2002–09/strange-world.html.

Pollack, A. 2000. Selling evolution in ways Darwin never imagined. *New York Times*, Oct. 28, 2000. http://www.nytimes.com/2000/10/28/business/28EVOL.html.

Polyak, V.J., W.C. McIntosh, N. Güven, and P. Provencio. 1998. Age and origin of Carlsbad Cavern and related caves from $^{40}Ar/^{39}Ar$ of alunite. *Science* 279: 1919–1922. See also p. 1874.

Ponnamperuma, C., and J. Chela-Flores (eds.). 1995. *Chemical Evolution: Structure and Model of the First Cell*. Dordrecht, Netherlands: Kluwer Academic Publishers.

Poole, A.M., D.C. Jeffares, and D. Penny. 1998. The path from the RNA world. *Journal of Molecular Evolution* 46: 1–17.

Porter, S.M., and A.H. Knoll. 2000. Testate amoebae in the Neoproterozoic Era: Evidence from vase-shaped microfossils in the Chuar Group, Grand Canyon. *Paleobiology* 26: 360–385.

POSH (Parents for Objective Science and History). n.d. Biology text review. http://posh .roundearth.net/biology.htm.

Powell, J.A. 1992. Interrelationships of yuccas and yucca moths. *Trends in Ecology and Evolution* 7: 10–15.

Prantzos, N. 1999. Gamma-ray line astrophysics and stellar nucleosynthesis: Perspectives for INTEGRAL. http://xxx.lanl.gov/abs/astro-ph/9901373.

Price, G.M. 1913. *The Fundamentals of Geology*. Mountain View, CA: Pacific Press.

*Prigogine, I. 1977. Time, structure, and fluctuations. http://www.nobel.se/chemistry/laureates/197-7/prigogine-lecture.pdf.

Prijambada, I.D., S. Negoro, T. Yomo, and I. Urabe. 1995. Emergence of nylon oligomer degradation enzymes in *Pseudomonas aeruginosa* PAO through experimental evolution. *Applied and Environmental Microbiology* 61: 2020–2022.

Prophecy Fulfilled. n.d. http://www.bci.org/prophecy-fulfilled.

Pudritz, R.E. 2002. Clustered star formation and the origin of stellar masses. *Science* 295: 68–76.

Qiu, Y.-L., Y. Cho, J.C. Cox, and J.D. Palmer. 1998. The gain of three mitochondrial introns identifies liverworts as the earliest land plants. *Nature* 394: 671–674.

*Radford, B. 2000. Ten-percent myth. http://66.165.133.65/science/stats/10percnt.htm.

*Rainey, P. 2003. Evolution: Five big questions: 4. Is evolution predictable? *New Scientist* 178(2399): 37–38.

Rainey, P.B., and K. Rainey. 2003. Evolution of cooperation and conflict in experimental bacterial populations. *Nature* 425: 72–74.

Randi, J. 1982. Australian Skeptics 1980 divining test. *The Skeptic* 2(1): 2–6. http://www.skeptics
.com.au/journal/divining.htm.

Rasmussen, B., S. Bengtson, I. R. Fletcher, and N. J. McNaughton. 2002. Discoidal impressions and
trace-like fossils more than 1200 million years old. *Science* 296: 1112–1115.

*Ratzsch, D. 2001. *Nature, Design and Science*. Albany: State University of New York Press.

Raymond, J., O. Zhaxybayeva, J. P. Gogarten, and R. E. Blankenship. 2003. Evolution of photosynthetic prokaryotes: A maximum-likelihood mapping approach. *Philosophical Transactions,
Biological Sciences* 358: 223–230.

*RBH. 2003. Untitled. http://www.iscid.org/boards/ubb-get_topic-f-6-t-000384.html#000013.

Reader, J. S., and G. F. Joyce. 2002. A ribozyme composed of only two different nucleotides. *Nature* 420: 841–844.

Redecker, D., R. Kodner, and L. E. Graham. 2000. Glomalean fungi from the Ordovician. *Science*
289: 1920–1921. See also pp. 1884–1885.

Reed, F. 2004. Spare me. http://fredoneverything.net/EvolutionAgain.shtml.

Reineck, H.-E, and I. B. Singh. 1980. *Depositional Sedimentary Environments*. 2nd ed. Berlin:
Spinger-Verlag.

ReMine, W. J. 1993. *The Biotic Message*. St. Paul, MN: St. Paul Science.

Ren, D. 1998. Flower-associated Brachycera flies as fossil evidence for Jurassic angiosperm origins.
Science 280: 85–88. See also pp. 57–59.

Rendle-Short, J. 1980. What should a Christian think about evolution? *Creation* 3(1): 15–17.
http://www.answersingenesis.org/home/area/magazines/docs/v3n1_prof.asp.

Renne, P. R., et al. 1997. ^{40}Ar/^{39}Ar dating into the historical realm: Calibration against Pliny the
Younger. *Science* 277: 1279–1280.

*RESA. n.d. Origins of life. http://www.resa.net/nasa/origins_life.htm.

Retallack, G. 1981. Comment and reply on "Reinterpretation of the depositional environment of
the Yellowstone fossil forests." *Geology* 9: 52–53.

*Richardson, M. K., and G. Keuck. 2002. Haeckel's ABC of evolution and deveolopment. *Biological
Reviews* 77: 495–528. (technical)

*Richardson, M. K., et al. 1998. Haeckel, embryos, and evolution. *Science* 280: 983–986.

Richmond B. G., and D. S. Strait. 2000. Evidence that humans evolved from a knuckle-walking
ancestor. *Nature* 404: 382–385. See also pp. 339–340.

Ricklefs, R. 1993. *The Economy of Nature*. New York: Freeman.

Riggs, A. C. 1984. Major carbon-14 deficiency in modern snail shells from southern Nevada springs.
Science 224: 58–61.

*Ritland, R. 1982. Historical development of the current understanding of the geologic column:
Part II. *Origins* 9: 28–50. http://www.grisda.org/origins/09028.htm.

Rittenour, T. M., J. Brigham-Grette, and M. E. Mann. 2000. El Niño-like climate teleconnections
in New England during the Late Pleistocene. *Science* 288: 1039–1042. See also p. 945.

Robinson, B. A. 1995. Public beliefs about evolution and creation. http://www.religioustolerance.
org/ev_publi.htm, citing *Newsweek*, Jun. 29, 1987, p. 23.

Robinson, B. A. 2000. Bible passages that are immoral by today's standards. http://www.religious
tolerance.org/imm_bibl.htm.

*Robinson, B. A. 2002. A failed attempt to dialog with creation scientists. http://www.religious
tolerance.org/ev_dialog.htm.

Robinson, B. A. 2003. 17 indicators that evolution didn't happen (with rebuttals). http://www.re
ligioustolerance.org/ev_noway.htm#11.

Rode, B. M., H. L. Son, and Y. Suwannachot. 1999. The combination of salt induced peptide formation reaction and clay catalysis: A way to higher peptides under primitive earth conditions.
Origins of Life and Evolution of the Biosphere 29: 273–286.

Rogerson, J. W. 1992. Interpretation, history of. In *The Anchor Bible Dictionary*, vol. 3, ed. D. N.
Freedman, 425–433. New York: Doubleday.

Rokas, A., B. L. Williams, N. King, and S. B. Carroll. 2003. Genome-scale approaches to resolving incongruence in molecular phylogenies. *Nature* 425: 798–804.

*Rosenberg, G. D., and S. K. Runcorn (eds.). 1975. *Growth Rhythms and the History of the Earth's Rotation.* New York: Wiley. (technical)

Ross, C. P., and R. Rezak. 1959. The rocks and fossils of Glacier National Park: The story of their origin and history. U.S. Geological Survey Professional Paper 294-K.

Ross, H. 1994. Astronomical evidences for a personal, transcendent God. In *The Creation Hypothesis*, ed. J. P. Moreland, 141–172. Downers Grove, IL: InterVarsity Press.

Rotblat, J. 1999. A Hippocratic Oath for scientists. *Science* 286: 1475.

Rowe, T., et al. 2001. The *Archaeoraptor* forgery. *Nature* 410: 539–540.

Rubin, K. 2001. The formation of the Hawaiian Islands. http://www.soest.hawaii.edu/GG/HCV/haw_formation.html.

*Rudge, D. W. 1999. Taking the peppered moth with a grain of salt. *Biology and Philosophy* 14: 9–37.

*Rudge, D. W. 2000. Does being wrong make Kettlewell wrong for science teaching? *Journal of Biological Education* 35: 5–11.

Ruse, M. 2000. Creationists correct?: Darwinians wrongly mix science with morality, politics. *National Post*, May 13, 2000. http://www.members.shaw.ca/mschindler/A/eyring_2_2.htm.

*Ruse, M. 2001. *Can a Darwinian Be a Christian?* Cambridge: Cambridge University Press.

Russeau, D. D., and N. Wu. 1997. A new molluscan record of the monsoon variability over the past 130,000 yr in the Luochuan loess sequence, China. *Geology* 25: 275–278.

*Russell, M. 2003. Evolution: Five big questions: 1. How did life begin? *New Scientist* 178(2399): 33–34.

Russell, M. J., and A. J. Hall. 1997. The emergence of life from iron monosulphide bubbles at a submarine hydrothermal redox and pH front. *Journal of the Geological Society of London* 154: 377–402. http://www.gla.ac.uk/Project/originoflife/html/2001/pdf_articles.htm.

*Ryan, R. 2003. Anatomy and evolution of the woodpecker's tongue. http://www.talkorigins.org/faqs/woodpecker/woodpecker.html.

*Sacks, O. 1970. *The Man Who Mistook His Wife for a Hat.* New York: HarperCollins.

Sacks, O. 1995. *An Anthropologist on Mars.* New York: Vintage Books.

Saghatelian, A., Y. Yokobayashi, K. Soltani, and M. R. Ghadiri. 2001. A chiroselective peptide replicator. *Nature* 409: 797–801.

Saladino, R., et al. 2001. A possible prebiotic synthesis of purine, adenine, cytosine, and 4(3H)-pyrimidinone from formamide: Implications for the origin of life. *Bioorganic and Medicinal Chemistry* 9(5): 1249–1253.

Saladino, R., et al. 2003. One-pot TiO_2-catalyzed synthesis of nucleic bases and acyclonucleosides from formamide: Implications for the origin of life. *ChemBioChem* 4(6): 514–521.

Sá Martins, J. S. 2000. Simulated coevolution in a mutating ecology. *Physical Review E* 61: R2212–R2215.

Sanderson, M. J., and H. B. Shaffer. 2002. Troubleshooting molecular phylogenetic analyses. *Annual Review of Ecology and Systematics* 33: 49–72.

Sansom, I. J., M. P. Smith, H. A. Armstrong, and M. M. Smith. 1992. Presence of the earliest vertebrate hard tissues in conodonts. *Science* 256: 1308–1311.

Santorum, R. 2002. Illiberal education in Ohio schools. *Washington Times*, March 14, 2002.

Sarfati, J. 1999. The Yellowstone petrified forests. *Creation Ex Nihilo* 21(2): 18–21. http://www.answersingenesis.org/docs/4109.asp.

Sarfati, J. 2002. *Refuting Evolution 2.* Green Forest, AR: Master Books. http://www.answersingenesis.org/home/area/re2/index.asp.

Sarfati, J. 2004. *Refuting Compromise.* Green Forest, AR: Master Books.

Sasada, M. 1989. Fluid inclusion evidence for recent temperature increases at Fenton Hill Hot Dry Rock Test Site west of the Valles Caldera, New Mexico, U.S.A. *Journal of Volcanology and Geothermal Research* 36: 257–266.

*Schacter, D.L. 2001. *The Seven Sins of Memory.* New York: Houghton Mifflin.

Schadewald, R.J. 1982. Six 'Flood' arguments creationists can't answer. *Creation/Evolution* 3(3): 12–17.

Schadewald, R.J. 1986. Scientific creationism and error. *Creation/Evolution* 6(1): 1–9. http://www.talkorigins.org/faqs/cre-error.html.

Schadewald, R.J. 1987. The Flat-earth Bible. *The Bulletin of the Tychonian Society* 44 (July). http://www.lhup.edu/~dsimanek/febible.htm.

*Schafersman, S. 2003. Texas Citizens for Science responds to latest Discovery Institute challenge. http://texscience.org/files/faqs.htm.

*Scheckler, S.E. 1999. Progymnosperms. In *Encyclopedia of Paleontology*, vol. 2, ed. R. Singer, 992–995. Chicago: Fitzroy Dearborn.

**Schilthuizen, M. 2001. *Frogs, Flies, and Dandelions: The making of species.* Oxford: Oxford University Press.

*Schimmrich, S.H. 1998. Geochronology *kata* John Woodmorappe. http://www.talkorigins.org/faqs/woodmorappe-geochronology.html.

Schlesinger, G., and S.L. Miller. 1983. Prebiotic synthesis in atmospheres containing CH_4, CO, and CO_2. I. Amino acids. *Journal of Molecular Evolution* 19: 376–382.

Schmidt-Dannert, C. 2001. Directed evolution of single proteins, metabolic pathways, and viruses. *Biochemistry* 40: 13125–13136.

Schmitz, B., B. Peucker-Ehrenbrink, M. Lindstrom, and M. Tassinari. 1997. Accretion rates of meteorites and cosmic dust in the Early Ordovician. *Science* 278: 88–90.

Schmitz, B., T. Häggström, and M. Tassinari. 2003. Sediment-dispersed extraterrestrial chromite traces a major asteroid disruption event. *Science* 300: 961–964. See also Martel, L.M.V. 2004. Tiny traces of a big asteroid breakup. http://www.psrd.hawaii.edu/Mar04/fossilMeteorites.html.

*Schneer, C.J. n.d. William "Strata" Smith on the web. http://www.unh.edu/esci/wmsmith.html.

Schneider, E.D., and J.J. Kay. 1994. Life as a manifestation of the second law of thermodynamics. *Mathematical and Computer Modelling* 19(6–8): 25–48. http://www.fes.uwaterloo.ca/u/jjkay/pubs/Life_as/lifeas.pdf.

Schneider, T.D. 2000. Evolution of biological information. *Nucleic Acids Research* 28: 2794–2799. http://www-lecb.ncifcrf.gov.~toms/paper/ev.

*Schopf, J.W. (ed.). 1983. *Earth's Earliest Biosphere. Its Origin and Evolution.* Princeton, NJ: Princeton University Press. (technical)

Schopf, J.W. 1993. Microfossils of the Early Archean Apex Chert: New evidence of the antiquity of life. *Science* 260: 640–646.

*Schopf, J.W. 2000. Solution to Darwin's dilemma: Discovery of the missing Precambrian record of life. *PNAS* 97: 6947–6953. http://www.pnas.org/cgi/content/full/97/13/6947.

Schueller, G. 1998. Stuff of life. *New Scientist* 159(2151): 31–35. http://www.newscientist.com/hot topics/astrobiology/stuffof.jsp.

*Schur, C. 2000. Trace fossils and sedimentary structures: The Permian Coconino sandstone. http://www.psiaz.com/Schur/azpaleo/cocotr.html.

Schwartz, G.E.R., L.G.S. Russek, L.A. Nelson, and C. Barentsen. 2001. Accuracy and replicability of anomalous after-death communication across highly skilled mediums. *Journal of the Society for Psychical Research* 65: 1–25.

Schweitzer, M.H., and J.R. Horner. 1999. Intravascular microstructures in trabecular bone tissues of *Tyrannosaurus rex*. *Annales de Paléontologie* 85: 179–192.

Schweitzer, M.H., and T. Staedter. 1997. The real Jurassic Park. *Earth* (June): 55–57.

Schweitzer, M.H., et al. 1997a. Heme compounds in dinosaur trabecular bone. *PNAS* 94: 6291–6296.

Schweitzer, M.H., et al. 1997b. Preservation of biomolecules in cancellous bone of *Tyrannosaurus rex*. *Journal of Vertebrate Paleontology* 17: 349–359.

Scientific American. 1849. Young Mamalia. *Scientific American* 4(51): 403. Reprinted in *Scientific American* 281(3): 10.

Scott, E.C. 1994. Debates and the Globetrotters. http://www.talkorigins.org/faqs/debating/globetrotters.html.

*Scott, E.C., and G. Branch. 2003. Evolution: What's wrong with 'teaching the controversy.' *Trends in Ecology and Evolution* 18: 499–502.

Scott, E.C., and H.P. Cole. 1985. The elusive scientific basis of creation "science." *Quarterly Review of Biology* 60: 21–30.

Scrutton, C.T. 1964. Periodicity in Devonian coral growth. *Palaeontology* 7(4): 552–558.

Scrutton, C.T. 1970. Evidence for a monthly periodicity in the growth of some corals. In *Palaeogeophysics*, ed. S.K. Runcorn, 11–16. London: Academic Press.

Scrutton, C.T. 1978. Periodic growth features in fossil organisms and the length of the day and month. In *Tidal Friction and the Earth's Rotation*, eds. P. Brosche and J. Sundermann, 154–196. Berlin: Springer-Verlag.

Searls, D. 2003. Pharmacophylogenomics: Genes, evolution and drug targets. *Nature Reviews Drug Discovery* 2: 613–623. http://www.nature.com/nature/view/030731.html.

*Seghers, J. 1998. Sola Scriptura. http://totustuus.com/solascri.htm.

Seife, C. 2002. Eternal-universe idea comes full circle. *Science* 296: 639.

Seilacher, A., P.K. Bose, and F. Pflüger. 1998. Triploblastic animals more than 1 billion years ago: Trace fossil evidence from India. *Science* 282: 80–83.

Sereno, P., and C. Rao. 1992. Early evolution of avian flight and perching: New evidence from the Lower Creates of China. *Science* 255: 845–848.

Service, R.F. 1999. Does life's handedness come from within? *Science* 286: 1282–1283.

SETI Institute. n.d. Frequently asked questions. http://www.seti.org/faq.html.

Setterfield, B. 1998. Birds, beetles, and life. http://www.setterfield.org/essays/giraffe.html.

Shackleton, N.J. 2000. The 100,000-year ice-age cycle identified and found to lag temperature, carbon dioxide, and orbital eccentricity. *Science* 289: 1897–1902.

*Shanks, N., and K.H. Joplin. 1999. Redundant complexity: A critical analysis of intelligent design in biochemistry. *Philosophy of Science* 66: 268–298. http://www.asa3.org/ASA/topics/Apologetics/POS6-99ShenksJoplin.html.

Shapiro, M.D., et al. 2004. Genetic and developmental basis of evolutionary pelvic reduction in threespine sticklebacks. *Nature* 428: 717–723. See also p. 703.

Shea, M. 1997. Five myths about seven books. *Envoy* (Mar./Apr.). http://www.envoymagazine.com/backissues/1.2/marapril_story2.html.

Shen, Y., R. Buick, and D.E. Canfield. 2001. Isotopic evidence for microbial sulphate reduction in the early Archaean era. *Nature* 410: 77–81.

Sherwin, F. 2001. Just how simple are bacteria? *Back to Genesis* 146 (Feb.).

Shlyakhter, A.I. 1976. Direct test of the constancy of fundamental nuclear constants. *Nature* 264: 340. http://sdg.lcs.mit.edu/~ilya_shl/alex/76a_oklo_fundamental_nuclear_constants.pdf.

Shu, D.-G., S. Conway Morris, and X.-L. Zhang. 1996. A *Pikaia*-like chordate from the Lower Cambrian of China. *Nature* 384: 157–158.

Shu, D.-G., et al. 2004. Ancestral echinoderms from the Chengjiang deposits of China. *Nature* 430: 422–428.

Shubin, N.H., E.B. Daeschler, and M.I. Coates. 2004. The early evolution of the tetrapod humerus. *Science* 304: 90–93. See also pp. 57–58.

Sievers, D., and G. von Kiedrowski. 1994. Self-replication of complementary nucleotide-based oligomers. *Nature* 369(6477): 221–224.

*Sigmund, K., E. Fehr, and M.A. Nowak. 2002. The economics of fair play. *Scientific American* 286(1): 82–87.

Simberloff, D. 1988. The contribution of population and community biology to conservation science. *Annual Review of Ecology and Systematics* 19: 473–511.

Simon, C., et al. 2000. Genetic evidence for assortative mating between 13-year cicadas and sympatric "17-year cicadas with 13-year life cycles" provides support for allochronic speciation. *Evolution* 54: 1326.

*Simons, L. M. 2000. Archaeoraptor fossil trail. *National Geographic* 198(4): 128–132.

*Singer, P. 2000. *A Darwinian Left: Politics, Evolution, and Cooperation.* New Haven, CT: Yale University Press.

Skeptic's Annotated Bible. n.d.a. False prophecies, broken promises, and misquotes in the Bible. http://www.skepticsannotatedbible.com/proph/long.html.

Skeptic's Annotated Bible. n.d.b. Slavery. http://www.skepticsannotatedbible.com/topics/slavery.html.

Skjaerlund, D. n.d. Creationism explains human diversity. http://www.forerunner.com/forerunner/X0722_Creationism_explains.html.

Sloan, C. P. 1999. Feathers for T. Rex? *National Geographic* 196(5): 98–107.

Smith, J. V., F. P. Arnold Jr., I. Parsons, and M. R. Lee. 1999. Biochemical evolution III: Polymerization on organophilic silica-rich surfaces, crystal-chemical modeling, formation of first cells, and geological clues. *PNAS* 96: 3479–3485. http://www.pnas.org/cgi/content/full/96/7/3479.

Snelling, A. A. 1997. Sedimentation experiments: Nature finally catches up! *Creation Ex Nihilo Technical Journal* 11(2): 125–126. http://www.answersingenesis.org/docs/456.asp.

Snelling, A. A. 1999. Dating dilemma: Fossil wood in 'ancient' sandstone. *Creation Ex Nihilo* 21(3): 39–41. http://www.answersingenesis.org/home/area/magazines/docs/v21n3_date-dilemma.asp.

Snelling, A. A. 2000. Polonium radiohaloes: Still "a very tiny mystery." *Impact* 326 (Aug.).

Snelling, A. A., and S. A. Austin. 1992. Grand Canyon: Startling evidence for Noah's Flood! *Creation Ex Nihilo* 15(1): 47. http://www.answersingenesis.org/home/area/magazines/docs/v15n1_grandcanyon.asp.

*Sobel, D. 1994. Secrets of the rings. *Discover* 15 (Apr.): 86–91.

Sober, E., and D. S. Wilson. 1998. *Unto Others: The Evolution and Psychology of Unselfish Behavior.* Cambridge, MA: Harvard University Press.

"Socrates." (Oct. 5, 2003). TheologyWeb forum: Too many fossils for a global flood. http://theologyweb.com/forum/showthread.php?postid=233947#post233947.

Sonett, C. P., et al. 1996. Late Proterozoic and Paleozoic tides, retreat of the Moon, and rotation of the Earth. *Science* 273: 100–104.

Song, X., and P. G. Richards. 1996. Seismological evidence for differential rotation of the earth's inner core. *Nature* 382: 221–224.

*Sonnert, G. 2002. *Ivory Bridges: Connecting Science and Society.* Cambridge, MA: MIT Press.

*South, D. 1995. A whale of a tale. http://www.talkorigins.org/faqs/polystrate/whale.html.

*SpaceDaily. 2004. Quasar studies keep fundamental physical constant—constant. http://www.spacedaily.com/news/cosmology-04i.html.

*Speer, B. R. 2000. Introduction to the Deuterostomia. http://www.ucmp.berkeley.edu/phyla/deuterostomia.html.

*Speer, B. R., and N. C. Arens. 1996. Introduction to the progymnosperms. http://www.ucmp.berkeley.edu/seedplants/progymnosperms.html.

Spotts, P. N. 2001. Raw materials for life may predate Earth's formation. *The Christian Science Monitor*, Jan. 30, 2001. http://search.csmonitor.com/durable/2001/01/30/fp2s2-csm.shtml.

*Sproul, B. 1991. *Primal Myths.* New York: HarperCollins.

Stanley, M. 2003. "An expedition to heal the wounds of war": The 1919 eclipse and Eddington as Quaker adventurer. *Isis* 94: 57–89.

Stanley, S. M. 1974. Relative growth of the titanothere horn: A new approach to an old problem. *Evolution* 28: 447–457.

*Stassen, C. 1997. The age of the earth. http://www.talkorigins.org/faqs/faq-age-of-earth.html.

*Stassen, C. 1998. Isochron dating. http://www.talkorigins.org/faqs/isochron-dating.html.

*Stassen, C. 1999. Feedback response (Jan.). http://www.talkorigins.org/origins/feedback/jan99.html (4th response down).

*Stassen, C. 2003. A criticism of the ICR's Grand Canyon Dating Project. http://www.talkorigins.org/faqs/icr-science.html.

*Stassen, C., et al. 1997. Some more observed speciation events. http://www.talkorigins.org/faqs/speciation.html.

State of Oklahoma. 2003. House Bill HB1504: Schools; requiring all textbooks to have an evolution disclaimer; codification; effective date; emergency. http://www2.lsb.state.ok.us/2003-04hb/hb1504_int.rtf.

Steele, D. 2000. Scientists search for secrets of robust systems. *Dallas Morning News*, Sep. 18, 2000. http://nasw.org/users/dsteele/Stories/Robust.html.

Steinhardt, P. J., and N. Turok. 2002. A cyclic model of the universe. *Science* 296: 1436–1439.

Stenger, V. J. 1995. *The Unconscious Quantum*. Amherst, NY: Prometheus Books.

*Stenger, V. J. 1997. Intelligent design: Humans, cockroaches, and the laws of physics. http://www.talkorigins.org/faqs/cosmo.html.

Stenger, V. J. 1999. The anthropic coincidences: A natural explanation. *The Skeptical Intelligencer* 3 (July): 2–17. http://www.stephenjaygould.org/ctrl/stenger_intel.html.

*Stenger, V. J. 2000. The emperor's new designer clothes. *Skeptical Briefs* (Dec.). http://www.csicop.org/sb/2000-12/reality-check.html.

Steward, R. C. 1977. Industrial and non-industrial melanism in the peppered moth, *Biston betularia* (L.). *Ecological Entomology* 2: 231–243.

*Stockwell, J. 2002. Borel's law and the origin of many creationist probability assertions. http://www.talkorigins.org/faqs/abioprob/borelfaq.html.

Stokes, W. L. 1986. Alleged human footprint from Middle Cambrian strata, Milliard County, Utah. *Journal of Geological Education* 34: 187–190.

**Strahler, A. N. 1987. *Science and Earth History: The Evolution/Creation Controversy*. Buffalo, NY: Prometheus Books.

*Straight Dope. 2002. Who wrote the Bible? (Part 5). http://www.straightdope.com/mailbag/mbible5.html.

Straus, W. L., Jr., and A. J. E. Cave. 1957. Pathology and the posture of Neanderthal man. *Quarterly Review of Biology* 32: 348–363.

Stribling, R., and S. L. Miller. 1987. Energy yields for hydrogen cyanide and formaldehyde syntheses: The HCN and amino acid concentrations in the primitive ocean. *Origins of Life and Evolution of the Biosphere* 17: 261–273.

Stricherz, V. 1998. Burke displays fossil of toothless whale. *University Week*, Oct. 10, 1998. http://depts.washington.edu/uweek/archives/1998.10.OCT_29/_article2.html.

Stuiver, M., et al. 1986. Radiocarbon age calibration back to 13,300 years BP and the 14_C age matching of the German oak and US bristlecone pine chronologies. *Radiocarbon* 28(2B): 969–979.

Sun, D., et al. 1998. Magnetostratigraphy and paleoclimatic interpretation of a continuous 7.2Ma Late Cenozoic eolian sediments from the Chinese Loess Plateau. *Geophysical Research Letters* 25: 85–88.

Sun, G., et al. 2002. *Archefructaceae*, a new basal angiosperm family. *Science* 296: 899–904. See also p. 821.

*Sungenis, R. A. 1997. *Not by Scripture Alone*. Santa Barbara: Queenship.

*Sutera, R. 2001. The origin of whales and the power of independent evidence. *RNCSE* 20(5): 33–41. http://www.talkorigins.org/features/whales.

Swanson, R. 1978. A (recently) living plesiosaur found? *Creation Research Society Quarterly* 15: 8.

Swift, D., et al. n.d. The dinosaur figurines of Acambaro, Mexico. http://www.bible.ca/tracks/tracks-acambaro.htm.

*Sykes, B. 2001. *The Seven Daughters of Eve*. New York: W. W. Norton.

Sykes, G. 1937. The Colorado River Delta. *American Geographical Society Special Publication 19*. New York: American Geographical Society.

*Talk.Origins Archive. n.d.a. Debates and gatherings. http://www.talkorigins.org/origins/faqs-debates.html.

*Talk.Origins Archive. n.d.b. Irreducible complexity and Michael Behe. http://www.talkorigins .org/faqs/behe.html.

Tamura, K., and P. Schimmel. 2001. Oligonucleotide-directed peptide synthesis in a ribosome- and ribozyme-free system. *PNAS* 98: 1393–1397.

*Tamzek, N. 2002. Icon of obfuscation. http://www.talkorigins.org/faqs/wells/iconob.html.

Tao, H., and V.W. Cornish. 2002. Milestones in directed enzyme evolution. *Current Opinion in Chemical Biology* 6: 858–864.

*Tattersall, I. 1995. *The Fossil Trail*. New York: Oxford University Press.

*Tattersall, I. 2001. How we came to be human. *Scientific American* 285(6): 56–63. Excerpted from *The Monkey in the Mirror*. New York: Harcourt, 2002, 138–168.

Taylor, D.M. 2003. Alignment of Chimp_rp43-42n4 against human chromosome 15. http://www-personal.umich.edu/~lilyth/erv.

Taylor, P. 1998. *The Great Dinosaur Mystery and the Bible*. http://www.christiananswers.net/dinosaurs/ j-dragon1.html.

*Taylor, R.E. 1987. *Radiocarbon Dating: An Archaeological Perspective*. Orlando, FL: Academic Press.

TCCOP. 1998. Evolutionism vs. creationism: Evidence, extinction, and the evolutionarian "spin doctors." http://www.geocities.com/deke1942/tccop/evolution.htm.

Tchernov, E., et al. 2000. A fossil snake with limbs. *Science* 287: 2010–2012. See also pp. 1939–1941.

Tegmark, M., A. de Oliveira-Costa, and A.J.S. Hamilton. 2003. A high resolution foreground cleaned CMB map from WMAP. *Physical Review D* 68: 123523. http://cul.arxiv.org/abs/astro-ph/0302496.

*Theobald, D. 2003. The vestigiality of the human vermiform appendix: A modern reappraisal. http://www.talkorigins.org/faqs/vestiges/appendix.html.

*Theobald, D. 2004. 29+ evidences for macroevolution: The scientific case for common descent. http://www.talkorigins.org/faqs/comdesc.

*Thewissen, J.G.M. (ed.). 1998. *The Emergence of Whales: Evolutionary patterns in the origin of Cetacea*. New York: Plenum Press. (technical)

Thewissen, J.G.M., and S.T. Hussain. 1993. Origin of underwater hearing in whales. *Nature* 361: 444–445.

Thewissen, J.G.M., S.T. Hussain, and M. Arif. 1994. Fossil evidence for the origin of aquatic lo-comotion in archaeocete whales. *Science* 263: 210–212. See also pp. 180–181.

*Thewissen, J.G.M., S.I. Madar, and S.T. Hussain. 1998. Whale ankles and evolutionary rela-tionships. *Nature* 395: 452. See also Wong, K. 1999. Cetacean creation. *Scientific American* 280(1): 26, 30.

*Thewissen, J.G.M., and E.M. Williams. 2002. The early radiations of Cetacea (Mammalia): Evo-lutionary pattern and developmental correlations. *Annual Review of Ecology and Systematics* 33: 73–90. (technical)

Thielemann, F.-K., et al. 1998. Nucleosynthesis basics and applications to supernovae. In *Nuclear and Particle Astrophysics*, eds. J. Hirsch and D. Page, 27. Cambridge: Cambridge University Press. http://xxx.lanl.gov/abs/astro-ph/9802077.

"Thomas." n.d. "Dr." Kent Hovind. http://www.geocities.com/odonate/hovind.htm.

Thomas, A.L.R. 1997. The breath of life: Did increased oxygen levels trigger the Cambrian Ex-plosion? *Trends in Ecology and Evolution* 12: 44–45.

*Thomas, D. 1998. "Creation physicist" D. Russell Humphreys, and his questionable "evidence for a young world." http://www.cesame-nm.org/Viewpoint/contributions/Hump.html.

*Thomas, D. 2003. The C-Files: The smoking gun—"intelligent design" IS creationism! http://www.nmsr.org/smkg-gun.htm.

*Thomas, D. n.d. Evolution and information: The nylon bug. http://www.nmsr.org/nylon.htm.

Thompson, A. 1996. An evolved circuit, intrinsic in silicon, entwined with physics. http://www.cogs.susx.ac.uk/users/adrianth/ices96/paper.html.

Thompson, B., and K. Butt. 2001. Creation vs. evolution—[Part II], Lesson 6. Montgomery, AL: Apologetics Press. http://www.apologeticspress.org/rr/reprints/hsc0106.pdf.

*Thompson, T. 1994. Is the planet Venus young? http://www.talkorigins.org/faqs/venus-young.html.

*Thompson, T. 1996. Meteorite dust and the age of the earth. http://www.talkorigins.org/faqs/moon-dust.html.

*Thompson, T. 1997. On creation science and the alleged decay of the earth's magnetic field. http://www.talkorigins.org/faqs/magfields.html.

*Thompson, T. 1999. On creation science and transitional fossils. http://www.tim-thompson.com/trans-fossils.html.

*Thompson, T. 2000. The recession of the Moon and the age of the Earth–Moon system. http://www.talkorigins.org/faqs/moonrec.html.

*Thompson, T. 2003. A radiometric dating resource list. http://www.tim-thompson.com/radio metric.html.

*Thompson, T. n.d.a. Answers in Genesis and Saturn's rings. http://home.austarnet.com.au/stear/aig_and_saturn's_rings.htm.

*Thompson, T. n.d.b. Is the Earth young? http://www.tim-thompson.com/young-earth.html.

*Thompson, T. n.d.c. Luminescence and radiometric dating. http://www.tim-thompson.com/radio metric.html.

Thomson, W. 1871. Address of Sir William Thomson, Knt., LL.D., F.R.S, President. *Report of the Forty-First Meeting of the British Association for the Advancement of Science*, lxxxiv–cv.

Thorne, A., et al. 1999. Australia's oldest human remains: Age of the Lake Mungo 3 skeleton. *Journal of Human Evolution* 36: 591–612.

Thwaites, W.M. 1985. New proteins without God's help. *Creation/Evolution* 5(2): 1–3. http://www.ncseweb.org/resources/articles/4661_issue_16_volume_5_number_2__4_10_2003.asp.

*Thwaites, W.M., and F.T. Awbrey. 1982. As the world turns: Can creationists keep time? *Creation/Evolution* 3(3): 18–22. http://www.natcenscied.org/resources/articles/9626_issue_09__volume_3_number_3__1_3_2003.asp.

*Till, F. 1990. What about scientific foreknowledge in the Bible? http://www.infidels.org/library/magazines/tsr/1990/4/4scien90.html.

*Toland, J. 1976. *Adolf Hitler*. Garden City, NY: Doubleday.

Tparents. n.d. Dr. Jonathan Wells returns to UTS. http://www.tparents.org/library/unification/publications/cornerst/cs970506/cst_dr-jonathan.html.

Travis, D.J., A.M. Carleton, and R.G. Lauritsen. 2002. Contrails reduce daily temperature range. *Nature* 418: 601.

Travis, J. 1998. The accidental immune system. *Science News* 154: 302–303.

Trendall, A.F., et al. (eds.). 1990. *Geology and Mineral Resources of Western Australia, Memoir 3*. Geological Survey of Western Australia. State Printing Division, Perth.

*Trott, R., and J. Lippard. 2003. Creationism implies racism? http://www.talkorigins.org/faqs/racism.html.

*True, H.L., and S.L. Lindquist. 2000. A yeast prion provides a mechanism for genetic variation and phenotypic diversity. *Nature* 407: 477–483. (technical)

Truman, J.W., and L.M. Riddiford. 1999. The origins of insect metamorphosis. *Nature* 401: 447–452.

Tryon, E.P. 1973. Is the universe a vacuum fluctuation? *Nature* 246: 396–397.

TSRI. 2001. New study by scientists at the Scripps Research Institute suggests an answer for one of the oldest questions in biology. (press release, Feb. 15). http://www.scripps.edu/news/press/021401.html.

Turner, G. 1981. The development of the atmosphere. In *The Evolving Earth*, ed. L.R.M. Cocks, 121–136. London: British Museum.

Tuttle, R.H. 1990. The pitted pattern of Laetoli feet. *Natural History* (March): 60–64.

Ukraintseva, V. V. 1993. *Vegetation Cover and Environment of the "Mammoth Epoch" in Siberia*. Hot Springs, SD: Mammoth Site of Hot Springs of South Dakota.

Underhill, P. A., et al. 2000. Y chromosome sequence variation and the history of human populations. *Nature Genetics* 26: 358–361.

*University of Oregon. n.d. Uniformitarianism. http://zebu.uoregon.edu/2003/glossary/uniformitarianism.html.

*Ussery, D. 1999. A biochemist's response to "The biochemical challenge to evolution." *Bios* 70: 40–45. http://www.cbs.dtu.dk/staff/dave/Behe.html.

Valentine, J. W., A. G. Collins, and C. P. Meyer. 1994. Morphological complexity increase in metazoans. *Paleobiology* 20: 131–142.

van Arnhem, C. 2002. The Genesis site: Chinese characters. http://ourworld.compuserve.com/homepages/CW_Arnhem/chinchar/chinchar5.html.

Vandenberghe, J., et al. 1997. New absolute time scale for the Quaternary climate in the Chinese loess region by grain-size analysis. *Geology* 25: 35–38.

van Gent, R. H. 1984. Red Sirius. *Nature* 312: 302.

*Van Till, H. J. 2002. Is the Creation a 'right stuff' universe? *Perspectives on Science and Christian Faith* 54(4): 232.

Van Valen, L. M., and V. C. Maiorana. 1991. HeLa, a new microbial species. *Evolutionary Theory* 10: 71–74.

Vardiman, L. 1992. Ice cores and the age of the earth. *Impact* 226 (Apr.). http://www.icr.org/pubs/imp/imp-226.htm.

Vecsey, C. 1991. *Imagine Ourselves Richly*. San Francisco: HarperCollins.

Velicer, G. J., and Y. N. Yu. 2003. Evolution of novel cooperative swarming in the bacterium *Myxococcus xanthus*. *Nature* 425: 75–78.

Velikovsky, I. 1955. *Earth in Upheaval*. New York: Pocket Books.

Veneziano, G. 2004. The myth of the beginning of time. *Scientific American* (May): 54–65.

*Vickers, B. 1998. Some questionable creationist credentials. http://www.talkorigins.org/faqs/credentials.html.

Vickers-Rich, P., L. M. Chiappe, and S. Kurzanov. 2002. The enigmatic birdlike dinosaur *Avimimus portentosus*. In *Mesozoic Birds: Above the Heads of Dinosaurs*, eds. Chiappe and L. M. Witmer, 65–86. Berkeley: University of California Press.

Vlaardingerbroek, B. n.d. Kent Hovind's bogus challenge. http://home.austarnet.com.au/stear/kent_hovind's_bogus_challenge.htm.

Volpicelli, J., et al. 1999. The role of uncontrollable trauma in the development of PTSD and alcohol addiction. *Alcohol Research and Health* 23(4): 256–262. http://www.niaaa.nih.gov/publications/arh23-4/256-262.pdf.

*Voltaire. 1759. *Candide*. http://www.literature.org/authors/voltaire/candide.

*VonRoeschlaub, W. K. 1998. God and evolution. http://www.talkorigins.org/faqs/faq-god.html.

*Wächtershäuser, G. 2000. Life as we don't know it. *Science* 289: 1307–1308.

*Waggoner, B. 1996. Introduction to the Cephalochordata. http://www.ucmp.berkeley.edu/chordata/cephalo.html.

Waitt, R. B., Jr. 1985. Case for periodic, colossal jökulhlaups from Pleistocene glacial Lake Missoula. *Geological Society of America Bulletin* 96: 1271–1286.

Wake, D. B. 1997. Incipient species formation in salamanders of the *Ensatina* complex. *PNAS* 94: 7761–7767.

Wakefield, D., Jr. 1971. Mummified seals of southern Victoria Land. *Antarctic Journal* 6: 210–211.

*Wakefield, J. R. 1998. The geology of Gentry's "tiny mystery." *Journal of Geological Education* 36: 161–175. http://www.csun.edu~vcgeo005/gentry/tiny.htm.

Walker, G. 2003. The collector. *New Scientist* 179 (2405): 38–41.

Walker, T. 2000. Dating dilemma deepens: Moore on ancient radiocarbon. http://www.answersingenesis.org/home/area/feedback/negative6-26-2000.asp.

Wallace, B. 1991. *Fifty Years of Genetic Load: An Odyssey*. Ithaca, NY: Cornell University Press.

Wallace, T. 2002. Five major *evolutionist* misconceptions about evolution. http://www.trueorigins .org/isakrbtl.asp.

Wang, D.Y.-C., S. Kumar, and S. B. Hedges. 1999. Divergence time estimates for the early history of animal phyla and the origin of plants, animals and fungi. *Proceedings of the Royal Society of London, Series B, Biological Sciences* 266: 163–171.

*Wang, J. 1998. Scientists flock to explore China's 'site of the century.' *Science* 279: 1626–1627.

Wang, Y.J., et al. 2001. A high-resolution absolute-dated Late Pleistocene monsoon record from Hulu Cave, China. *Science* 294: 2345–2348.

Ward, L.W., and B.W. Blackwelder. 1975. *Chesapecten*, a new genus of Pectinidae (Mollusca: Bivalvia) from the Miocene and Pliocene of eastern North America. U.S. Geological Survey Professional Paper 861.

Ward-Thompson, D. 2002. Isolated star formation: From cloud formation to core collapse. *Science* 295: 76–81. See also related articles in the same issue.

Watchtower Bible and Tract Society. 1985. *Life—How Did It Get Here?* Brooklyn, NY.

Watchtower Bible and Tract Society. 1989. *The Bible—God's Word or Man's?* Brooklyn, NY.

Watson, K. 2001. Radiometric time scale. http://pubs.usgs.gov/gip/geotime/radiometric.html.

Webb J.K., et al. 1999. Search for time variation of the fine structure constant. *Physical Review Letters* 82: 884–887. http://xxx.lanl.gov/abs/astro-ph/?9803165.

*Weber, C.G. 1980. Common creationist attacks on geology. *Creation/Evolution* 1(2): 10–25.

*Weber, C.G. 1981. The bombardier beetle myth exploded. *Creation/Evolution* 2(1): 1–5.

Webster, A.J., R.J.H. Payne, and M. Pagel. 2003. Molecular phylogenies link rates of evolution and speciation. *Science* 301: 478.

Wedekind, C., and V.A. Braithwaite. 2002. The long-term benefits of human generosity in indirect reciprocity. *Current Biology* 12: 1012–1015.

Wedekind, C., and M. Milinski. 2000. Cooperation through image scoring in humans. *Science* 288: 850–852. See also pp. 819–820.

*Wein, R. 2002. Not a free lunch but a box of chocolates: A critique of William Dembski's book *No Free Lunch*. http://www.talkorigins.org/design/faqs/nfl.

*Weinberg, S. 1999. A designer universe? http://www.physlink.com/Education/essay_weinberg.cfm.

**Weiner, J. 1994. *The Beak of the Finch: A Story of Evolution in Our Time*. New York: Knopf.

*Weiner, J. 1999. *Time, Love, Memory: A Great Biologist and His Quest for the Origins of Behavior*. New York: Knopf.

Weisgraber K.H., et al. 1983. Apolipoprotein A-I$_{Milano}$. Detection of normal A-I in affected subjects and evidence for a cysteine for arginine substitution in the variant A-I. *Journal of Biological Chemistry* 258: 2508–2513.

Wells, J. 1991. Marriage and the family: The Unification blessing. http://www.tparents.org/ Library/Unification/Talks/Wells/MARRGE3.htm.

Wells, J. 1999. Second thoughts about peppered moths. *The Scientist*, 13(11): 13. http://www.arn .org/docs/wells/jw_pepmoth.htm.

Wells, J. 2000. *Icons of Evolution*. Washington, DC: Regnery.

Wells, J. n.d. Darwinism: Why I went for a second Ph.D. http://www.tparents.org/library/ unification/talks/wells/darwin.htm.

Wells, J.W. 1963. Coral growth and geochronometry. *Nature* 197: 948–950.

Wells, J.W. 1970. Problems of annual and daily growth-rings in corals. In *Palaeogeophysics*, ed. S.K. Runcorn, 3–9. London: Academic Press.

West, J.G., Jr. 2003. Intelligent design and creationism just aren't the same. http://www.arn .org/docs2/news/idandcreationismnotsame011503.htm.

Weston, P. 1998. The fallacy of racism. *Creation Ex Nihilo* 20(1): 52–53, http://www .answersingenesis.org/docs/384.asp.

Weston-Broome, S. 2001. Louisiana House Concurrent Resolution no. 74: Civil Rights: Provides relative to racism and education about racism. HLS 01-2652 ORIGINAL.

Whalley, P., and E. A. Jarzembowski. 1981. A new assessment of *Rhyniella*, the earliest known insect, from the Devonian of Rhynie, Scotland. *Nature* 291: 317.

Wheeler, M. R. 1987. Drosophilidae. In *Manual of Nearctic Diptera*, vol. 2, 1011. Agriculture Canada, Hull, Quebec: Canadian Government Publishing Centre.

Whitcomb, J. C., Jr., and H. M. Morris. 1961. *The Genesis Flood*. Philadelphia, PA: Presbyterian and Reformed Publishing.

Whitehouse, D. 2001. Songbird shows how evolution works. BBC News Online, Jan. 18, 2001. http://news.bbc.co.uk/1/hi/sci/tech/1123973.stm.

White, T. 2002a. Palaeos Vertebrates 420.500: Cynodontia: Probainognathia. http://www.palaeos .com/Vertebrates/Units/410Cynodontia/410.500.html.

White, T. 2002b. Palaeos Vertebrates 420.300: Mammaliformes: Symmetrodonta. http://www .palaeos.com/Vertebrates/Units/Unit420/420.300.html.

White, T. D., et al. 2003. Pleistocene *Homo sapiens* from Middle Awash, Ethiopia. *Nature* 423: 742–747.

Whittet, D.C.B. 1999. A physical interpretation of the 'red Sirius' anomaly. *Monthly Notices of the Royal Astronomical Society* 310: 355–359.

Wichman, H. A., et al. 1999. Different trajectories of parallel evolution during viral adaptation. *Science* 285: 422–424.

Wicken, J. S. 1979. The generation of complexity in evolution: A thermodynamic and information-theoretical discussion. *Journal of Theoretical Biology* 77: 349–365.

Wieland, C. 1997. Sensational dinosaur blood report. *Creation* 19(4): 42–43. http://www.an swersingenesis.org/docs/4232cen_s1997.asp.

Wieland, C. 1998. A shrinking date for 'Eve.' *Creation Ex Nihilo Technical Journal* 12(1): 1–3. http://www.answersingenesis.org/docs/4055.asp.

*Wiens, R. C. 2002. Radiometric dating: A Christian perspective. http://www.asa3.org/ASA/ resources/Wiens.html.

Wiker, B. D. 2003. Does science point to God? Part II: The Christian critics. http://www.crisis magazine.com/julaug2003/feature1.htm.

Wilder-Smith, A. E. 1970. *The Creation of Life: A Cybernetic Approach to Evolution*. Wheaton, IL: Harold Shaw.

*Wilkins, J. 1996. Darwin's precursors and influences: 1. Transmutationism. http://www.talkorigins .org/faqs/precursors/precurstrans.html.

*Wilkins, J. 1997a. Evolution and philosophy: Does evolution make might right? http://www.talk origins.org/faqs/evolphil/social.html.

*Wilkins, J. 1997b. Evolution and philosophy: A good tautology is hard to find. http://www.talk origins.org/faqs/evolphil/tautology.html.

*Wilkins, J. 1997c. Evolution and philosophy: Is evolution science, and what does 'science' mean? http://www.talkorigins.org/faqs/evolphil/falsify.html.

*Wilkins, J. 1997d. Evolution and philosophy: Predictions and explanations. http://www .talkorigins.org/faqs/evolphil/predict.html.

*Wilkins, J. 1997e. Macroevolution. http://www.talkorigins.org/faqs/macroevolution.html.

*Wilkins, J. 1998. So you want to be an anti-Darwinian: Varieties of opposition to Darwinism. http://www.talkorigins.org/faqs/anti-darwin.html.

*Wilkins, J. 2000. Evolutionists against eugenics. http://www.talkorigins.org/origins/postmonth/ nov00.html.

*Wilkins, J. 2001. Defining evolution. *RNCSE* 21(1–2): 29–37. http://www.ncseweb.org/resources/ rncse_content/vol21/9925_defining_evolution_12_30_1899.asp.

*Wilkins, J. S. 2003. How to be a chaste species pluralist–realist: The origins of species modes and the Synapomorphic Species Concept. *Biology and Philosophy* 18: 621–638.

*Wilkins, J. S., and W. R. Elsberry. 2001. The advantages of theft over toil: The design inference and arguing from ignorance. *Biology and Philosophy* 16: 711–724. http://www.talkdesign.org/ faqs/theftovertoil/theftovertoil.html.

Williams, A. 2002. Kingdom of the plants. *Creation* 24(1): 46–48. http://www.answersingenesis.org/home/area/magazines/docs/v24n1_kingdom.asp.

Williams, A. 2003. Copying confusion. *Creation* 25(4): 15. http://www.answersingenesis.org/creation/v25/i4/DNAduplication.asp.

*Williams, B. 1998. Creationist deception exposed. *The Skeptic* 18(3): 7–10. http://www.tccsa.tc/video/creationist_deception_exposed.pdf.

Williams, D. F., et al. 1997. Lake Baikal record of continental climate response to orbital insolation during the past 5 million years. *Science* 278: 1114–1117.

Williams, G. E. 1997. Precambrian length of day and the validity of tidal rhythmite paleotidal values. *Geophysical Research Letters* 24: 421–424.

*Williams, R. n.d.a. Examples of beneficial mutations and natural selection. http://www.gate.net~rwms/EvoMutations.html.

*Williams, R. n.d.b. Examples of beneficial mutations in humans. http://www.gate.net~rwms/EvoHumBenMutations.html.

*Williams, R. n.d.c. Haldane's dilemma. http://www.gate.net~rwms/haldane1.html.

*Willis, P. 1997. Turning a corner in the search for the origin of life. *Santa Fe Institute Bulletin* 12(2). http://www.santafe.edu/sfi/publications/Bulletins/bulletin-summer97/turning.html.

Willis, T. 1997. Dinosaurs: Incredible new evidence of their VERY recent life. *CSA News* (Nov/Dec.), http://www.csama.org/199711NL.HTM.

Willis, T. 2000. The laws of cause and effect, and the 1st and 2nd laws of thermodynamics have been invalidated by modern science, part 2. *CSA News* 17(2) (Mar/Apr): 1–2.

Winker, C. D., and S. M. Kidwell. 1986. Paleocurrent evidence for lateral displacement of the Pliocene Colorado River delta by the San Andreas fault system, southeastern California. *Geology* 14: 788–791.

Winn, P. 2003. A new day, some new science. *Citizen Link* (Aug. 15) http://www.family.org/cforum/feature/A0027372.cfm.

Wise, K. P. 1986. The way geologists date! In *Proceedings of the First International Conference on Creationism*, eds. R. E. Walsh, C. L. Brooks, and R. S. Crowell, 1: 135–138.

Woese, C. R. 2000. Interpreting the universal phylogenetic tree. *PNAS* 97: 8392–8396. http://www.pnas.org/cgi/content/abstract/97/15/8392.

*Wolpert, D. 2002. William Dembski's treatment of the No Free Lunch theorems is written in jello. *Mathematical Reviews* (Feb.). http://www.talkreason.org/articles/jello.cfm.

Wolpert, D. H., and W. G. Macready. 1997. No Free Lunch theorems for optimization. *IEEE Transactions on Evolutionary Computation* 1(1): 67–82. http://citeseer.nj.nec.com/wolpert96no.html.

*Wong, M. 2001. Young-earth creationism: Pseudoscience. http://www.stardestroyer.net/Creationism/YoungEarth/Hartman-5.shtml.

Wood, I. n.d. Is Kent Hovind a liar too? http://home.austarnet.com.au/stear/kent_hovind's_lies.htm.

*Wood, J. S. 2003. Io: Jupiter's volcanic moon: tidal heating. http://www.planetaryexploration.net/jupiter/io/tidal_heating.html.

Wood, T. C., and D. P. Cavanaugh. 2003. An evaluation of lineages and trajectories as baraminological membership criteria. *Occasional Papers of the Baraminology Study Group* 2: 1–6. http://www.bryancore.org/bsg/opbsg/002.html.

Woodmorappe, J. 1979. Radiometric geochronology reappraised. *Creation Research Society Quarterly* 16: 102–129.

Woodmorappe, J. 1982. Anomalously occurring fossils. *Creation Research Society Quarterly* 18 (March). http://www.nwcreation.net/anomalies.html.

Woodmorappe, J. 1996. *Noah's Ark: A Feasibility Study*. Santee, CA: Institute for Creation Research.

*Woolf, J. 1999. Young-earth creationism and the geology of the Grand Canyon. http://my.erinet.com~jwoolf/gc_intro.html.

Wright, M. 1996. Re: Chinese characters as pornography. http://www.google.com/groups?as_umsgid=31C87C35.FF3%40redshift.com.

*Wright, M. n.d. Do Chinese characters tell us something about Genesis? http://www.coastalfog.net/languages/chinchar/chinchar.html.

Wright, M. C., and G. F. Joyce. 1997. Continuous in vitro evolution of catalytic function. *Science* 276: 614–617. See also pp. 546–547.

*Wright, R. 1994. *The Moral Animal: The New Science of Evolutionary Psychology.* New York: Pantheon Books.

*Wuethrich, B., 1998. Why sex? Putting theory to the test. *Science* 281: 1980–1982. See also several related articles in the same issue.

Wyatt, R. E. 1989. *Discovered—Noah's Ark.* Nashville, TN: World Bible Society.

Wysong, R. L. 1976. *The Creation–Evolution Controversy.* Midland, MI: Inquiry Press.

*Xiong, J., and C. E. Bauer. 2002. Complex evolution of photosynthesis. *Annual Review of Plant Biology* 53: 503–521. (technical)

Xu, X., M. A. Norell, X. Wang, P. J. Makovicky, and X. Wu. 2002. A basal troodontid from the Early Cretaceous of China. *Nature* 415: 780–784.

Xu, X., Z. Tang, and X. Wang. 1999. A therizinosaurid dinosaur with integumentary structures from China. *Nature* 399: 350–354.

Xu, X., and X. Wang. 2003. A new maniraptorian dinosaur from the Early Cretaceous Yixian Formation of Western Liaoning. *Vertebrate Palasiatica* 41(3): 195–202.

Xu, X., X.-L. Wang and X.-C. Wu. 1999. A dromaeosaur dinosaur with a filamentous integument from the Yixian Formation of China. *Nature* 401: 262–266.

Xu, X., and X.-C. Wu. 2001. Cranial morphology of *Sinornithosaurus millenii* Xu et al. 1999 (Dinosauria: Theropoda: Dromaeosauridae) from the Yixian Formation of Liaoning, China. *Canadian Journal of Earth Sciences* 38: 1739–1752.

Xu, X., Z. Zhou, and R. O. Prum. 2001. Branched integumental structures in *Sinornithosaurus* and the origin of feathers. *Nature* 410: 200–204.

Xu, X., Z. Zhou, X. Wang, X. Kuang, F. Zhang, and X. Du. 2003. Four-winged dinosaurs from China. *Nature* 421: 335–340.

Yahya, H. 1999. The evolution deceit: Thermodynamics falsifies evolution. http://www.evolutiondeceit.com/chapter12.php.

Yahya, H. 2003a. Darwinism Refuted, Evolution and thermodynamics. http://www.darwinismrefuted.com/thermodynamics_01.html.

Yahya, H. 2003b. Darwinism Refuted, The invalidity of punctuated equilibrium. http://www.darwinismrefuted.com/equilibrium.html.

Yahya, H. 2003c. Darwinism Refuted, The myth of homology. http://www.darwinismrefuted.com/myht_of_homology_05.html.

Yahya, H. 2003d. Darwinism Refuted, The true origin of species. http://www.darwinismrefuted.com/origin_of_species_02.html.

Yahya, H. 2004. Errors concerning human intelligence on the BBC's Horizon programme. http://www.harunyahya.net/V2/Lang/en/Pg/WorkDetail/Number/1905.

Yahya, H. n.d. *Miracles of the Qur'an.* http://www.harunyahya.com/miracles_of_the_quran_01.php.

Yamamoto, Y., and W. R. Jeffery. 2000. Central role for the lens in cave fish eye degeneration. *Science* 289: 631–633.

*Yates, S. 1994. The Lady Hope story: A widespread falsehood. http://www.talkorigins.org/faqs/hope.html.

Yockey, H. P. 1992. *Information Theory and Molecular Biology.* Cambridge: Cambridge University Press.

Yoshida, T., et al. 2003. Rapid evolution drives ecological dynamics in a predator–prey system. *Nature* 424: 303–306.

*Young, D. A. 1988. *Christianity and the Age of the Earth.* Thousand Oaks, CA: Artisan Sales.

Young, K. V., E. D. Brodie Jr., and E. D. Brodie III. 2004. How the horned lizard got its horns. *Science* 304: 65.

*Young, M. 2004. Moonshine: Why the peppered moth remains an icon of evolution. http://www.talkdesign.org/faqs/moonshine.htm.

Yuehai Ke, et al. 2001. African origin of modern humans in East Asia: A tale of 12,000 Y chromosomes. *Science* 292: 1151–1153. See also pp. 1051–1052.

Yunis, J. J., and O. Prakash. 1982. The origin of man: A chromosomal pictorial legacy. *Science* 215: 1525–1530.

Yuretich, R. F. 1984a. Comment and reply on "Yellowstone fossil forests: New evidence for burial in place." *Geology* 12: 639.

Yuretich, R. F. 1984b. Yellowstone fossil forests: New evidence for burial in place. *Geology* 12: 159–162.

Zepik, H., et al. 2002. Chiral amplification of oligopeptides in two-dimensional crystalline self-assemblies on water. *Science* 295: 1266–1269.

Zhang, F., and Z. Zhou. 2000. A primitive enantiornithine bird and the origin of feathers. *Science* 290: 1955–1959.

Zhang, J., Y.-P. Zhang. and H. F. Rosenberg. 2002. Adaptive evolution of a duplicated pancreatic ribonuclease gene in a leaf-eating monkey. *Nature Genetics* 30: 411–415.

Zhang, M., D. Yuan, Y. Lin, H. Cheng, J. Qin, and H. Zhang. 2004. The record of paleoclimatic change from stalagmites and the determination of termination II in the south of Guizhou Province, China. *Science in China Series D* 47(1): 1–12. http://www.karst.edu.cn/publication/Zhang%20Ml200401.pdf.

Zheng, Y.-F. 1989. Influences of the nature of the initial Rb-Sr system on isochron validity. *Chemical Geology* (Isotope Geoscience Section) 80: 1–16.

Zhou, Z., J. A. Clarke, and F. Zhang. 2002. *Archaeoraptor*'s better half. *Nature* 420: 285.

Zhou, Z., and F. Zhang. 2002. A long-tailed, seed-eating bird from the Early Cretaceous of China. *Nature* 418: 405–409.

Zhou, Z., and F. Zhang. 2003. Anatomy of the primitive bird *Sapeornis chaoyangensis* from the Early Cretaceous of Liaoning, China. *Canadian Journal of Earth Sciences* 40: 731–747.

*Zimmer, C. 1995. Back to the sea. *Discover* 16 (Jan.): 82–84.

Zimmer, C. 1996. The light at the bottom of the sea. *Discover* (Nov.): 62–66, 71–73.

*Zimmer, C. 1998. *At the Water's Edge*. New York: Touchstone.

*Zimmer, C. 2000a. In search of vertebrate origins: Beyond brain and bone. *Science* 287: 1576–1579.

**Zimmer, C. 2000b. *Parasite Rex: Inside the Bizarre World of Nature's Most Dangerous Creatures*. New York: Free Press.

Zorpette, G. 2000. Looking for Madam Tetrachromat. *Red Herring* (Dec.). http://www.redherring.com/mag/issue86/mag-mutant-86.html (registration required).

INDEX

Page numbers in bold are the most relevant of the listed alternatives.

About the Author

MARK ISAAK has written numerous articles on the creation/evolution debate. He is the editor of the "Index of Creationist Claims" on the acclaimed Web site www.talkorigins.org.